Hydraulics 2018

IOP Conference Series: Materials Science and Engineering
Volume 492

Moscow, Russia
27 November 2018

ISBN: 978-1-5108-8332-1
ISSN: 1757-8981

Printed from e-media with permission by:

Curran Associates, Inc.
57 Morehouse Lane
Red Hook, NY 12571

Some format issues inherent in the e-media version may also appear in this print version.

Printed by Curran Associates, Inc. (2019)

For permission requests, please contact the Institute of Physics
at the address below.

Institute of Physics
Dirac House, Temple Back
Bristol BS1 6BE UK

Phone: 44 1 17 929 7481
Fax: 44 1 17 920 0979

techtracking@iop.org

Additional copies of this publication are available from:

Curran Associates, Inc.
57 Morehouse Lane
Red Hook, NY 12571 USA
Phone: 845-758-0400
Fax: 845-758-2633
Email: curran@proceedings.com
Web: www.proceedings.com

Table of contents

Volume 492

Hydraulics

27 November 2018, Moscow, Russian Federation

Accepted papers received: 28 January 2019
Published online: 13 March 2019

Preface

Papers

IOP Conf. Series: Materials Science and Engineering **492** (2019) 011001 doi:10.1088/1757-899X/492/1/011001

Hydraulics 2018

The conference "Hydraulics" has been held annually in November since 2014 at the Bauman Moscow State Technical University. The Department of Fluid Mechanics, Hydraulic Machines and Hydraulic and Pneumatic Automation is the conference organizer. The best conference reports are published as articles in the journal "Hydraulics", indexed in the Russian Science Citation Index. Every year, representatives of various scientific, educational and industrial organizations of Russia and other countries come to us with reports and speeches. The main topics of the reports are hydromechanics, hydraulic machines, and hydraulic drives. A part of the reports is devoted to industrial ecology issues. We are pleased to see at our conference scientists from many countries and from many organizations devoting their work to industrial hydraulics, ecology and solving fundamental issues in these areas.

Sincerely, A. Protopopov

IOP Conf. Series: Materials Science and Engineering **492** (2019) 011002 doi:10.1088/1757-899X/492/1/011002

Peer review statement

All papers published in this volume of *IOP Conference Series: Materials Science and Engineering* have been peer reviewed through processes administered by the proceedings Editors. Reviews were conducted by expert referees to the professional and scientific standards expected of a proceedings journal published by IOP Publishing.

IOP Conf. Series: Materials Science and Engineering **492** (2019) 012001 doi:10.1088/1757-899X/492/1/012001

An ejector for mixing a reagent with discharge water

B Ksenofontov[1, 2], Ya Vasilieva[1] and S Kaptinova[1]

[1] Bauman Moscow State Technical University, 5 Second Baumanskaya Street, Moscow, 105005, Russian Federation

[2] E-mail: ksenofontov@bmstu.ru

Abstract. The paper presents a research on ejectors used to mix discharge waters with flocculant or coagulant solutions intensifying the flotational discharge water treatment. Results of modelling using ANSYS environment are provided, visualizing the processes going on in the mixing chamber of the ejector as well as estimating parameters of those processes. The velocity gradient values were estimated, which is the main parameter with regard to the mixing process, and therefore characterizing the intensity of discharge water treatment. The dependency between the value of the velocity gradient and the velocity of the water entering the ejector was analyzed. It was shown that the gradient value changes nearly proportionally to the input velocity, and this dependency holds in all directions. The ration between the X component of the gradient and its Y and Z components amounts to approximately 2:1. That means, turbulation does not significantly differ direction-wise, and the mixing conditions are favorable.

1. Introduction

Treatment of discharge water using chemicals is one of the most common methods for discharge water treatment [1-22]. The process of mixing reagents (in particular, coagulants) is as important for the final as the composition of the reagents. Coagulation process runs quite fast, and therefore it is crucial to introduce the coagulants into water evenly and fast. The intensity of the mixing process is characterized by the value of velocity gradient G (1/sec), as well as Camp criterion.

$$Ca = G \cdot \tau, (1)$$

where

τ denotes mixing time, in seconds.

Velocity gradient may be calculated using the following formula:

$$G = \sqrt{\frac{W}{\mu}}, (2)$$

where

W – ratio of power lost in the mixing process to the volume of water mixed;

μ – dynamic viscosity.

Mixing time may be calculated using the following formula:

$$T = \frac{V}{Q}, (3)$$

where
Q – discharge water flow, м^3/с;
V – effective volume of the mixing chamber, m^3.
The optimality condition for the mixing is:

$$T = \tau .(4)$$

Optimal mixing depends on the implementation of the conditions specified above.

Fast introduction of the chemicals into discharge water increases the efficiency of the water treatment process. However, the use of devices equipped with stirring paddles, considered the most efficient technology, sometimes lead to the destruction of the forming aggregates, and therefore it requires strict control over the mixing time. The ejector here works as a jet-pump and improves both dosage and mixing processes (figure 1).

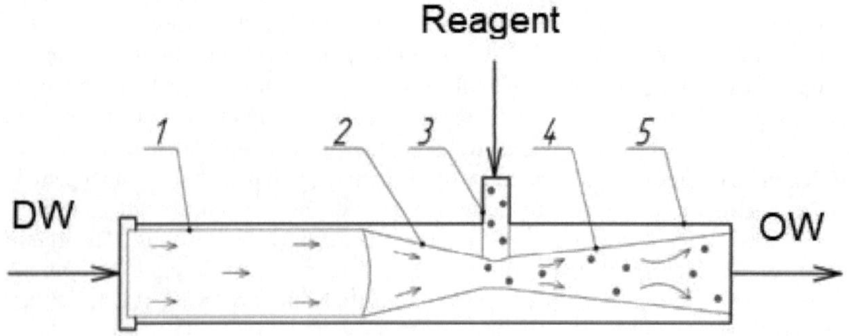

Figure 1. Ejector:*1* – intake chamber; *2* – nozzle; *3* – reagent intake pipe; *4*– suction pipe; *5* – ejector body.

2. Single phase model description and results

Single-phase model was used for the modelling, i.e. it was assumed that liquids flowing into both the intake reagent intake pipe and the water intake chamber have similar characteristics. Heat (energy) exchange was not considered for modelling. Turbulence was considered for modelling. Stationary operation mode was achieved after 60 sec of operation in the model.

The flow velocity (and its variation) is the main parameter characterizing the efficiency of an ejector as a mean of mixing reagents into the discharge waters.

Figure 2 shows velocity distributions for the given portion of the mixing chamber (length = 3 cm).

The X axis on the figure 2 represents the position of the ejector's cross-section (the section plane is Z0Y, i.e. the cross-section goes perpendicularly to the section depicted on figure 1) for which velocities are given on the graph, starting from the left end of the mixing chamber. The velocities are modelled for the axis of symmetry of the flow. The Y axis represents the velocities, measured in m/s.

However, XZ_053 (white) line depicts the magnitude of the velocity in X0Z cross-section (i.e. along the ejector's length), for X=53 mm (i.e. directly at the entrance to mixing chamber).

The velocities are then estimated with a step of 3 mm starting from the entrance to the mixing chamber (its left end). Figure 2 clearly shows that XZ_053 curve (red) does not stick to the ejector's wall, as it is the position of the reagent intake pipe.

However, the single-phase model does not fully represent the processes going on the mixing chamber when discharge waters are mixed with the reagent added through the ejector's pipe (which was originally designed for air intake), because in reality the reagent and discharge waters are distinct liquids with different characteristics, and therefore they should be represented as two different flows in

IOP Conf. Series: Materials Science and Engineering **492** (2019) 012001 doi:10.1088/1757-899X/492/1/012001

the ANSYS system. At this stage, there is no need to specify the reagent's properties (this would require focusing on the concrete composition of the reagent), all we need is to set the model into "two flows mode" or dual-phase mode.

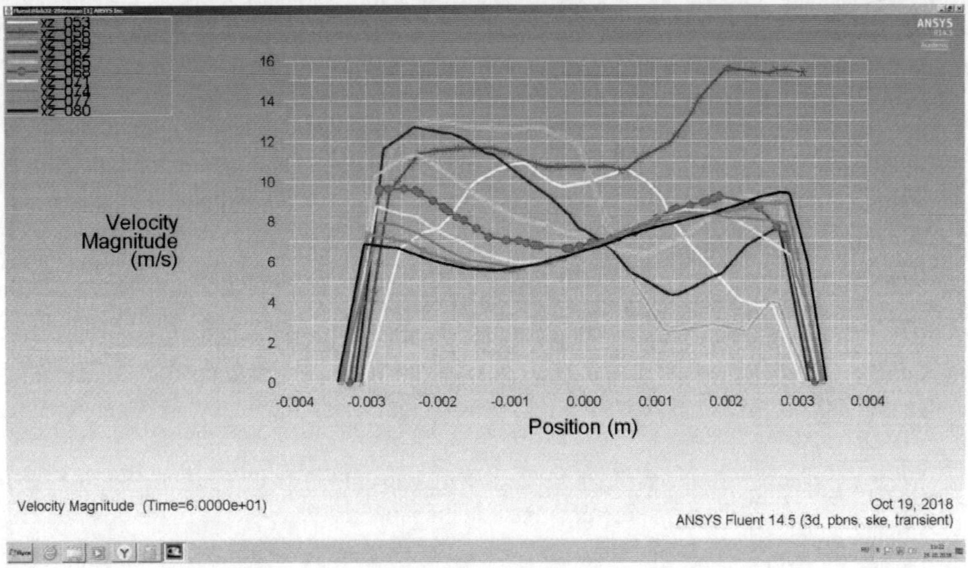

Figure 2. Velocity distribution along the mixing chamber.

3. Dual phase model description and results

The second model considered in this paper is dual phase model. Now we can analyze the homogeneity of the environment (liquid in our case). The homogeneity is directly related to the discharge water treatment efficiency, because more homogenic distribution of the flocculant in discharge water ensures better flotational treatment.

Figure 3 illustrates the volume distribution of liquids, i.e. phase composition of the flow.

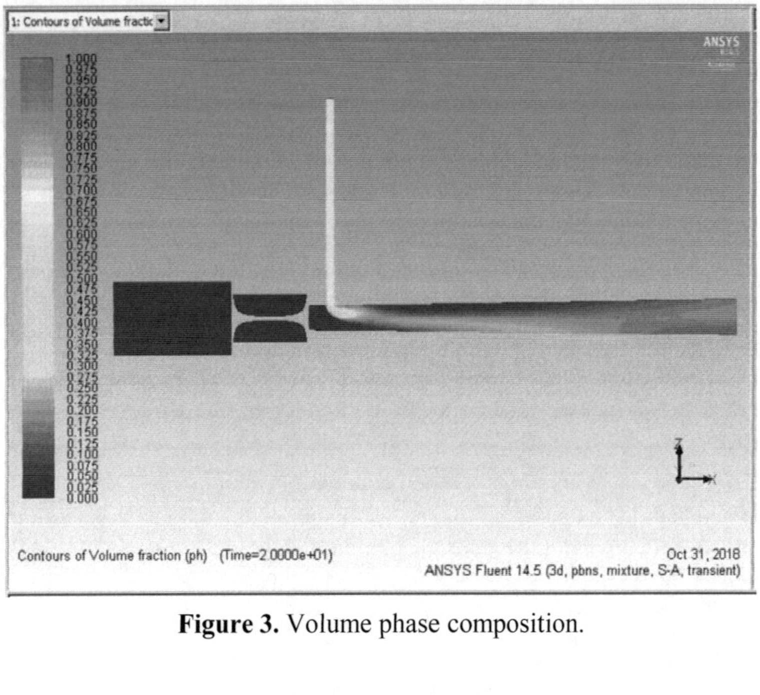

Figure 3. Volume phase composition.

The composition presented on Figure 3 shows that an increase in mixing process intensity can be achieved by more than one way: the flows of both phases may be twisted together, e.g. by installing a separator of specific form on the entrance of the intake chamber; the intake velocity may be increased alongside with the length of the ejector. Alternative solutions may be similar to those described in [23-26].

4. Discussion of the results

According to (2), velocity gradient is one of the main parameters characterizing the mixing process. Table 1 illustrates the dependency between the velocity gradient value and the discharge water velocity at the intake.

Table 1. Dependence between velocity gradient value and water intake velocity.

Water intake velocity, m/s	Flow, kg/s			Velocity, m/sec			N_{exit}, %	Gradient values for velocity components, 1/sec		
	Water	Reagent	Residual	Reagent intake	Mix, max.	Mix, exiting injector		$V_x$max	$V_y$max	$V_z$max
0,4 (7c)	0,102	0,003	-4,7e-06	0,919	5,829	1,645	2,76	12776	6188	7065
0,8 (14c)	0,203	0,007	-1,0e-05	2,115	11,634	3,304	3,16	24162	11931	14019
1,2 (21c)	0,305	0,010	-1,5e-05	3,310	17,452	4,962	3,29	36226	17920	20988
1,6 (28c)	0,406	0,014	-2,0e-05	4,487	23,273	6,620	3,35	48302	23906	27986
2,0 (35c)	0,508	0,018	-2,4e-05	5,655	29,094	8,277	3,37	60406	29897	34999
2,4 (42c)	0,609	0,021	-2,8e-05	6,811	34,920	9,934	3,39	72532	35896	42029

Data represented in Table 1 clearly show that the velocity gradient value (as well as the values of all its components) considerably depends on the water intake velocity. The ratio of the X component of the velocity gradient to its Y and Z components is approximately 2:1. It shows that the turbulation does not differ significantly direction-wise, what is likely an argument supporting good conditions for mixing of the reagent and the treated water.

5. Conclusion

The data obtained in this paper suggest the achievement of an efficient mixing process for introduction of the reagents to the treated water even at low water intake rates. Therefore, low-pressure pumps can be used; the mixing system could be simplified, and an enclosed mixing unit could be developed.

References

[1] Ksenofontov B S and Goncharenko E E 2016 Intensification of Purification of Surface Sewage by use a Bioflocculant*Herald of the Bauman Moscow State Technical University Series Natural Sciences*3 (66)

[2] Ksenofontov B S (2018) Engineering problems of dehydration of rainfall of sewage *Ecology and Industry* 22(9), 4–7.

[3] Kumar E, Bhatnagar A, Hogland W, Marques M and Sillanpää M 2014 Interaction of anionic pollutants with Al-based adsorbents in aqueous media–a review *Chemical Engineering Journal* **241** 443–456.

[4] Zhang W, Chi J, Ma L and Wang L 2017 Occluded Organic Nanofibers Template the Hierarchical Organization of Nanosized Particles in Calcium Oxalate Raphides of Musa spp. *Crystal Growth & Design* 18(2) 1155–1161

[5] Covaci E, Duca G and Sturza R 2015 Equilibrium and kinetic parameters for the sedimentation of tartaric salts in young wines *Chemistry Journal of Moldova* 10(1) 33–41

[6] Zhu H W, Xu C L, Wu D H, Wei B Q, Vajtai R and Ajayan P M 2002 Direct synthesis of long single-walled carbon nanotube strands *Science* 296(5569) 884–886

[7] Zanjani M R K, Yamini Y, Shariati S and Jönsson J Å 2007 A new liquid-phase microextraction method based on solidification of floating organic drop *Analytica Chimica Acta* 585(2) 286–293

[8] Fridman G, Peddinghaus M, Balasubramanian M, Ayan H, Fridman A, Gutsol A and Brooks A 2006 Blood coagulation and living tissue sterilization by floating-electrode dielectric barrier discharge in air *Plasma Chemistry and Plasma Processing* 26(4) 425–442

[9] Costa C A and Rubio J 2005 Deinking flotation: influence of calcium soap and surface-active substances *Minerals Engineering* 18(1) 59–64

[10] Banford A W and Aktas Z 2004 The effect of reagent addition strategy on the performance of coal flotation *Minerals Engineering* 17(6) 745–760

[11] Couper J R, Penney W R and Fair J R (2009) *Chemical Process Equipment-Selection and Design* (Revised 2nd Edition). Gulf Professional Publishing.

[12] Schmidt E. 2015 *US Patent No 8967390* Washington DC US Patent and Trademark Office. Majumder S K 2016 *Hydrodynamics and transport processes of inverse bubbly flow*. Elsevier.

[13] Richter R and Flachberger H 2010 September Physical/chemical treatment of mercury-contaminated wastes from former Chlorine-Alkali-Electrolysis Plants–Application of flotation technologies in soil washing plants *XXV International Mineral Processing Congress (IMPC)* pp 3881-3895

[14] Maselko J, Kiehl M, Couture J, Dyonizy A, Kaminker V, Nowak P and Pantaleone J (2014) Emergence of complex behavior in chemical cells: the system AlCl3–NaOH *Langmuir* 30(20) 5726–5731

[15] Diebold E R, Beaty T A and Buck Jr H B 2012 *US Patent No 8148164* Washington DC US Patent and Trademark Office

[16] Chan Y J, Chong M F, Law C L and Hassell D G 2009 A review on anaerobic–aerobic treatment of industrial and municipal wastewater *Chemical Engineering Journal* 155(1-2) 1–18

[17] Vymazal J (2014) Constructed wetlands for treatment of industrial wastewaters: a review *Ecological Engineering* 73 724–751

[18] Bautista P, Mohedano A F, Casas J A, Zazo J A and Rodriguez J J 2008 An overview of the application of Fenton oxidation to industrial wastewaters treatment. *Journal of Chemical Technology & Biotechnology: International Research in Process, Environmental & Clean Technology* 83(10) 1323–1338

[19] Bianco B, De Michelis I andVegliò F 2011 Fenton treatment of complex industrial wastewater: Optimization of process conditions by surface response method. *Journal of hazardous materials* 186(2-3) 1733–1738

[20] Comninellis C, Kapalka A, Malato S, Parsons S A, Poulios I and Mantzavinos D 2008 Advanced oxidation processes for water treatment: advances and trends for R&D. *Journal of Chemical Technology & Biotechnology: International Research in Process, Environmental & Clean Techn*

[21] Comninellis C, Kapalka A, Malato S, Parsons S A, Poulios I and Mantzavinos D 2008 Advanced oxidation processes for water treatment: advances and trends for R&D. *Journal of Chemical*

Technology & Biotechnology: International Research in Process, Environmental & Clean Techn
[22] Ksenofontov B S, Antonova E S, Ivanov M V, Kozodaev A S and Taranov R A 2015 The influence of oil contaminated soil on the quality of surface waste water. *Water Practice and Technology* 10(4) 814–822
[23] Dimitriyenko Yu I, Gubareva E A and Markevich M N 2016 S V Collectors Mathematical modeling of dielectric properties of the nanostructured composite materials by method of asymptotic averaging *Messenger of MSTU of N.E. Bauman. Series: Natural and technical science* no 1 (64) pp 76–89.
[24] Algazin O D and Kopayev A V 2015 The solution of the mixed regional task for Laplace's equation in a multidimensional infinite layer *the MSTU Bulletin of N.E. Bauman. Series: Natural and technical science* no 1 (58) page 3–13
[25] Algazin O D 2016 The exact solution of a task of Dirikhle for the elliptic equation degenerating on border like Trikomi — Keldysha in a half-space *the MSTU Bulletin of N.E. Bauman. Series: Natural sciences* no 5 pp 4–17 DOI: 10.18698/1812-3368-2016-5-4-17
[26] Algazin O D 2018 Construction by method of similarity of the fundamental solution of a task Dirikhle for the equation like Keldysh in a half-space *the MSTU Bulletin of N.E. Bauman. Series: Natural sciences* no 1 pp 4–15 DOI: 10.18698/1812-3368-2018-1-4-15

IOP Conf. Series: Materials Science and Engineering **492** (2019) 012002 doi:10.1088/1757-899X/492/1/012002

On the issue of starting-up overheating of electric motors of centrifugal pumps

A Protopopov[1,2] and D Bondareva[1]

[1]Bauman Moscow State Technical University, 5 Second Baumanskaya Street, Moscow, 105005, Russian Federation

[2]E-mail: proforg6@yandex.ru

Abstract. In various industries where centrifugal pumps are used, a common problem is the starting overheating of electric motors. Such overheating can lead to motor failure, especially in the case of starting-up the centrifugal pump on the open valve. It happens due to the fact that the starting current is many times greater than the rated current, and the rated current with an open valve usually makes is much more. In this case, complex methods of centrifugal pumps analysis and manuals do not contain any methods of evaluation the magnitude of the starting overheating of centrifugal pump electric motors. In order to fill this lacuna, a mathematical model of the centrifugal pump start-up, which allows to estimate the starting overheating value, is developed in this article.

1. Introduction

Centrifugal pumps are widely used in various industries: mining, chemical industry, municipal engineering, aviation, astronautics, agriculture, robotics, etc. In this case, one of the actual problems is the problem of starting such pumps [1]-[8].

The fact is that at the starting time of the electric motor centrifugal pump its starting overheating occurs. This happens due to the fact that the starting current is many times greater than the rated current. The starting overheating is usually stronger in the case of the pump starting on the open valve, than the rated current. It is known from the characteristics of any centrifugal pump, that the rated current is greater when the valve is open [9]-[15].

Due to this, the specifications for any modern centrifugal pump have recommendations about the desirability of starting it up on a closed valve [16]-[20].

However, there is nothing more detailed except such discrete recommendations like "it is better to start up on a closed valve than to open one" in the literature. It is not known from the literature how much this or that pump overheats and why and to what extent the amount of starting overheating depends. To fill this lacuna, this article proposes a mathematical model of starting up, which is able to estimate the amount of starting overheating.

2. Mathematical model of centrifugal pump starting-up

Based on the theorem of the change in the amount of motion, the moment equation can be written:

$$J \cdot \frac{d\omega}{dt} = M_{dv}(t) - \alpha \cdot M_{rk}(t), \quad (1)$$

where

IOP Conf. Series: Materials Science and Engineering **492** (2019) 012002 doi:10.1088/1757-899X/492/1/012002

J – the moment of inertia of the rotor relative to the axis;
ω – angular rotation speed of the pump shaft;
M_{dv} – motor moment without load;
M_{rk} – impeller moment at starting-up;
α – loss by coupling, bearings, pump seals when the moment is transmitted , α >1.

Let's consider the terms in equation (1).

The moment on the impeller is calculated:

$$M_{rk} = M_c + M_{dt}, \quad (2)$$

where
M_c – centrifugal moment;
M_{dt} – the moment of disk friction.

The impeller centrifugal moment:

$$M_c = \rho \cdot Q \cdot {R_2}^2 \cdot \omega(t), \quad (3)$$

where
ρ – working fluid density;
Q can be determined by the formula

$$Q = \mu_p \cdot \pi \cdot D_1 \cdot a\sqrt{2 \cdot g \cdot H(t)}, \quad (4)$$

where
D_1 – impeller diameter on groove seal;
μ_p– the discharge coefficient in the front axial clearance between the impeller and the pump casing.

The moment of disk friction is equal to:

$$\mathrm{M}_{dt}(t) = \frac{\omega(t) \cdot \pi \cdot \mu \cdot {R_2}^4}{a}, \quad (5)$$

where
a – axial clearance between impeller and pump casing.

Then it turns out that the impeller moment is equal to:

$$M_{rk}(t) = \frac{\omega(t) \cdot \pi \cdot \mu \cdot {R_2}^4}{a} + \rho \cdot Q \cdot {R_2}^2 \cdot \omega(t) \quad (6)$$

The engine moment can be presented in the form of a linear dependence on the angular velocity:

$$M_{dv}(t) = K - K_1 \cdot \omega(t), \quad (7)$$

where
K and K_1 – coefficients of the torque-mechanical characteristics of the electric motor.

Let's make the balance equation for the required pressure:

$$H_H = H_{st} + H_{tr} + h_{in}, \quad (8)$$

where
H_H – pump head required to overcome losses;
H_{st} – static head between the tanks of the feeder and receiver;
H_{tr} – pressure loss in the pipeline;
h_{in} – inertial head.

Let's consider the terms in equation (8).

From the similarity of centrifugal pumps:

$$H_H(Q;\omega) = H_o \cdot \left(\frac{\omega_o}{\omega}\right)^2, \quad (9)$$

IOP Conf. Series: Materials Science and Engineering **492** (2019) 012002 doi:10.1088/1757-899X/492/1/012002

where

$H_н$, ω – head and angular velocity of the pump at starting-up;

H_o, ω_o – head and angular velocity of the pump at initial value.

The inertial head is determined by the acceleration or deceleration of the fluid flow, therefore, to find it, the second Newton law for the element of the stream of an ideal incompressible fluid, will be used and the desired equation will be obtained:

$$h_{in} = \frac{1}{g}\int_{l_1}^{l_2}\frac{\partial V}{\partial t}\cdot dl = \frac{j}{g}\cdot l, \quad (10)$$

where

j – the fluid flow acceleration;

l – the pipeline length.

The acceleration of the fluid flow will be obatined by differentiating the flow formula j in time t:

$$\frac{dV}{dt} = j = \frac{1}{F}\cdot Q' \quad (11)$$

The pressure losses in the pipeline are the sum of friction losses along the length and losses in local resistances, expressed in terms of flow. Based on these conditions, we'll write out the general loss formula:

$$H_{tr} = \left(\frac{\lambda\frac{l}{d}+\xi(t)}{2g\cdot F^2}\right)\cdot Q^2(t), \quad (12)$$

where

λ – friction resistance coefficient;

$\zeta(t)$ – the aggregated coefficient of local resistance;

l,d –the length and diameter of the pipeline;

$F = \frac{\pi d^2}{4}$ – the pipeline cross-sectional area;

$Q(t)$ – flow rate through the section;

$$\xi(t) = K_2 - K_3\cdot t, \quad (13)$$

where

K_2 and K_3–coefficientsdescribing the linear law of the resistance coefficient changing.

The initial conditions for the task are the following conditions:

$$\omega(0) = 0 \quad (14)$$

$$Q(0) = 0 \quad (15)$$

Thus, the mathematical model of the pump starting-up process is as follows:

$$\begin{cases} H_o\cdot\left(\frac{\omega_o}{\omega}\right)^2 = H_{st} + \left(\frac{\lambda\frac{l}{d}+\xi(t)}{2g\cdot F^2}\right)\cdot Q^2(t) + \frac{Q'(t)\cdot l}{F\cdot g} \\ J\cdot\frac{d\omega}{dt} = \left(K - K_1\cdot\omega(t)\right) - \alpha\cdot\left(\frac{\omega(t)\cdot\pi\cdot\mu\cdot R_2{}^4}{a} + \rho\cdot Q\cdot R_2{}^2\cdot\omega(t)\right) \\ \omega(0) = 0 \\ Q(0) = 0 \end{cases} \quad (16)$$

Let's solve the system of equations (16). To do this, we'll rewrite the system of equations (16) in the form:

IOP Conf. Series: Materials Science and Engineering **492** (2019) 012002 doi:10.1088/1757-899X/492/1/012002

$$\begin{cases} Q'(t) = \frac{F \cdot g}{l} \cdot \left(H_o \cdot \left(\frac{\omega_o}{\omega}\right)^2 - H_{st} - \left(\frac{\lambda \frac{l}{d} + \xi(t)}{2g \cdot F^2}\right) \cdot Q^2(t) \right) \\ \omega'(t) = \frac{(K - K_1 \cdot \omega(t)) - \alpha \cdot \left(\frac{\omega(t) \cdot \pi \cdot \mu \cdot R_2{}^4}{a} + \rho \cdot Q \cdot R_2{}^2 \cdot \omega(t)\right)}{J} \\ \omega(0) = 0 \\ Q(0) = 0 \end{cases} \quad (17)$$

3. The results of mathematical modeling

The system of equations (17) is solved in the Mathcad system using the 4th order Runge–Kutta method. The graph of the angular velocity versus time is obtained (figure 1).

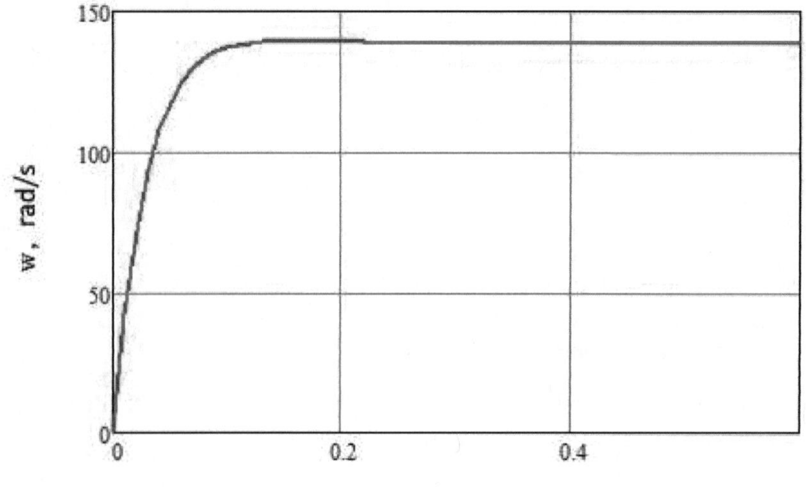

Figure 1. The graph of rotor angular velocity versus time.

The graph of flow versus time (figure 2):

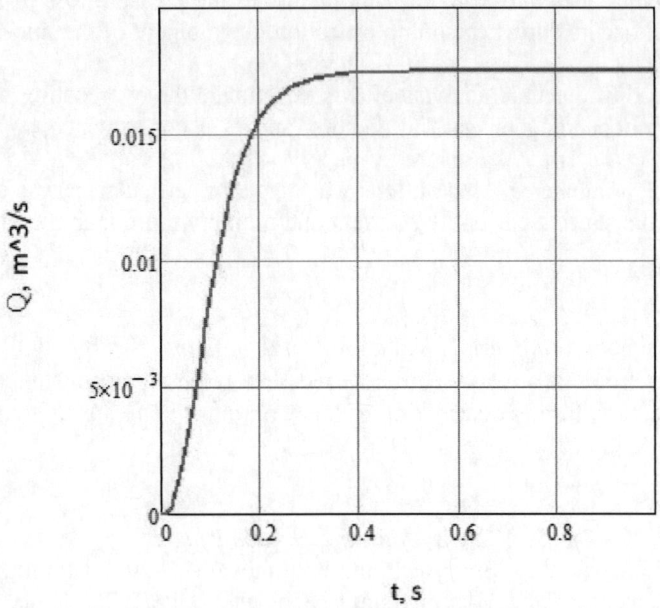

Figure 2. Graph of Q flow versus time.

We need both of these graphs to determine the time of the transition process, which we can substitute into the Joule-Lenz law.

Let's write the law of Joule–Lenz for the amount of heat:

$$\left(I_{nom} \cdot K_p\right) \cdot U \cdot t_{pp} = C_{medi} \cdot m_{prov} \cdot \Delta T, \quad (18)$$

where

I_{nom} – current intensity;
K_p – starting coefficient;
t_{pp} – transition time;
U – electric potential;
C_{medi} – copper specific thermal capacity;
m_{prov} – conductor mass;
ΔT – motor overheating temperature.

Electric current power in the motor winding:

$$P = I_{nom} \cdot U = \frac{\rho \cdot g \cdot Q \cdot H}{\eta}, \quad (19)$$

where

η – pump efficiency.

Let's express the value of the current I_{nom} from the equation (19)

$$I_{nom} = \frac{\rho \cdot g \cdot Q \cdot H}{\eta \cdot U}, \quad (20)$$

The general equations for temperature from equations (18) and (20)

$$\Delta T = \frac{(I_{nom} \cdot K_p) \cdot U \cdot t_{pp}}{C_{medi} \cdot m_{prov}} = \frac{\rho \cdot g \cdot Q \cdot H \cdot K_p \cdot t_{pp}}{C_{medi} \cdot m_{prov} \cdot \eta}, \quad (21)$$

Thus, we obtain the calculation of the temperature of the motor overheating during starting-up.

4. Conclusion

A mathematical model of starting-up of the centrifugal pump rotor with an asynchronous electric motor is obtained.

IOP Conf. Series: Materials Science and Engineering **492** (2019) 012002 doi:10.1088/1757-899X/492/1/012002

The described method takes into account such factors as the pressure loss in the pipeline H_{tr}, the inertia pressure h_{in} that occurs during the pump operation, the moment of the impeller M_{rk} and electric motor moment M_d.

The obtained method of dynamic analysis allows to estimate the overheating of the electric motor of a centrifugal pump depending on various factors, and as a consequence, to predict its possible failure.

In the above model, a number of assumptions was made, in particular, it was assumed that during the transition process the starting current is constant and multiply more than the rated current with the same parameters.

Reference

[1] Borovin G K and Lapshin V V 2018 *Mathematica Montisnigri*

[2] Guskov A M, Lomakin V O, Banin E P and Kuleshova M S 2017 Minimization of Hemolysis and Improvement of the Hydrodynamic Efficiency of a Circulatory Support Pump by Optimizing the Pump Flow path *Biomedical Engineering* **4** pp 229–233

[3] Lomakin V O, Chaburko P S and Kuleshova M S 2017 Multi-criteria Optimization of the Flow of a Centrifugal Pump on Energy and Vibroacoustic Characteristics *Procedia Engineering* **176** pp 476–482

[4] Gouskov A M, Lomakin V O, Banin E P and Kuleshova M S 2016 Assessment of Hemolysis in a Ventricular Assist Axial Flow Blood Pump *Biomedical Engineering* 4 pp 12–15

[5] Lomakin V O, Kuleshova M S and Bozh'eva S M 2016 Numerical Modeling of Liquid Flow in a Pump Station *Power Technology and Engineering* **5** pp 324–327

[6] Lomakin V O 2015 *Proceedings of 2015 International Conference on Fluid Power and Mechatronics*

[7] Lomakin V O, Kuleshova M S and Kraeva E A 2015 Fluid Flow in the Throttle Channel in the Presence of Cavitation *Procedia Engineering* **106** pp 27–35

[8] Borovin G K and Lapshin V V 2014 Optimal attitude control of two pivotally connected bodies in the supportless phase of motion *Journal of Computer and Systems Sciences International* **4** pp 610–622

[9] Borovin G K and Kostyuk A V 2002 Mathematical modeling of the hydraulic control system of a walking robot *Journal of Computer and Systems Sciences International*

[10] Borovin G K andKostyuk A V 2002 Mathematical modeling of the hydraulic control system of a walking robot *IzvestiyaAkademiiNauk. TeoriyaiSistemyUpravleniya*

[11] Shargatov V A, Gorkunov S V and Il'ichev A T 2019 Dynamics of front-like water evaporation phase transition interfaces *Communications in Nonlinear Science and Numerical Simulation* 67 pp 223–236 DOI: 10.1016/j.cnsns.2018.07.006

[12] Kraposhin M V, Banholzer M, Pfitzner M and Marchevsky I K 2018 A hybrid pressure-based solver for nonideal single-phase fluid flows at all speeds *International Journal for Numerical Methods in Fluids* 88 (2) pp 79–99 DOI: 10.1002/fld.4512

[13] Pelevin FV, Avraamov N I, Ir'yanov N Y, Orlin SA, Lozovetskii V V and Ponomarev A V 2018 Intensification of Heat Exchange in the Regenerative Cooling System of a Liquid-Propellant Rocket Engine *Journal of Engineering Physics and Thermophysics* 91 (3) pp 601–610 DOI: 10.1007/s10891-018-1781-4

[14] Ivanov A S and Zhanysbekova Z Z2018 Threaded Joints of Slewing Bearings *Russian Engineering Research* 38 (4) pp 245–250 DOI: 10.3103/S1068798X18040111

[15] KolesnikovA G, Cherepanov D S, Chekulaev A V and Mironova M O 2018 Analysis of Drive Mechanisms for the Working Stand in Periodic Cold-Rolled Pipe Mills *Metallurgist* 61 (11-12) pp 1102–1107DOI: 10.1007/s11015-018-0612-3

[16] Bondarenko V L, Valyakina A V, Borisenko A V, Trotsenko A V and Valyakin V N 2018 Vapor-Liquid Equilibrium of the Ethylene–Butane Mixture *Chemical and Petroleum Engineering* 53 (11-12) pp 778–787 DOI: 10.1007/s10556-018-0421-3

[17] Gouskov A, NikolaevS, Kuts V, Nizametdinov F, Korovaitseva E and Yuan S 2018 Analysis of displacement fields of particle shaping surface during nanoscale ductile mode cutting of brittle materials *International Journal of Advanced Manufacturing Technology*95 (5–8) pp 1911–1918 DOI: 10.1007/s00170-017-1233-x

[18] ZakharovM N, Laryushkin P A and Erastova K G 2018 Stable Geometry of a Plane Five-Link Mechanism *Russian Engineering Research* 38 (2) pp 72–76
DOI: 10.3103/S1068798X1802020X

[19] Arefyev K Y, Prokhorov A N and Saveliev A S 2018 Study of the breakup of liquid droplets in the vortex wake behind pylon at high airspeeds *Thermophysics and Aeromechanics* 25 (1) pp 55–66 DOI: 10.1134/S0869864318010055

[20] Serdyukov V I, Serdyukova N A and Shishkina S I 2018 Improving Operational Reliability by Means of Artificial Intelligence *Russian Engineering Research* 38 (1) pp 15–18
DOI: 10.3103/S1068798X1801015X

IOP Conf. Series: Materials Science and Engineering **492** (2019) 012003 doi:10.1088/1757-899X/492/1/012003

Development of methods for high-speed centrifugal pump analysis

A Protopopov[1,2] and V Vigovskij[1]

[1]Bauman Moscow State Technical University, 5 Second Baumanskaya Street, Moscow, 105005, Russian Federation

[2]E-mail: proforg6@yandex.ru

Abstract: The article deals with the problems associated with the analysis of high-speed low-flow centrifugal pumps. The existing methods for analysis of such pumps and their disadvantages were reviewed. A method of multi-criteria optimization of high-speed centrifugal pumps has been proposed.
The method is based on the search for a compromise between the cavitation characteristics and the hydraulic head of the centrifugal pump by plotting a compromise curve. The LP-tau searching method was used for finding the points of this curve in the article. The LP-tau search in the article was carried out by two design parameters - the rotor speed of the centrifugal pump and the diameter of the impeller blades.

1. Introduction

Nowadays low-flow centrifugal pumps are used in various industries. They have found their use as power sources in various closed systems, for example, in experimental stands of various purposes for supplying the working fluid. The following hydraulic characteristics are specific for low-flow centrifugal pumps: pump flow from 0 to 20 l/min, head - from 0 to 25 m. Most of the existing methods for calculating the main parameters of centrifugal pumps, which are described in the literature, for example ([1] - [7]), are aimed at relatively large values of the working fluid input. The use of such techniques for low-flow centrifugal pumps will lead to distorted pump parameters after calculation.

In the literature [8] - [17], one can also find a wide use of hydrodynamic modeling methods. However, the high costs of human and machine time for calculations can be attributed to the disadvantages of such methods.

Thus, all of the above listed necessitates the search of new methods for calculating low-flow centrifugal pumps free from the above-mentioned disadvantages.

One of the main parameters affecting the operation of the pump are the pressure and cavitation margin. As a rule, these are conflicting parameters: the second has to be sacrificed when the first is increasing. There is a problem in finding a compromise solution, when the second parameter is maximum at the selected value of the first parameter. This question is not ordinary, taking into account that many other parameters and nuances affect these characteristics.

The pressure and cavitational safety margin in low-flow centrifugal pumps, among other parameters, are influenced by the pump shaft speed and the impeller diameter at the inlet. The attempt to determine the final dependence creates new difficulties, since both criteria are significant and it is not possible to determine the weights for them, since a large error appears. Therefore, it was decided to use LP-tau search.

2. Generating LP-tau sequence points

The LP-tau search generates points in a quasi-random manner within the specified interval for two parameters. In our case, the diameter of the impeller at the inlet of the working cavity "D" can vary from 3.5 to 6.5 mm, the rotational speed of the shaft "*n*" - from 13000 to 17000 rpm. Then the field of generated work points will be a rectangle.

The resulting LP-tau sequence in the study region is shown at figure 1.

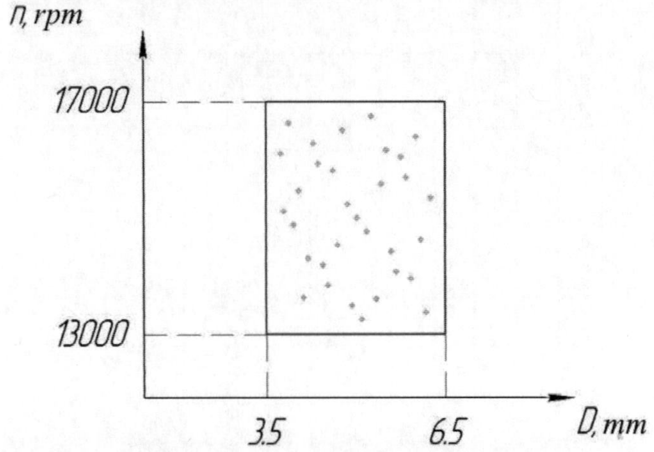

Figure 1. The point generation region.

3. The construction of a compromise curve

The values of the cavitation margin and pump head are found according to the values of the design parameters obtained above.

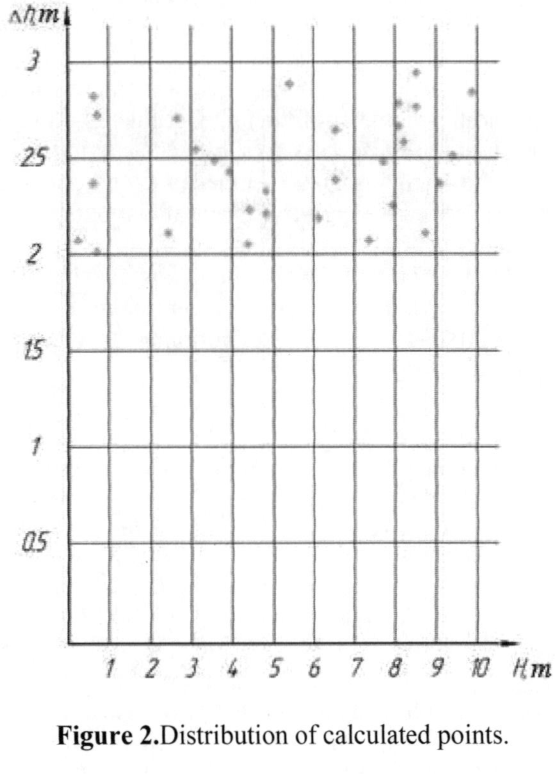

Figure 2. Distribution of calculated points.

The compromise curve is constructed for this distribution by the following reasons: for the same value of one parameter, the point with the highest value of the second is considered to be the "winning". It is clear that the compromise curve have to pass through these "winning" points.

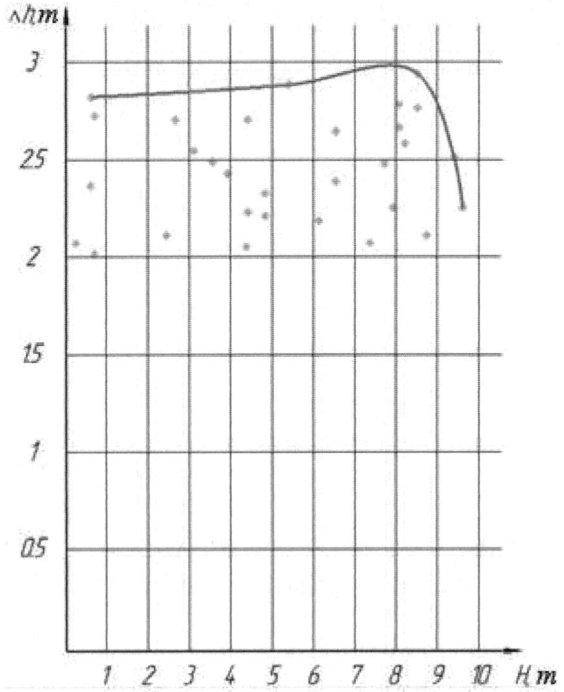

Figure 3. The compromise curve.

Conclusion

The use of LP-tau sequence for finding a compromise between the pressure and the cavitation margin when varying the rotor speed and the diameter of the impeller inlet is a very effective method. This is clearly seen from the obtained compromise curve - when considering the 32 points of the LP-tau sequence, 5 of them entered the Pareto set and formed a compromise curve. Thus, the above-mentioned method can be recommended for high-speed centrifugal pump analysis.

References

[1] Lomakin V O, Petrov A I and Kuleshova M S 2014 Issledovanie dvuhfaznogo techenija v osecentrobezhnom kolese metodami gidrodinamicheskogo modelirovanija *Nauka i obrazovanie: jelektronnoe nauchno-tehnicheskoe izdanie* no 9 pp 45–64

[2] Lomakin A A1966 *Centrobezhnye i osevye nasosy* Moscow Mashinostroenie Publ. 354 p

[3] Lomakin V O, Artemov A V and Petrov A I 2012 Opredelenie vlijanija osnovnyh geometricheskih parametrov otvoda nasosa NM 10000-210 na ego harakteristiki *Nauka i obrazovanie: jelektronnoe nauchno-tehnicheskoe izdanie* no 8 pp 5

[4] Lomakin V O and Petrov A I 2012 Verifikacija rezul'tatov rascheta v pakete gidrodinamicheskogo modelirovanija STAR-CCM+ protochnoj chaste centrobezhnogo nasosa AH 50-32-200 *Izvestija vysshih uchebnyh zavedenij. Mashinostroenie* no C p 6

[5] Lomakin V O, Petrov A I and Shherbachev P S 2012 Razrabotka bokovogo poluspiral'nogo

IOP Conf. Series: Materials Science and Engineering **492** (2019) 012003 doi:10.1088/1757-899X/492/1/012003

podvoda s uvelichennym momentum skorosti na vhode v rabochee koleso. *Izvestija vysshih uchebnyh zavedenij. Mashinostroenie* no S pp 3–5.

[6] Petrov A I, Martynov N D, Pokrovskij P A, Pashhenko V I, Ustjuzhanin P Ju, Korolev P V and Artemov A V 2010 Opyt razrabotki stenda dlja ispytanij krupnyh centrobezhnyh nasosov *Nauka I obrazovanie: jelektronnoe nauchno-tehnicheskoe izdanie* no 11 2 p

[7] Borovin G K and Lapshin V V 2018 *Mathematica Montisnigri*

[8] Guskov A M, Lomakin V O, Banin E P and Kuleshova M S 2017 Minimization of Hemolysis and Improvement of the Hydrodynamic Efficiency of a Circulatory Support Pump by Optimizing the Pump Flowpath *Biomedical Engineering* **4** pp 229–233

[9] Lomakin V O, Chaburko P S and Kuleshova M S 2017 Multi-criteria Optimization of the Flow of a Centrifugal Pump on Energy and Vibroacoustic Characteristics *Procedia Engineering* **176** pp 476–482

[10] Gouskov A M, Lomakin V O, Banin E P and Kuleshova M S 2016 Assessment of Hemolysis in a Ventricular Assist Axial Flow Blood Pump *Biomedical Engineering* 4 pp 12–15

[11] Lomakin V O, Kuleshova M S, Bozh'eva S M 2016 Numerical Modeling of Liquid Flow in a Pump Station *Power Technology and Engineering* **5** pp 324–327

[12] Lomakin V O 2015 *Proceedings of 2015International Conference on Fluid Power and Mechatronics*

[13] Lomakin V O, Kuleshova M S and Kraeva E A 2015 Fluid Flow in the Throttle Channel in the Presence of Cavitation *Procedia Engineering* **106** pp 27–35

[14] Pugachev P V, Svoboda D G and Zharkovsky A A 2016 *Calculation and design of blade hydraulic machines. Calculation of viscous flow in blade hydraulic machines using the ANSYS CFX package*. St. Petersburg Publishing house of Polytechnic University 120 p.

[15] Petrov A I 2016 Technique of continuous obtaining of characteristics of a vane pump for variable temperature and viscosity of a working fluid during tests in a thermal pressure chamber *Engineering Bulletin Electronic Journal* no 10 Available at: http://technomag.edu.ru/doc/850931.html

[16] Petrov A I and Isaev N Ju 2017 Gidrodinamicheskoe modelirovanie raboty centrobezhnogo nasosa v zone otricatel'nyhpodach *Gidravlika* no 3 pp 91–101

[17] Petrov A I and Isaev N Ju 2017 Issledovanie raboty lopastnogo nasosa v zone otricatel'nyh podach metodami gidrodinamicheskogo modelirovanija *Nauchnoe obozrenie* no 13 pp 75–80

Numerical investigation of two-phase media separation in the hydrodynamic filter

V Devisilov[1] **and E Sharai**[1, 2]

[1]Bauman Moscow State Technical University, 5 Second Baumanskaya Street, Moscow, 105005, Russian Federation

[2]E-mail: e9.sharay@yandex.ru

Abstract. Today lots of industries require filtration of two-phase flows with high viscosity non-Newtonian dispersion medium and solid dispersion phase in order to reduce energy consumption for cleaning applications. Specific feature of the separation process for such flows is the limited service life of the filtering equipment. These problems can be solved by modernized tangential filtration with complex flow organization. In the working convergent annular channel are implemented two mechanisms of purification: centrifugal separation and filtration. This work investigates the behavior of solids in non-Newtonian fluids in a context of complex force impacts on flow, as implemented in hydrodynamic filters. The numerical investigation of the flow structure in the convergent annular channel with rotating inner permeable wall was performed. The optimal ratios of the operating parameters are obtained, which ensure steady flow and minimum separation load on the permeable baffle. The obtained results can be used in the design of systems that have high requirements for the purity of the liquid.

1. Introduction

Today lots of industrial applications require separation of two-phase flows with high viscosity non-Newtonian dispersion medium and solid dispersion phase. The fineness of fluid filtration depends on the reliability of technical systems, the probability of occurrence of accidents in production lines, the degree of environmental pollution. The special importance of the mechanical impurities removal is in the systems of waste liquids regeneration, in particular, of waste oils. Hydraulic systems of hydraulic machines and mechanisms often use oils. Hydraulic oils can be contaminated with solid particles when the equipment is worn out. Conversion and recycling of used oil is economically justified and contributes to reducing the anthropogenic load on the environment.

Physical methods for cleaning liquids from solid impurities are widely used, based on the processing of oil in a force field under the action of gravitational, centrifugal force, as well as the method of filtration. Filtration allows guaranteeing the required fineness of purification in contrast to the other methods.

A large number of liquid media used in the industry, exhibit non-Newtonian pseudoplastic properties. These mediums include polymer solutions, heavily polluted waste water, mineral and synthetic oils. For these media can be applied rheological equation of non-Newtonian fluid state, written in the form of a power-law de Waele-Ostwald within the certain range of shear rates [1-3]. Such liquids are characterized by an inverse relationship between the effective viscosity and deformation rate.

IOP Conf. Series: Materials Science and Engineering **492** (2019) 012004 doi:10.1088/1757-899X/492/1/012004

Specific feature of the separation process of such flows by the method of filtration is the limited service life of the filtering equipment due to the need for frequent regeneration of the filtering material because of the clogging of its pores by solid particles and plugging of pore channels because of obliteration processes [4]. The high viscosity of dispersion medium results in the necessity of creating a large pressure drop across the filtering baffle, which requires considerable energy costs. To solve these problems, it is advisable to use combined action devices connecting filtration and other physical separation methods. This allows to provide the required fineness of purification, to increase cleaning efficiency, to reduce the size of production areas, and decrease cleansing process costs by reducing the effective viscosity of the liquid to be cleaned with the appearance of additional shear rates [5, 6].

Among the combined action devices can be distinguished the self-cleaning hydrodynamic filter (figure 1), which allows to considerably reduce energy costs in comparison with traditional filtration [7-10]. The principle of hydrodynamic filtering is that the liquid is passed through rotating porous baffle, wherein a portion of the fluid flow to be cleansed (10–15% of total volume) is bypassed along the filtering baffle [11, 12].

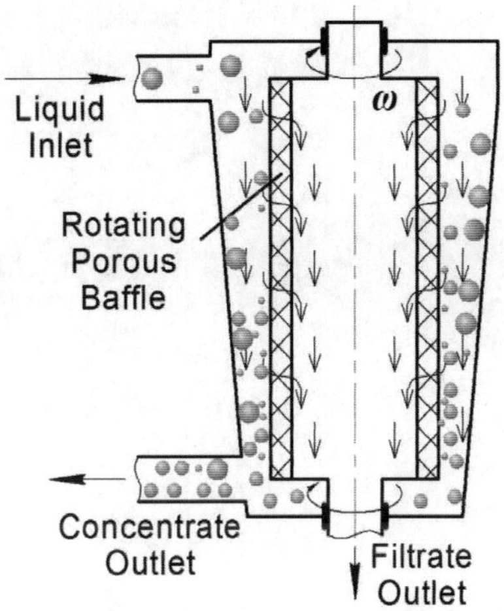

Figure 1. Hydrodynamic filter.

The hydrodynamic force acting on the suspended solids from the pumped flow, contributes to their washout from the baffle surface. This ensures constant cleaning of the filtering surface. By rotating the filtering baffle, it becomes possible to separate under the influence of centrifugal force the solids, having a higher density than the liquid phase. Convergent annular gap between the housing wall and the filtering baffle provides uniform degree of self-cleaning of the baffle over its entire length because of the constancy of the longitudinal fluid flow rate. Reducing the load on the filtering material can be achieved by increasing the proportion of solids separated by the action of centrifugal forces from the total amount of contaminants trapped in the filter by all cleaning mechanisms. This can be achieved by increasing the rotation of the flow, namely the presence of a tangential inlet flow and the rotation of the filtering baffle.

Hydrodynamic filter combines a tangential filtration and centrifugal separation in one device. This allows to increase service life (time between regenerations) and the energy costs of cleaning highly viscous media.

2. Problem statement

The pattern of medium flow in the hydrodynamic filter is rather complex. It is therefore advisable to successively consider the individual components of the separation process, driven by the influence of main force factors occurring in its working area. The working area is an annular channel formed by the stationary outer conical surface and the inner rotating cylinder. For the purpose of this study, let's consider the three component of fluid flow in the working area. The first velocity component of the flow u_r (radial velocity component at coordinate r, m s^{-1}) is caused by filtration through the permeable inner cylinder, determined by the velocity u_R (radial velocity component at filtering baffle at $r = R$, m s^{-1}, where R – radius of filtering baffle) of the fluid suction through its surface (figure 2). The second component is connected with the bypassing of the portion of liquid flow at the velocity w_r, axial velocity component, by means of which hydrodynamic flush of solids from the surface of the permeable cylinder is provided. The third component is caused by the rotation of the permeable cylinder with radius R and the tangential inlet, which ensures centrifugal separation of solid particles, thereby reducing the load on the filtering material. The defining parameters of separation process are tangential velocity v_r and initial swirl velocity, described by the average velocity in the inlet pipe u_{in}.

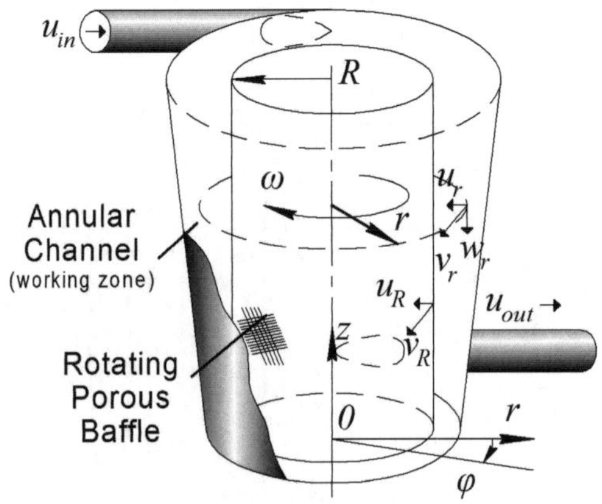

Figure 2. Flow diagram.

And the last component of the flow v_r is a special case of a known flow between two coaxial cylinders, where the outer one is fixed and the inner one is rotating. It is known, that in case of such flow, unstable liquid separation may occur due to the fact, which the liquid particles in the vicinity of the rotating cylinder tend to move to the outer cylinder. Thereby, starting from a certain Taylor number, regularly alternating leftward and rightward rotating toroidal vortices, called Taylor vortexes, appear between the cylinders [13, 14]. However, known conditions for the appearances of Taylor vortices are not the flow stability condition in the hydrodynamic filter, since the outer surface is conical and because there are certain factors, that has a stabilizing effect. In the flow being studied, these factors are the viscosity of the fluid and the suction of the fluid through the permeable baffle [15-20].When filtering high-viscosity media, buckling will occur at a higher rotation speed, as well as by reducing the thickness of the boundary layer due to the suction of the liquid through the rotating cylinder. Therefore, for proper selection of filtering operating parameters ensuring effective separation of particles by the centrifugal mechanism, it is necessary to know the boundaries of appearance vortex structures in view of the aforementioned factors.

Thus, the objective of this study is to determine the effect of the vortexes, which occur between the stationary outer conical wall and the inner permeable rotating cylinder, on the efficiency of separation of solid particles using centrifugal mechanism.

3. Method of research

The system of equations for hydrodynamics two-phase medium is the system of equations, based on the Euler model and Navier-Stokes equations [21], and can be expressed as follows:

- the law of conservation of momentum for a solid phase

$$\frac{\partial}{\partial t}(\alpha_s\rho_s V_{j,s})+\frac{\partial}{\partial x_j}(\alpha_s\rho_s V_{j,s}V_{i,s})=-\alpha_s\frac{\partial P}{\partial x_i}+\frac{\partial\tau_{ij,s}}{\partial x_j}+\alpha_s\rho_s g_i+K_{sf}(V_{i,f}-V_{i,s}); \quad (1)$$

- the law of conservation of momentum for fluid phase

$$\frac{\partial}{\partial t}(\alpha_f\rho_f w_{j,f})+\frac{\partial}{\partial x_j}(\alpha_f\rho_f V_{j,f}V_{i,f})=-\alpha_f\frac{\partial P}{\partial x_j}+\frac{\partial\tau_{ij,f}}{\partial x_j}+\alpha_f\rho_f g_i+K_{fs}(V_{i,s}-V_{i,f}); \quad (2)$$

- continuity equation for solid phase

$$\frac{\partial}{\partial t}(\alpha_s\rho_s)+\nabla(\alpha_s\rho_s V_s)=0; \quad (3)$$

- continuity equation for fluid phase

$$\frac{\partial}{\partial t}(\alpha_f\rho_f)+\nabla(\alpha_f\rho_f V_f)=0; \quad (4)$$

- the volume fractions of the phases

$$\alpha_f+\alpha_s=1; \quad (5)$$

where

α_f, α_s – relative volumetric concentrations of fluid and solid phase;

ρ_f, ρ_s – density of fluid phase and solid phase;

V_f, V_s – velocity of fluid phase and solid phase;

P – pressure within the medium;

x – coordinate;

g – mass forces;

K_{fs} – phase interaction factor;

i – index numbers of the coordinate axis, i = 1, 2, 3.

Stress tensors:

- for fluid phase

$$\tau_{ij,f}=\frac{2}{3}\alpha_f\rho_f\kappa_f\delta_{ij}+\alpha_f\mu_f^{eff}\left[\left(\frac{\partial V_{i,f}}{\partial x_j}+\frac{\partial V_{j,f}}{\partial x_i}\right)-\frac{2}{3}\delta_{ij}\frac{\partial V_{k,f}}{\partial x_k}\right];$$

- for solid phase

$$\tau_{ij,s}=\alpha_s\mu_s\left(\frac{\partial V_{i,s}}{\partial x_j}+\frac{\partial V_{j,s}}{\partial x_i}\right)+\left(\alpha_s\nu_s-\frac{2}{3}\alpha_s\mu_s\right)\delta_{ij}\frac{\partial V_{k,s}}{\partial x_k},$$

where

κ_f – turbulence kinetic energy;

$\delta_{i,j}$ – Kronecker delta;

μ_f^{eff} – effective viscosity of the fluid;

μ_s – solids shear viscosity.

The equation of state for non-Newtonian fluid, which is written as power-law de Waele-Ostwald dependence, has the form of [1]:

$$\tau = m\left|\gamma\right|^{n-1}\gamma, \tag{6}$$

where flow index or the intensity of non-Newtonian properties $n< 1$,
m – consistency index;
γ – shear rate.

As a model of two-phase flow turbulence, we adopted the $\kappa-\varepsilon$ turbulence model, where the transport equations are written for the turbulence kinetic energy κ and the turbulent energy dissipation rate ε for each of the phases separately.In this paper, the system (1) – (5) is solved numerically by finite volume method under stationary conditions.

4. Research results

Calculations have been made for two-phase flow with the following parameters:

1. The continuous fluid. Density $\rho_f = 900\ \text{kg}\cdot\text{m}^{-3}$; consistency index $m = 0.15\ \text{Pa}\cdot\text{s}^n$, the degree of intensity of non-Newtonian properties n was varying from 0.8 to 1.0.
2. The solid phase. Density $\rho_S = 2250\ \text{kg}\cdot\text{m}^{-3}$. At the first iteration the solid phase monodispercity condition was adopted. Diameter of particles d_s was varying from 20 to 100 μm.
3. The geometrical parameters. The radius of the filtering baffle R = 0.02 m;the length of the channel L = 0.1 m; the slant angle of 12°; little radius cone of 0.025 m.

The following are the results of the numerical solution of the system equations (1) – (5) of the analytical model. The effective viscosity in the working area was determined. For this type of complex flow the influence of dispersed phase particle diameters d_s, circumferential rotation velocity of the inner cylindrical surface and the degree of intensity of non-Newtonian properties of particles on efficiency of particle separation using centrifugal separation mechanism were examined.

4.1. The effective viscosity in working zone of the hydrodynamic filter

The rotation of the filtering baffle and the tangential input of the cleaned flow into the working zone makes possible to reduce the effective viscosity of pseudoplastic fluid in the channel by creating additional shear rates. Figures 3a-3b shows the distribution of the relative effective viscosity $\overline{\mu}_f^{eff}$ along the relative radial distance $\overline{r}$ at z/L equal to 0.1 (figure3a), 0.5 (figure3b):

$$\overline{\mu}_f^{eff} = \frac{\mu_f^{eff}}{\mu_f^{n=1}};\ \overline{r} = \frac{r}{R},$$

where

$\mu_f^{n=1}$ – effective viscosity of the fluid with $n = 1$.

As can be seen from the graphs, the introduction of rotation of the filtering baffle really allows reducing the effective viscosity in the annular channel. For instance, the rotation of the filteringbaffle with ω = 100 rad·s^{-1} allows to reduce the effective viscosity by an average of 15% for fluid with the degree n = 0.9, and with n = 0.8 by 25%.

However, there is the limit to reducing the effective viscosity. For considered continuous fluid it is not advisable to use rotation more than ω = 400 rad·s^{-1}. In this case the decrease in effective viscosity will be less than 1% when rotational speed of filtering baffle increases by 100 rad·s^{-1} (figure 4). This effect is due to the presence of pseudoplastic properties in the liquid phase, which are subordinate topower-law de Waele-Ostwald. In this case the dependence between shear stress and shear rate conventionally has three regions: the first Newtonian region (the viscosity does not depend on the shear rate); region of viscosity decrease due to orientation of molecules or particles; the second

Newtonian region (the viscosity remains constant, independent of the further shear rate growth). Therefore, it is necessary to select the operating parameters so that they correspond to the boundary of the second and third regions of the relationship between the shear stress and the shear rate for the regenerated pseudoplastic liquid.

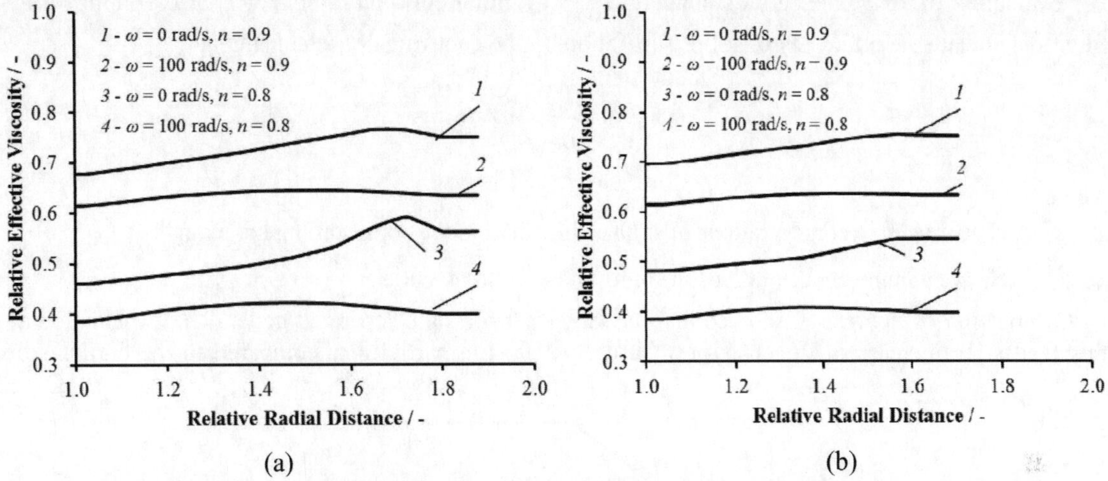

Figure3. The effective viscosity in working zone of the hydrodynamic filter at z/L equal 0.1 (a) and 0.5 (b).

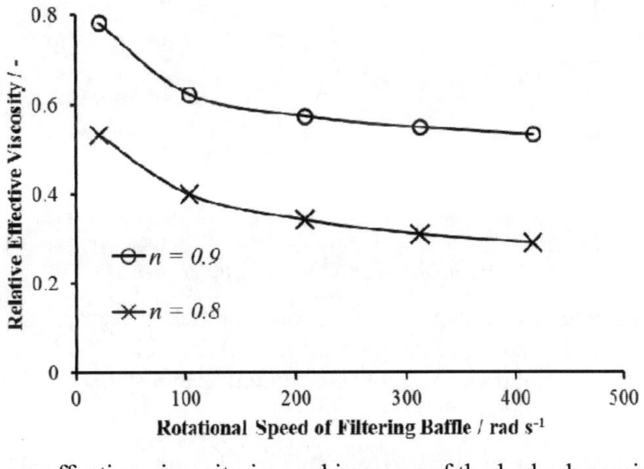

Figure4. The average effective viscosity in working zoneof the hydrodynamic filter at $z/L = 0.5$.

4.2. The efficiency of solid particles separation by the centrifugal mechanism

In general, the efficiency of liquid cleaning in hydrodynamic filter will be determined by the pore size of the filtering material. The simplest version is when the pore size is smaller than the particle size, and the solid particle settles on the surface of the baffle, forming a layer of sediment. If the particle size is commensurate with the pore size, it can penetrate the filtering baffle and get into cleaned flow (filtrate), violating the guarantee of the purity of the liquid.When the sediment layer accumulates on the surface of the filtering baffle, the pressure drop rapidly increases and regeneration is required. To reduce the rate of sediment accumulation in the hydrodynamic filter the rotation of the filtering baffle is implemented, what prevents accumulation of sediment layer. The more particles are separated by centrifugal mechanism and withdrawn from the filter along with the concentrate, the longer the time between regenerations of the filtering material. Therefore, the results of calculating the efficiency of solid particles separation using centrifugal separation mechanism, the main component that affects the efficiency of the filter, are given below.

The efficiency of solid particles separation using centrifugal separation mechanism was estimated by the relative volume concentration of solid particles $\alpha_{s,out}$ at the exit of the concentrate branch pipe. For evaluation of convenience let's convert $\alpha_{s,out}$ to initial concentration $\alpha_{s,inlet}$ at the input of the filter to calculate the efficiency of solids separation by the centrifugal mechanism:

$$\eta = \frac{\alpha_{s,out}}{\alpha_{s,inlet}},$$

where

$\alpha_{s,inlet}$ – relative volume concentrations of solids at the inlet to the computational domain;

$\alpha_{s,out}$ – relative volume concentrations of solids on concentrate outlet.

Comparison of figures5, 6 showed that the larger particles are separated better. The explanation of this fact is the increase of the centrifugal force for 100 μm particles as compared to the particles of 20 μm.

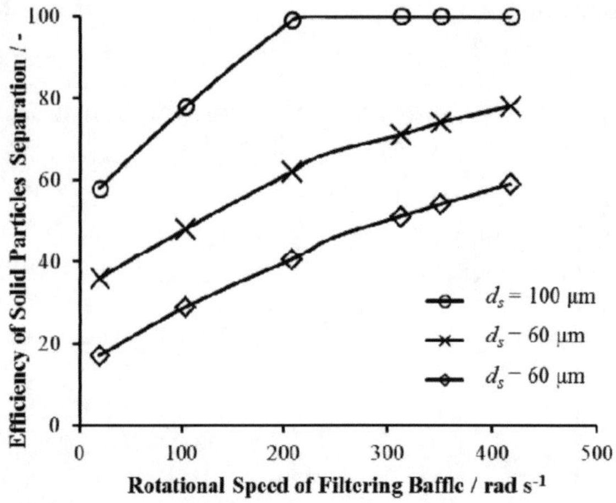

Figure5. Dependence of solid particles separation efficiency using centrifugal separation mechanism η on rotational speed of filtering baffle at $n = 0.8$.

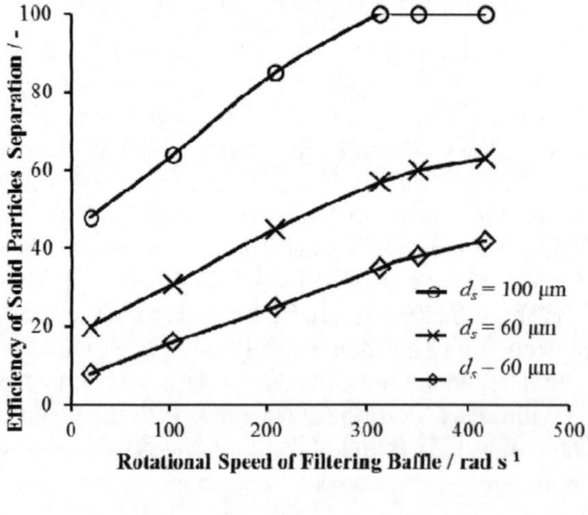

Figure6. Dependence of solid particles separation efficiency using centrifugal separation mechanism η on rotational speed of filtering baffle at $n = 0.9$.

As can be seen from the curves on figures5, 6, particle separation efficiency using centrifugal mechanism increases as the flow index n decreases. This is due to the effect of "liquefaction" of power-law fluid in case of a complex shift, when the effective viscosity decreases with the decreasing of the flow index n. Therefore, with the increased n a solid particle is affected by greater drag force that results in a decreasing of separation efficiency using centrifugal mechanism.

Increasing efficiency η due to increasing of circumferential speed of the inner permeable cylinder rotation is logical, because centrifugal force, that affects a solid particle, is directly proportional to the square of the circumferential speed. However, this increasing is limited and observed at the steady state for the existence of the circulation vortex in the working zone. Further increase in the speed of rotation leads to a sharp increase in hydraulic resistance.

5. Inference

Based on the deterministic approach, we have developed the model of the medium flow process in the convergent channel formed by the stationary conical outer surface and the inner cylindrical rotating permeable surface. We have examined the process of movement of the two-phase flow containing dispersed phase, consisting of solid particles and liquid disperse medium, the rheological properties of which are described by the power-law de Waele-Ostwald dependence, within an annular channel with rotating permeable inner surface. We have identified key parameters of the solid particles separation process in a complex flow. The results of the study can be used to develop an engineering method for calculating hydrodynamic filters, the principle of operation of which is based on filtering in condition of complex impacts on the flow.The obtained results can be used in the design of systems for cleaning high-viscosity fluids from solid particles.

Acknowledgments

This work was performed under the state order of The Ministry of Education and Science of the Russian Federation (No. 10.7766.2017/8.9).

References

[1] Gupta A, Misra J and Reza M 2003 Effects of suction or blowing on the velocity and temperature distribution in the flow past a porous flat plate of a power-law fluid *Fluid Dynamics Research*vol 32 (iss 6) pp 283–294

[2] Yi Xia, Jianzhong Lin and Xiaoke Ku 2018 Flow-induced rotation of circular cylinder in Poiseuille flow of power–law fluids *Journal of Non-Newtonian Fluid Mechanics*vol 260 pp 120–132

[3] Thakur P, Tiwari N. and Chhabra R 2018 Flow of a power–law fluid across a rotating cylinder in a confinement *Journal of Non-Newtonian Fluid Mechanics* vol 251 pp 145–161

[4] Murugesan C and Rudramoorthy D 2016 Numerical And Experimental Study Of Single stage And Multistage Centrifugal Mixed Flow Submersible Borewell Pumps *International Journal of Fluid Machinery and Systems*vol9 (iss 2) pp 107–118

[5] Aleksandrov A, Devisilov V, Sharai E and Kiselyova D 2018 Effect of geometric parameters of working channel of hydrodynamic filter with protective baffle on medium flow structure *Herald of the Bauman Moscow State Technical University, Series Natural Sciences*iss 2 (77) pp 23–78 DOI: 10.18698/1812-3368-2018-2-23-38

IOP Conf. Series: Materials Science and Engineering **492** (2019) 012004 doi:10.1088/1757-899X/492/1/012004

[6] Devisilov V and Sharai E 2012 Hydrodynamics of a rheologically complicated liquid in a self-cleaning filter*Theoretical Foundations of Chemical Engineering*vol 46 (no 6) pp 594–600

[7] Lee J, Bang K, Cho Y and Joh S 2006 The hydrodynamic filter separator for removal of urban storm runoff *Water science and technology* vol 53 (7) pp 243–252

[8] Pandit H, Shakya N, Stole H and Garg N 2009 Hydraulic and sediment removal performance of a modified hydrocyclone*Minerals Engineering*vol 22 (4) pp 412–414

[9] Kroner K and Nissinen V 1988 Dynamic filtration of microbial suspensions using an axially rotating filter*Journal of Membrane Science* vol 36 pp85-100

[10] Fernandez-Cerezo L, Rayat A, Chatel A, Pollard J, Lye G and Hoare M 2018 An ultra scale-down method to investigate monoclonal antibody processing during tangential flow filtration (TFF) using ultrafiltration membranes*Biotechnology and bioengineering*

[11] Devisilov V and Sharai E 2018 Particle Separation in an Annular Converging Channel with an Inner Rotating Permeable Baffle *High Temperature*vol 56 (iss 4) pp 576–580 DOI: 10.1134/S0018151X18040053

[12] Devisilov V and Sharai E 2016 Numerical study of the flow structure in a hydrodynamic filter*Theoretical Foundations of Chemical Engineering*vol 50 (no 2) pp 209–216 DOI: 10.1134/S0040579516020044

[13] Varaksin A, Protasov M, Marinichev D and Vasil'ev N 2015 An Analysis of the Parameters of Flare Particles for Optical Diagnostics of Vortex Flows *Measurement Techniques* vol 58 no 6 pp 655–660

[14] Singh J and Bajaj R 2008 Parametric modulation in the Taylor–Couetteferrofluid flow *Fluid Dynamics Research*vol 40 (iss 10) pp 737–752

[15] Noui-Mehidi M2011 Design optimization of a conical annular centrifugal contractor *Fluid Dynamics & Materials Processing*vol 7 (no 2) pp 141–152 DOI: 10.3970/fdmp.2011.007.141

[16] Zhang Y, Xu L and Li D 2012 Numerical computation of end plate effect on Taylor vortices between rotating conical cylinder *Communications in Nonlinear Science and Numerical Simulation* vol 17 (iss 1) pp 235–241 DOI: 10.1016/j.cnsns.2011.05.021

[17] Lalaoua A and Chaieb Z2016 Flow patterns in a combined Taylor — Couette geometry*Topical Problems of Fluid Mechanics* pp 109–118 DOI: 10.14311/TPFM.2016.016 Available at: http://www.it.cas.cz/fm/im/im/proceeding/2016/16

[18] Xu X, Wen P, Xu L and Cao D 2010 Occurrence of Taylor vortices in the flow between two rotating conical cylinders*Communications in Nonlinear Science and Numerical Simulation* vol 15 (iss 5) pp 1228–1239 DOI: 10.1016/j.cnsns.2009.05.061

[19] Dou H-S, Khoo B and Yeo K 2008 Instability of Taylor — Couette flow between concentric rotating cylinders *Int. J. of Thermal Science*vol 47 (no 11) pp 1422–1435 DOI: 10.1016/j.ijthermalsci.2007.12.012

[20] Li Q, Wen P and Xu L 2010 Transition to Taylor vortex flow between rotating conical cylinders *Journal of Hydrodynamics* vol 22 (iss 2) pp 241–245 DOI: 10.1016/S1001-6058(09)60050-0

[21] Kamali R, Emdad H and Alishahi M 2008 Multicomponent fluid flow analysis using a new set of conservation equations *Fluid Dynamics Research* vol 40(iss 5) pp 343–363

IOP Conf. Series: Materials Science and Engineering **492** (2019) 012005 doi:10.1088/1757-899X/492/1/012005

The application of dynamic risk analysis methods for safety improvement on highways located near or on the territory of hydraulic engineering dams

A Rozhok[1, 2, 3] and V Tatarinov[1, 4]

[1] Bauman Moscow State Technical University, 5 Second Baumanskaya Street, Moscow, 105005, Russian Federation
[2]University of Genoa, Genoa, 16145 Italy

[3]E-mail: rozhok_anastasiya@mail.ru
[4]E-mail: tatavictor@bmstu.ru

Abstract. A theoretical basis for the analysis of the hazards of industrial systems such as "Man-Machine-Environment system", including methods for qualitative and quantitative analysis, physical criteria and characteristics of hazard assessment, is proposed on the basis of fuzzy sets. A mathematical model for improving safety on highways located near or on the territory of hydraulic structures dams using the methods of dynamic risk analysis has been proposed. A mathematical model of a fuzzy system, taken as the basis of the method of dynamic risk analysis, is proposed.
An approach to the analysis of hazards is proposed, which consists in the fact that risk research and hazard ranking have to be carried out on the basis of the theory of fuzzy sets and fuzzy logic under conditions of strong uncertainty and the presence of a human factor.
A methodology for classifying the systems of the "machine" component by hazard classes and the ranking method for the "Man-Machine-Environment system" by degree of danger are proposed on the basis of fuzzy sets. The proposed fuzzy criteria for assessing the risk allow to select the following types of the "Man-Machine-Environment system" ranking: in the presence of a single spectrum of requirements; in the presence of "m" spectra requirements; by the criterion of the total hazard assessment; by the degree of compliance with regulatory requirements; in terms of proximity to various regulatory requirements; by degree of satisfaction with all regulatory requirements.

1. Introduction

The percentage of emergency situations (ES) occurring at the Russian Federation territory, is more than 60% of the total number of emergencies. In the regions of possible impact of damaging factors in the emergency situations a large population lives on motorways, there are fire and explosive objects, bridges and tunnels, hydraulic structures. Dangerous natural phenomena in intensity and duration can have a negative impact, exceeding in scale technogenic emergencies. The proportion of damage caused by natural phenomena is more than 80% of the total damage in an emergency on the territory of the Russian Federation for the year [1-5]. All this predetermines the necessity of long-term forecasting of emergency situations, monitoring the main sources of technogenic and natural hazards, conducting risk analysis at sites and territories, and developing measures to reduce risk [6].

The risks are integral indicators that determine the degree of technogenic and natural hazards within the territory [7, 8, 9]. A consistent scientific and methodological approach to risk analysis will increase the level of safety of the population and territories from anthropogenic and natural emergency situations based on optimizing population protection measures in response to large-scale accidents and disasters [10].

The science of risks in the technogenic and natural spheres has been formed in the last quarter of the XX century due to the practical needs of ensuring safety in the technosphere, particularly, in nuclear power engineering, as well as the development of space technologies [11, 12].

For the most accurate predictions of hazards on highways located near or on the territory of hydraulic engineering dams, it is recommended to apply the method of dynamic risk analysis [13]. The methodology is based on a statistical analysis of the risk of natural and technogenic hazards. The development of the methodology is based on the integration of statistical data, emergency information and meteorological information.

The method of meteorological forecasting of the hazard level is conducted taking into account seasonal changes in climatic conditions, as well as predicted meteorological data [14, 15]. Dynamic risk analysis involves the prediction of risk indicators based on the statistical information processing. The basis for calculating risk indicators is the trend identification, which determines the increase or decrease in risk indicators for previous years or decades, then the seasonality factor is calculated, which is analyzed by months, taking into account changes in indicators [16]. The emergencies can be predicted and prevented, based on meteorological information. Meteorological conditions include air temperature, rainfall, wind speed, taking into account the specific territorial belonging and a certain period of time. The coefficient of anomalies is calculated with their help.

For the most accurate prediction, taking into account the lack of information, it is possible to use a mathematical model [17], based on the application of the method of fuzzy sets [18, 19].

A distinctive feature of this article is the application of a consistent scientific and methodological approach to the implementation of dynamic risk analysis methods to improve safety on highways located near or on the territory of hydraulic engineering dams [20].

2. The proposed methodology

The use of fuzzy sets is the basic for classifying hazards and ranking the "Man-Machine-Environment system".

The analysis of the "Man-Machine-Environment system" allows to select the main individual indicators and characteristics for the state of the "Man-Machine-Environment system". Analysis of the system also allows to reflect emerging hazards.

The classification of the hazards of the "Man-Machine-Environment system" is also be carried out for certain types of harmful and dangerous factors.

These factors affect directly the highway workers, including drivers who organize and carry out road works, loading and unloading of cargo, as well as the carriage of dangerous goods.

Four classes of hazards are listed according to the degree of health effect of harmful substances, depending on the concentration of this substance. They are introduced in table 1.

Table 1.

Classofhazard	Title
1-stclass	Extra-hazardous
2-ndclass	Highly-hazardous
3-rdclass	Moderatelyhazardous
4-thclass	Lowhazardous

The established hygienic criteria for assessing the working conditions of workers of the highway and hydraulic structures are determined by the hygienic assessment of the existing conditions and the sort of work, respectively, in terms of the hazards of the working environment factors, the severity and

IOP Conf. Series: Materials Science and Engineering **492** (2019) 012005 doi:10.1088/1757-899X/492/1/012005

intensity of the working process. The severity coefficients of work and the injury frequency rate are used to account injuries and allow a comparison of the work performed between them.

The variety and multiplicity of hazard factors create a problem of complex danger in the “Man-Machine-Environment system”. A hazard assessment system can become the solution of this problem. With its help one can rank the “Man-Machine-Environment system” by hazards, and, if it is necessary, by other significant parameters that can characterize this system and are relevant to the assessment problem under consideration. Such characteristics may be the population in the areas adjacent to the highway, the number of engineering workers on hydraulic engineering structures that are close to the roadway, the amount of consumed energy, the number of road workers, the novelty and complexity of the system. The solution of the multicriteria problem under consideration with deterministic fuzzy data can be obtained on the basis of the fuzzy sets theory.

Hazard identification begins with the definition of a certain contour as a set or vector:

$$E = \{e_1, e_2, \ldots, e_n\}, (1)$$

where

e_n – the contour element, which is one of the hazard or dangerous factors of many other factors - chemical, physical, biological, economic, ergonomic, organizational, social, environmental, anthropometric, aesthetic.

Ensuring safety on a highway located near or on the territory of hydraulic structures dams is also determined by the combination of many presented factors, particularly physical and organizational. These factors are assessed quantitatively and qualitatively by expert judgment and measurements.

It is necessary to consider systems relating specifically to this particular element, during evaluating each of the contour elements. Evaluation of system elements can be made on the basis of expert opinion and to represent any evaluation system that is convenient for use. Table 2, developed by the authors, shows an example of assessing system elements by systems and categories on a five-point scale for highways located on the territory of hydraulic structures or near the it, taking into account the preparedness of evacuation services to eliminate emergencies, and also taking into account the work of highway services.

Table 2. The evaluation table for highway systems.

System	Mark
EmergencyResponseSystem	**4.25**
EmergencyAlertSystem	5
EvacuationSystem	3
Medical Assistance in Emergency Situations	4
Rescue System in case of Emergency Situations	5
System for providing training and prevention of emergency situations	**4**
HazardAnalysisSystem	4
ProfessionalSelection	3
StaffTraining	4
Deployment of services responsible for events in Emergency Situations	4
Development of an action plan in case of Emergency Situations	5
DisasterRecoveryFramework	**3.8**
Post-accidentDamageAssessment	5
WorkRecovery	3
Restitution	3
Determining the causes of Emergency Situation	4
Evaluation of the state of the “Man-Machine-Environment system”	4

IOP Conf. Series: Materials Science and Engineering **492** (2019) 012005 doi:10.1088/1757-899X/492/1/012005

Each element of the circuit has a quantitative characteristic, which allows to obtain one of the three main estimation spectra.

1) Regulatory spectrum – the assessment of elements is performed using regulatory documentation, in some cases the assessment is based on common sense.

2) Objective spectrum – the assessment of elements is in the form of target values, which are the goal at the present moment, the assessment is made on the basis of the available possibilities.

3) The real spectrum – the real state of the elements in the present moment is displayed with the spectrum.

Evaluation spectra are converted to the following formulas.

Regulatory fuzzy set $T = \int_E \mu_T(e)/e$;(2)

Objective fuzzy set $C = \int_E \mu_C(e)/e$;(3)

Real fuzzy set $R = \int_E \mu_R(e)/e$,(4)

where

$\mu_i(e)$ – the membership function of the element *e* to the *i*-th set, $i = T, C, R$.

When considering possible dangers appearing on motorways, the most applicable is to consider the estimated spectra of the normative and real fuzzy sets. However, their incompatibility with each other may indicate the presence or possibility of the occurrence of hazards. In this way, it is possible to obtain appropriate hazard assessment criteria.

For convenience of the hazard assessment it is proposed to consider the following cases:

- Assigning a no-hazard condition in the form $Q \leq [Q]$; $[Q]$ is a valid value of Q, for which the inequality $\mu\_R (e) \leq \mu\_T (e)$ is satisfied, which can also be expressed as: $1 + \mu_T(e) - \mu_R(e) \geq 1$. Under this condition, the hazard criterion is characterized by the presence of danger

$$K_e(T,R) = 1 - \{1 \wedge [1 + \mu_T(e) - \mu_R(e)]\}. (5)$$

- Setting the no-hazard condition in the form of $Q \geq [Q]$; Under this condition, the hazard criterion is characterized by the presence of hazard

$$K_e(T,R) = 1 - \{1 \wedge [1 + \mu_R(e) - \mu_T(e)]\}. (6)$$

- Setting a no-hazard condition in the form of an interval $[Q^-, Q^+]$, the following inequality is necessary

$$\mu_T^-(e) \leq \mu_R(e) \leq \mu_T^+(e). (7)$$

In this inequality, the indices are specified and characterized by the minimum and maximum values of the membership function, respectively. Based on this inequality, the hazard criterion can be expressed as follows:

$$K_e(T,R) = 1 - min\{1 \wedge [1 + \mu_T^+(e) - \mu_R^+(e)]; 1 \wedge [1 + \mu_R^-(e) - \mu_T^-(e)]\}. (8)$$

Using all three cases of hazard assessment it is possible to form a criterion for the overall hazard assessment in the "Man-Machine-Environment system" in the following form

$$K_S(T,R) = \frac{1}{n}\sum_{i=1}^{n} K_{e_i}(T,R). (9)$$

Thus, the hazard assessment criteria proposed in this paper, calculated using the fuzzy sets $K_S(T,R)$ and $K_E(T,R)$, allow defining hazard indicators and classifying by the identified contour of elements.

The ranking of the identified contour elements can be carried out as follows:

- ranking due to the presence of a single range of requirements;
- ranking due to the presence of a certain range of requirements - m;
- ranking is based on the total hazard rating criterion;
- ranking is made according to the degree of compliance with regulatory requirements;
- ranking is made according to the degree of proximity to various regulatory requirements;
- ranking is made according to the degree of satisfaction with all regulatory requirements.

3. Conclusion

Thus, the paper proposed an idealized mathematical model based on fuzzy sets and risk analysis technique using fuzzy sets.

A model that can be used for ranking the systems of the "machine" component by hazard classes based on fuzzy sets has been proposed in this paper. Criteria for ranking the "Man-Machine-Environment system" by types of hazardous factors based on fuzzy sets have been also proposed.

The proposed hazard analysis model will allow:

– to predict the dangers of simple and complex technical systems, using established and systematic methods;

– to carry out the ranking of the component "machine" systems by hazard classes in the conditions of strong uncertainty;

– to rank all types of the "Man-Machine-Environment system" according to hazard criteria and compliance with regulatory documentation.

References

[1] Guskov S J and Levin V V2016 Model estimates of the probability of risk events in the system *2016IEEE Conference on Quality Management, Transport and Information Security, Information Technologies, IT and MQ and IS* 7751902 pp 61–64

[2] Pereezdchikov I V 2015 Ranking industrial plant hazards *Chemical and Petroleum Engineering* Volume 41 Issue 5-6 May 2005 pp 340–343

[3] Guskov S Y and Levin V V 2017 Model estimates of the probability of risk events in the system *Proceedings of the 2017 International Conference"Quality Management, Transport and Information Security, Information Technologies",* IT and QM and IS 8085876 pp 525–527

[4] Pavlenkov M N, Larionov V G, Voronin P M and Pavlenkov I M 2017 Methodology of Complicated System Hazard Analysis by the Use of Sharp and Fuzzy Sets Translated from*KhimicheskoeiNeftegazovoeMashinostroenie* 2005 no 7 pp 39–41 *Enterprise development factors' control European Research Studies Journal* 20 (2) pp 309–319

[5] Alfimtsev A N and Sakulin S A 2017 Data fusion based on the fuzzy integral *Model, methods and applications (Book Chapter) Data Fusion: Methods, Applications and Research* pp 1–64

[6] Kim H-S and Lee S-W 2018 Dependability-enhanced unified modeling and simulation methodology for Critical Infrastructures *Information and Software Technology* 102 pp 175–192

[7] Jiang Y, Yin S and Kaynak O 2018 Data-driven monitoring and safety control of industrial cyber-physical systems: Basics and beyond *IEEE Access* 6,8444966 pp 47374–47384

IOP Conf. Series: Materials Science and Engineering **492** (2019) 012005 doi:10.1088/1757-899X/492/1/012005

[8] Soussi A, Bouchta D, Amarti A E, (...), Trotta A and Zero E 2018 Risk analysis for hazardous material transport by road: Case study on Tangier-Tetouan region, Morocco *13th System of Systems Engineering Conference, SoSE* 2018 8428700, pp 464–470

[9] Kim H and Song Y 2018 An integrated measure of accessibility and reliability of mass transit systems *Transportation* 45(4) pp 1075–1100

[10] Beven K J, Almeida S, Aspinall W P, (...), Watson M and Wilkins K L 2018 Epistemic uncertainties and natural hazard risk assessment Part 1 A review of different natural hazard areas *Natural Hazards and Earth System Sciences* 18(10) pp 2741–2768

[11] Zio E 2018 The future of risk assessment *Reliability Engineering and System Safety* 177 pp 176–190

[12] Huang X, Wang X, Pei J, (...), Huang X and Luo Y 2018 Risk assessment of the areas along the highway due to hazardous material transportation accidents *Natural Hazards* 93(3) pp 1181–1202

[13] Kaur R K, Pandey B and Singh L K 2018 Dependability analysis of safety critical systems: Issues and challenges *Annals of Nuclear Energy* 120, pp 127–154

[14] Haasnoot M, van 't Klooster S and van Alphen J 2018 Designing a monitoring system to detect signals to adapt to uncertain climate change *Global Environmental Change* 52 pp 273–285

[15] Bruno Soares M, Daly M and Dessai S 2018 Assessing the value of seasonal climate forecasts for decision-making *Wiley Interdisciplinary Reviews: Climate Change* 9(4) p 523

[16] Jain P, Rogers W J, Pasman H J and Mannan M S 2018 A resilience-based integrated process systems hazard analysis (RIPSHA) approach Part II management system layer *Process Safety and Environmental Protection* 118 pp 115–124

[17] Pérez-González C J, Colebrook M, Roda-García JL and Rosa-Remedios C B 2019 Developing a data analytics platform to support decision making in emergency and security management *Expert Systems with Applications* 120 pp 167–184

[18] Zarei E, Khakzad N, Cozzani V and Reniers G 2019 Safety analysis of process systems using Fuzzy Bayesian Network (FBN) *Journal of Loss Prevention in the Process Industries* 57 pp 7–16

[19] Zhao R, Liu S, Liu Y, Zhang L and Li Y 2018 A safety vulnerability assessment for chemical enterprises: A hybrid of a data envelopment analysis and fuzzy decision-making *Journal of Loss Prevention in the Process Industries* 56 pp 95–103

[20] Torre-Bastida A I, Del Ser J, Laña I, (...), Bilbao M N, Campos-Cordobés S 2018 Big Data for transportation and mobility: Recent advances, trends and challenges *IET Intelligent Transport Systems* 12(8) pp 742–755

IOP Conf. Series: Materials Science and Engineering **492** (2019) 012006 doi:10.1088/1757-899X/492/1/012006

Information support for safety insurance of road transport of dangerous goods

V Tatarinov[1,3] **and A Kirsanov**[2]

[1] Bauman Moscow State Technical University, 5 Second Baumanskaya Street, Moscow, 105005, Russian Federation

[2] All-Russian center for monitoring and forecasting emergency situations of EMERCOM of Russia ("Antistihiya" center)

[3]E-mail: tatavictor@bmstu.ru

Abstract. In this article, the methods of system analysis are used to establish the presence of two open systems, one of which is intended for the carriage of dangerous goods by road and the other – for liquidation of consequences of accidents. It is established that in the case of a car accident with dangerous goods, part of the information necessary to make a decision on the elimination of the consequences of the accident is not available to the second system because only the first system that carries out the transportation has access to it. The integration of these systems on the basis of new information and communication technologies ensures the emergence of the developed system, eliminates the identified drawback and creates conditions for information support management.

1. Introduction

An accident with dangerous goods can be characterized by a number of features: the location of the emergency; the type of damaging factors and consequences that threaten the life and health of people and pose a threat to the economy and the environment [1, 2].

An even more dangerous emergency occurs in a car accident with dangerous goods [3]. Firstly, such an emergency is characterized by uncertainty of the location of occurrence due to the movement of dangerous goods in time and space. Based on this, it should be expected that an emergency can occur both in places of large concentrations of people and vehicles and in completely deserted places during long-distance transportation [4].

Secondly, the peculiarity of this type of emergency is the threat of chain development of the accident with the release of damaging factors abroad the place of occurrence as a result of the release of a dangerous substance into the surrounding space or the spread of fire due to the ignition of combustible substances, etc. Taking all of the above into consideration, the scale of this emergency can reach up to the regional level.

The purpose of the article is to build a system of information support for the management of transportation of dangerous goods by road that would include existing systems based on new information and communication technologies [5–7].

2. Theoretical analysis

In accordance with the existing practice in the Russian Federation of transportation of dangerous goods by road transport, the transport company is obliged to: obtain a permit for the transportation of dangerous goods; prepare the vehicle for transportation; teach the driver the rules of transportation and actions in case of emergency. At the same time, a system of dispatching services on duty (DSD) designed to eliminate emergencies including accidents with dangerous goods exists.

The decision to eliminate the consequences of a car accident with dangerous goods is based on information about the characteristics and quantity of the cargo, the location and time of the accident, weather conditions in the area of the accident. Analysis of the composition and sources of information allows to decompose it by time and location of its origin.

First component of information is regular information which is generated at the stage of preparation of dangerous goods for shipment, i.e. before the onset of the emergency. Its content is determined by the requirements of the guidance documents and is formed in the process of obtaining a transport company permit for the transport of dangerous goods and registration of accompanying documents.

The second component of the information is the operational information which is transmitted directly from the scene of the accident by the driver, if he is able to do it, or an eyewitness.

The third component is the calculated part. It is obtained in the process of evaluation calculations of the possible consequences of the accident to decide on the composition of forces and means to eliminate the damaging factors of the accident.

In the conditions of an emergency [8], the time factor in the vast majority of cases is crucial in making management decisions. Inconsistency and incompleteness, as well as sometimes a complete lack of information in the first minutes of emergency does not allow one to quickly make the required decision, appropriate to the circumstances. The main problem faced by the decision maker (DM) is the lack or complete lack of reliable information.

One of the possible ways to improve the level of safety of dangerous goods transportation by road is to create a software and hardware complex that would provide constant monitoring and notification of emergencies in real time. We define constant monitoring as the continuous monitoring of the movement of a vehicle with dangerous goods, monitoring the state of the "vehicle-cargo" system.

Constant monitoring [9, 10] would allow to assess the operational situation on the roads as well as the risks of possible emergencies on the road sections. This, in turn, will lead to the appropriate level of combat readiness of emergency services headquarters, to which certain sections of the roads would be assigned.

Any management is provided with information and analytical support [11], which include regulations, supporting documentation, reports, statistical data, etc. General control scheme of road transport of dangerous goods is represented in figure 1 as the IDEF0 functional model.

Organization of road transport can be divided into 5 main stages:

1. Preparation for transportation;
2. Development and coordination of transportation route;
3. Collection and registration of accompanying documentation;
4. Additional actions: cargo insurance, escort organization, etc.;
5. Delivery of the cargo.

The stage of preparation for transportation includes the following actions: the main parameters of the cargo being prepared for transportation (hazard class, quantity, dimensions, identification of the cargo according to the UN list) are established, the tank and the vehicle are selected and checked for their technical condition and the availability of the necessary safety equipment is checked. All information obtained at the stage of registration of the permit for the carriage of dangerous goods and filling in the accompanying documents is held by the dispatcher of the transport company and is not available to the person making the decision to eliminate the consequences of a car accident with dangerous goods. Elimination of this deficiency can be realized at introduction of system of electronic document flow taking into account differentiation of information access levels.

After the confirmation of the route and schedule of delivery of the goods, the next stage is carried out: the registration of the transport and accompanying documentation. At this stage, the most important thing in terms of obtaining operational information for emergency services in case of emergency are the following documents: bill of lading, waybill and the route of transportation of dangerous goods.

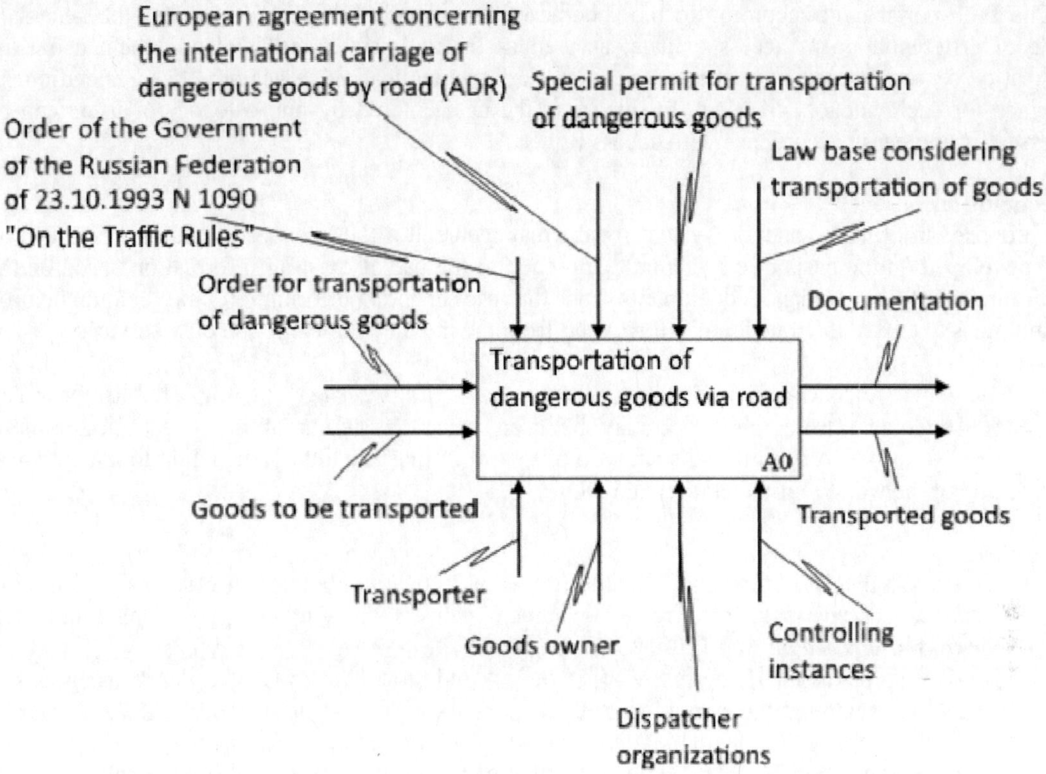

Figure 1.Functional model of organization of road transport of dangerous goods.

These documents contain information that can be used in information systems that provide control and monitoring of the safety of transportation of dangerous goods. Based on the analysis of these documents, a list of operational information is compiled.

Currently, there are no effective mechanisms for automatic transmission of operational information [12, 13] contained in the accompanying documents from the carriers to the dispatching services on duty. Information about the cargo and the location of the emergency can be transmitted by eyewitnesses or the driver, if he is not injured after the emergency. Thus, at present there is no communication system between a moving vehicle (truck) with dangerous goods and EMERCOM dispatching services on duty (DSD).

3. Practical realization

To implement the information transfer it is necessary to create a multi-level network system based on information and communication technologies [14]. Such a system is a Web-compatible integrated tool for access to search, collection and analysis of information that allows to process heterogeneous sets of text and graphic information more quickly. This system is called an electronic document management system (EDMS). A number of scientific publications have proved the improvement of management efficiency through the use of EDMS [15, 16].

In order to organize monitoring and security, the implementation of the EDMS in accordance with the proposed classification should be focused on supporting the management of the organization and the accumulation of information, as well as providing additional support for joint work. This in turn will ensure the performance of work by the staff of duty services even if they are geographically separated and the preservation of the results of this work. A web portal provides entering of information, control and provision [17] of it through the Internet networks on the communication channels protected from various types of attacks [18, 19].

IOP Conf. Series: Materials Science and Engineering **492** (2019) 012006 doi:10.1088/1757-899X/492/1/012006

The Web portal built according to the scheme of three-level architecture provides the necessary mode of differentiation of access to data, as well as the authority to fill and edit them using the application server. According to the set access policy, the application server provides a specific user interface for each category of users. In this case, the data entered by one category of users can be accessed by another in accordance with access rights.

4. Conclusion

The proposed information support system for the management of the transport of dangerous goods by road developed on the basis of a systematic approach to the use of existing information based on the requirements of the guidance documents and the use of new information and communication technologies to ensure the transfer of information from the moving vehicle to the duty services.

References

[1] Kozaev A, Saleh H M, Alexandov D and Buhvalov I 2017 Application of case-based method to choose scenarios to resolve emergencies on main gas pipeline*CEUR WorkshopProceedings*pp 120–126

[2] Korotkov S, Makuev V, Lopatnikov M, Sirotov A and Stonozhenko L 2016 Forest-use issues in Moscow Region at the beginning of the 21st century *Bulletin of the Transilvania University of Brasov Series II Forestry, Wood Industry, Agricultural Food Engineering* 9(2) pp 17–24

[3] Guskov S and Levin V 2016 Model estimates of the probability of risk events in the system 2016 *IEEE Conference on Quality Management, Transport and Information Security, Information Technologies, IT and MQ and IS* 7751902 pp 61–64

[4] Nikolaev A B, Sapego Y S, Ivakhnenko A, Mel'nikova T and Stroganov V Y 2017 Analysis of the incident detection technologies and algorithms in intelligent transport systems *International Journal of Applied Engineering Research* 12(15) pp 4765–4774

[5] Samimi P, Mukhtar M and Teimouri Y 2016 A combinatorial double auction resource allocation model in cloud computing Information *Sciences* 357 pp 201–216

[6] Kato N 2016 On cloud computing [Editor's Note] *IEEE Network* 30(1) pp 2

[7] Clarke R 2016 Big data, big risks *Information Systems Journal* 26(1) pp 77–90

[8] Nikolaev A B, Sapego Y S, Jakubovich A N, Berner L I and Stroganov V Y 2016 Fuzzy algorithm for the detection of incidents in the transport system *International Journal of Environmental and Science Education* 11(16) pp 9039–9059

[9] Shaytura S V, Kozhaev Y P, Ordov K V, Minitaeva A M and Feoktistova V M 2018 Geoinformation services in a spatial economy *International Journal of Civil Engineering and Technology* 9(2) pp 829–841

[10] Kovacs A A, Hartl R F, Parragh S N and Golden B L 2014 Vehicle routing problems in which consistency considerations are important: A survey *Networks* 64(3) pp 192–213

[11] Karanasios S., Allen D and Finnegan P 2015 Information systems journal special issue on: Activity theory in information systems research *Information Systems Journal* 25(3) pp 309–313

[12] Yu C, Wang Q G, Wang L, Huang J and Zhang D 2016 System identification in presence of outliers *IEEE Transactions on Cybernetics* 46(5) pp 1202–1216

[13] Gørtz I L and Nagarajan V 2016 Locating depots for capacitated vehicle routing *Networks* 68(2) pp 94–103

[14] Engemann K J and Miller H E 2015 Risk strategy and attitudinal sensitivity *Cybernetics and Systems* 46(3-4) pp 188–206

[15] Chen W, Delgrossi L, Kosch T and Saito T 2016 Automotive Networking and Applications *IEEE Communications Magazine* 54(12) pp 158
[16] Grudzień Ł and Hamrol A 2016 Information quality in design process documentation of quality management systems *International Journal of Information Management* 36(4) pp 599–606
[17] Cui Y, Wang L, Wang Y, Xia S and Wang X 2016 End-to-end coding for TCP *IEEE Network* 30(2) pp 68–73
[18] Wang S, Bie R, Zhao F, Cheng X and Choi H A 2016 Security in wearable communications *IEEE Network* 30(5) pp 61–67
[19] Barabanov A V, Markov A S and Tsirlov V L 2018 Information Security Controls against Cross-Site Request Forgery Attacks on Software Applications of Automated Systems *Journal of Physics: Conference Series* 1015(4) 042034
[20] August T, Shin H and Niculescu M F 2014 Cloud implications on software network structure and security risks *Information Systems Research* 25(3) pp 489–510

IOP Conf. Series: Materials Science and Engineering **492** (2019) 012007 doi:10.1088/1757-899X/492/1/012007

Study of the influence of the design parameters of the wet end of the gear pump on the force factors, acting in the gear pump with external gearing, using the software package MATLAB 2018a

A Sinitsyna[1,2], B Borisov[1] and O Gavryushina[1,3],

[1]Bauman Moscow State Technical University, 5 Second Baumanskaya Street, Moscow, 105005, Russian Federation

[2]E-mail: sin.anastasia23@gmail.com
[3]E-mail: gavros7@rambler.ru

Abstract. The main advantages of gear pumps with external gearing in comparison with other types of positive displacement pumps are presented. The problem of calculating the loads, acting in gear pumps with external gearing, is considered. Examples of the application of calculation methods for solving a wide range of engineering problems in the energy, shipbuilding and aircraft building, shipbuilding and automobile industries are given. The advantages of using modern calculation methods for scientific research are considered. An example of introducing a virtual laboratory workshop for conducting experiments with electrical equipment is presented. The description of the program made in the numerical computing environment for the implementation of algorithms MATLAB R2018a for the calculation of the scalar and the direction of the force acting on the driven gear is given. The advantages of the developed program are presented. The results of the study of the effect of the presence of chamfer on the sleeve and eccentricity on the scalar and the direction of the acting force are presented. The results of research obtained using the developed program have been analyzed.

1. Introduction

External gearing gear pumps are widely used in various sectors of the economy due to their indisputable advantages in comparison with other types of pumps [1, 2]. To all their major advantages the ease of use, good weight and dimensional characteristics, low cost and reliability are given. Among all the tasks of designing pumps, the task of calculating the loads acting on the driven gear stands apart, being the starting point for designers to choose the dimensions, weight and pump parameters. In that matter when referring to the pump design optimization, it is necessary to know the load as precisely as possible in order to get the optimum choice for bearing type and size. Therefore, it highlights the relevance of the study of the parameters affecting the amount and direction of the load, in order to "predict" the direction and module of the acting loads during the pump operation. Classical approaches to the study of the influence of factors changing the module and the direction of the load

IOP Conf. Series: Materials Science and Engineering **492** (2019) 012007 doi:10.1088/1757-899X/492/1/012007

during the pump operation are relatively labor-intensive and involve the manufacture of prototype models and stands, what leads to additional material costs and increasing the complexity of the research process. Furthermore, the development and manufacture of stands and the testing process itself take some time.

A modern engineering approach to the problem of studying loads and parameters affecting them involves the use of numerical modeling and techniques of programming. Examples of the application of modern computing methods for calculating the loads, acting in displacement pumps are presented in [3-4], in the flow parts of dynamic pumps in [5-7]. However, modern methods of calculation and programming technics, described in [8-10], are being actively implemented in all areas of engineering and industries - in the energy sector [11], in the automobile sector [12], in the shipbuilding and aircraft industries [13-15]. Furthermore, numerical methods of calculation and programming technics can greatly reduce the research time and difficulty, increasing their efficiency and allowing the research results demonstration more visibly. A study for a two-phase fluid in the flow part of an axial pump using the STAR CCM + software package is given in [16], a study of pump operation at various levels of cavitation in [17], a study of vortex and hysteresis effects in the intake device of a centrifugal multi-stage pump in [18]. Numerical modelling techniques are also actively used to calculate the parameters and operating conditions, dimensions and masses of diverse machines and mechanisms. Examples of the calculation and flow optimization in impeller pumps are given in [19-20]. The numerical modelling techniques have been actively disseminated in the field of experiments, allowing to reduce the material costs of the purchase and maintenance of experimental stands, simplify and expedite the process of experimentation and, in case of need, conduct it again or repeat. An example of comparing the results of a physical experiment with the results of a three-dimensional hydrodynamic simulation of the cavitation process in a centrifugal pump is given in [21]. For conducting experiments with electricity and electronics, LabVIEW virtual laboratory practice is actively applied. The LabVIEW software package provides a wide range of elements, a sort of operation modes for the studied electrical circuits; so the visual perception of information on the monitor screen is identical with laboratory installation. Due to the possibility of conducting virtual electric experiments, the risk of injury caused by failed experiments is entirely excluded, which, of course, ensures the full safety of such experiments. However, the lack of sensation of real physical experience is the significant shortage for these calculation software packages, which may negatively affect the training of specialists for working with electrical equipment in the real production conditions. Examples of the application and the description of the calculable software packages LabVIEW are given in [22-25]. Nowadays, it is safe to say that numerical modelling techniques and programming methods are crowding out the traditional methods of calculation "step by step" in all areas of science and technology.

2. Modern software for research

In the case of study, the programming method was chosen as the one for studying the influence of the design parameters for the flow part of a gear pump on the direction and magnitude of the fluid force. For this purpose, two subprograms have been developed in MATLAB language in the MATLAB R2018a software package: «For One Position of Teeth» and «For All Positions of Teeth». The «For One Position of Teeth» subprogram calculates the resulting force P_{rez}, the resulting force projections P_x and P_y (output is an array 3 × z size, where z is the number of gear teeth) for *one gear position*. The geometrical characteristics of gears (number of teeth z, radius of outside circle R_a, radius of base circle R_b, face width of tooth B), eccentricity e, angles, assigning the gear sleeve geometry, and nominal pressure P_{nom}. are given as the inputs. All angles are laid off from the horizontal (180°) clockwise, the horizontal – 0°. The «For All Positions of Teeth» subprogram doesn't work without the «For One Position of Teeth» subprogram, which is the internal loop of this subprogram. The inputs for the «For All Positions of Teeth» subprogram are the outputs of the «For One Placing of Teeth» subprogram (an array of forces' scalars 3 × z size). The corner segment $\left[\frac{-t_b}{2};\frac{t_b}{2}\right]$ is split into n segments, n+1 points of division. The parameter u, which determines the angle of rotation of the gear, is variable (figure 1):

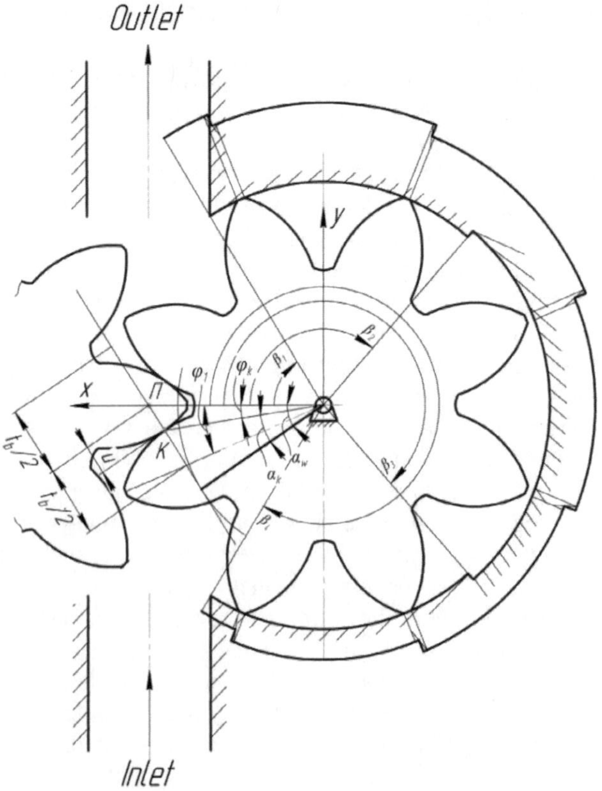

Figure 1. The input parameters

Consequently, inside the «For All Positions of Teeth» subprogram, the simulation of gears rotation on the angular segment $\left[\frac{-t_b}{2};\frac{t_b}{2}\right]$ is running, which allows to determine intermediate load values and calculating the angle of action of the maximum load on this gear with a given flow path geometry. The 2 arrays of $[1 \times (n+1)]$ size each are created inside the subprogram, which contain the angle and the scalars of the resultant force are written.

The program has the following advantages:

1) Versatility - the ability to change the inputs and to calculate the forces for pumps with different characteristics and geometry of the flow part.

2) The program is a platform for the study of the dependence of the magnitude and direction of eccentricity and chamfer on the direction and magnitude of the force acting on the gear.

3) A significant reducing the time and labour-intensity for the load's calculating.

As for the design parameters for research were chosen the chamfer on bushing and eccentricity in the pump body. It was decided to research the influence of the presence of chamfer on bushing and eccentricity in the pump body on the scalar and the direction of the fluid force. 4 variants of the geometry of the flow part were considered:

1) Without including chamfer on the bushing with eccentricity $e = 0$ µm;

2) Without including chamfer on the bushing with eccentricity $e = 16$ µm;

3) With including chamfer on the bushing with eccentricity $e = 0$ µm;

4) With including chamfer on the bushing with eccentricity $e = 16$ µm;

A variant of the bushing with chamfer is shown in the figure2:

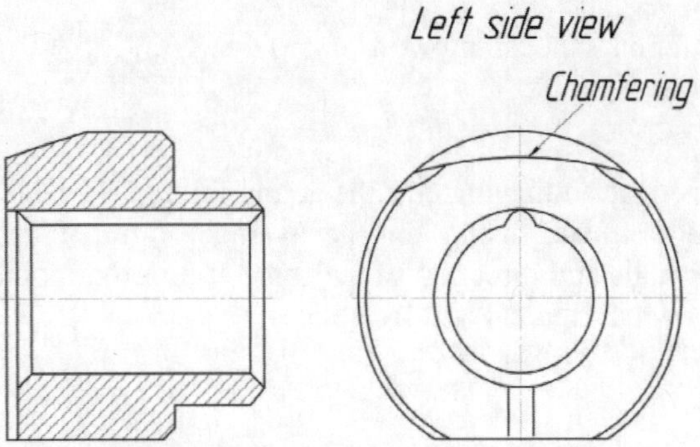

Figure 2. The bushing with chamfer

The variant of the flow part, with including chamfer on the bushing, assumes that not only the first and last teeth form an «infinite» drift between the gear and the pump body, but also an intermediate one (figure3):

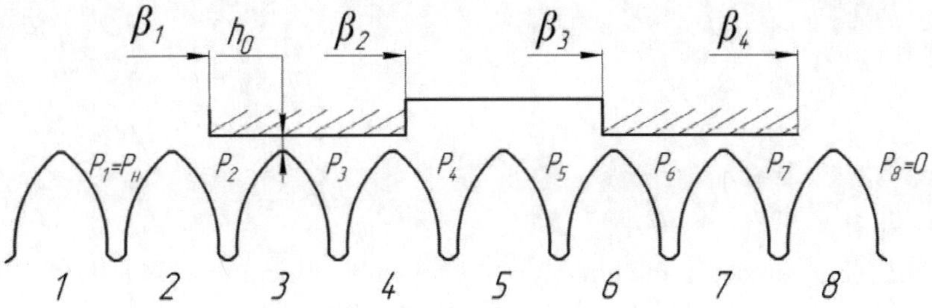

Figure 3. To the notion of «infinite» drift

The drift $h_0$between 3,4,6,7 teeth and the bushing is considered as «small», others drifts are considered as "infinitely great".

All dimensions and parameters required for inputs to the program were taken from the NSh32U pump.

The magnitude of the eccentricity e = 16 μm was taken from practice (displacement of the center of the gear during the pump operation).

The drift size $h_0 = 40$ μm was also taken from practice.

3. Results of the research

The results of the calculation of the direction and scalars of the force acting on the driven gear without including chamfer and with $e = 0$ мкмμm (var 1), without including chamfer on the bushing with eccentricity $e = 16$ μm (var 2), with including chamfer on the bushing with eccentricity $e = 0$ μm (var 3), with including chamfer on the bushing with eccentricity $e = 16$ μm (option 4) are presented in Table 1:

Table 1. The results of research

	Variant of flow part	Average angle of acting force β_{cp},°	The scalar of acting force P_{cp}, N
1	Without including chamfer on the bushing with eccentricity $e = 0$ μm	259,2	12 656
2	Without including chamfer on the bushing with eccentricity $e = 16$ μm	283,5	15 215
3	With including chamfer on the bushing with eccentricity $e = 0$ μm	264,3	11 290

IOP Conf. Series: Materials Science and Engineering **492** (2019) 012007 doi:10.1088/1757-899X/492/1/012007

4	With including chamfer on the bushing with eccentricity $e = 16$ μm	252,6	11 375

The averaging was carried out to the angle of the acting force, the scalar of the resulting force was calculated by the 2 average scalars, stand for the average angle of acting load. Diagrams with information about the magnitude and direction of the force at different points of the pitch arc, and the averaged load scalar are shown in figure 4 and figure 5:

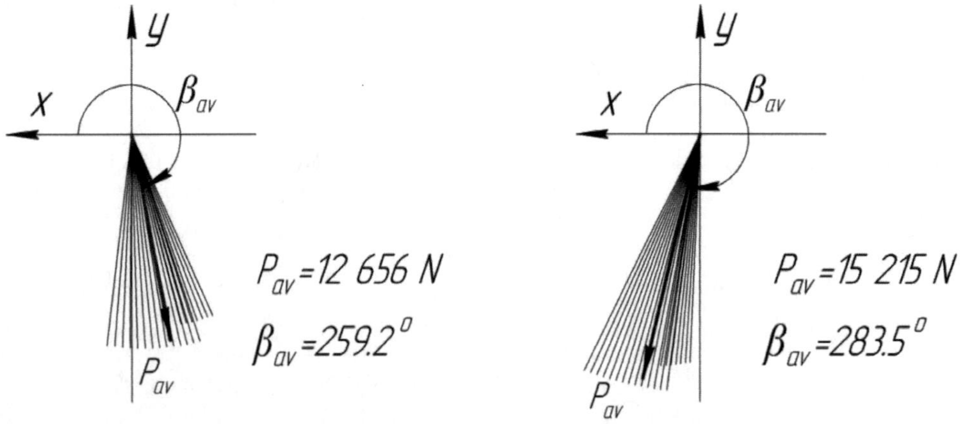

a) b)

Figure 4 . The average module and direction of load without including chamfer diagram with a) $e = 0$ μm and b)$e = 16$ μm

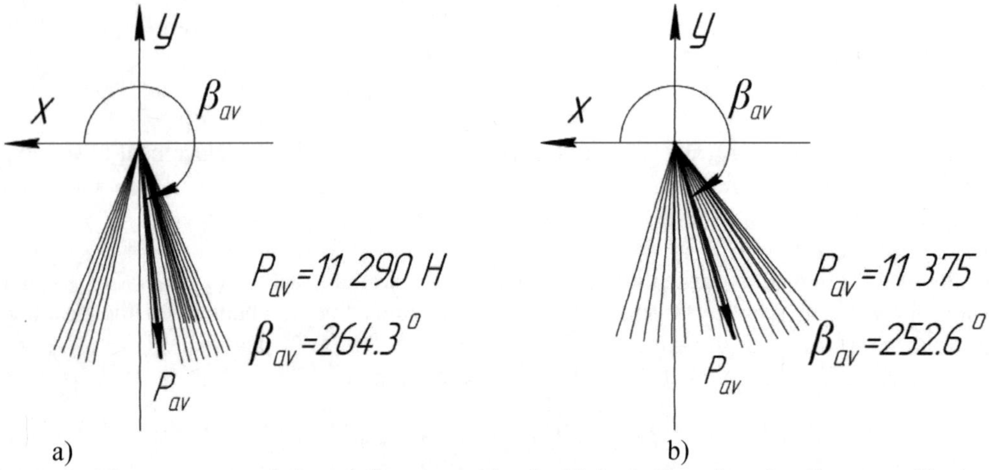

a) b)

Figure 5 . The average module and direction of load with including chamfer diagram with a) $e = 0$ μm and b)$e = 16$ μm

4. Conclusions

Based on the research results, we can make following conclusions:

1) In a radial arrangement of gears ($e = 0$ μm), the presence of unloading chamfers on the bushings reduces the fluid force acting on the driven gear by ≈ 11% .

With an eccentric arrangement of gears ($e = 16$ μm), the presence of unloading chamfers on the bushings leads to a more significant reduction by ≈ 25%

Consequently, the use of chamfers to reduce the acting force is more economically reasonable with an eccentric arrangement of gears.

2) Without chamfer on the bushing the presence of eccentricity significantly affects the scalars of force - with the eccentricity $e = 16$ μm, the force increases by $\approx 17\%$.

With chamfer on the bushing, the eccentricity has practically no effect on the changing of the scalars of force (with eccentricity, the acting force changes by $\approx 1 - 3\%$).

3) With a radial arrangement of gears, it is possible to "predict" the direction and scalar of the average acting force. With an eccentric arrangement of gears, it is very difficult to "predict" the direction of action of the average force, because the magnitude of eccentricity has a significant impact on the direction of acting force.

References

[1] Smith E C 1981Pumps *Proc. Conf. Pumping station design for the practicing engineer, (bozeman, u.s.a.: aug. 31-sep. 2, 1981)*, 1 - Fundamentals, Bozeman, U.S.A., Montana StateUniv 1981Chapter 1.4p23

[2] Yudin EMGear pumps1964*Engineering*p236

[3] BergadaJM, KumarS, WattonJ2012Axial piston pumps, new trends and development*Hydraulics: Fluid Dynamics, Mechanical Applications and Role in Engineering* pp1-158

[4] NorhirniMZ, Hamdi M, Nurmaya Musa S, Saw LH, Mardi NA, Hilman N2011Load and stress analysis for the swash plate of an axial piston pump/motor *Journal of Dynamic Systems, Measurement and Control, Transactions of the ASME* 133 (6)

[5] TanL, Shi W, Zhang D, Zhou L, Wang C2018Experiment on Fluid Force Induced Vibration for Centrifugal Pumps[Article@离心泵水力诱导激振试验研究]*NongyeJixieXuebao/Transactions of the Chinese Society for Agricultural Machinery* 49 (5)pp 179-185

[6] IsametovaME, Absadykov BN, Batyrgaliyev MK, Borovik II2018Centrifugal pump rotor dynamics study *News of the National Academy of Sciences of the Republic of Kazakhstan, Series of Geologyand Technical Sciences* 5 (431)pp226-233

[7] YanP, Chu N, WuD, Cao L, Yang S, Wu P2017Computational Fluid Dynamics- Based Pump Redesign to Improve Efficiency and Decrease UnsteadyRadial Forces*Journal of Fluids Engineering, Transactions of the ASME*139 (1)№ 011101

[8] LeeJS, JungJH, Kim BS2018Study on shape design and power performance analysis of small horizontal axis tidal turbine with duct using BEM and CFD methods*Transactions of the Korean Society of Mechanical Engineers*B 42 (3)pp 197-205

[9] Bennett E, Ivashchenko A2016Analysing a complex pump hydraulic system using STAR-CCM+® [Article@Analyser un systèmehydrauliquecomplexe de pompe à l'aide de STAR-CCM+®]*Eau, l'Industrie, les Nuisances* (393) pp61-65

[10] KerboucheK, CardosoAJM, YahiaK, MellitA2019Mathematical modelling and simulation of a directly coupled water pumping system, using a photovoltaic-fed synchronous reluctance motor*Lecture Notes in Electrical Engineering*519pp 431-437

[11] Sanchez Parra M, De Lara JaymeS, Re LopezV2012Guzman FloresRTurbogas power units retrofit: A technique to ensure improved efficiency, reliability and sustainability for smart grids*44th International Conference on Large High Voltage Electric Systems* 2012 p8

[12] Merino DE, Dubray-DemolA, GodoyE, DumurD2013Advanced control structure for an automotive heat pump system with security constraints *IFAC Proceedings Volumes (IFAC-PapersOnline)* pp725-730

[13] ZhangZ, ChengD, Meng F, He W2017Ship compartment recognition algorithm based on CAD model*HuazhongKejiDaxueXuebao (ZiranKexue Ban)/Journal of Huazhong University of Scienceand Technology (Natural Science Edition* 45 (12) pp 101-106

[14] FerreiraLA, Figueira YL, Iglesias IF, SoutoMÁ2017Offline CAD-based Robot Programming and Welding Parametrization of a Flexible and Adaptive Robotic Cell Using Enriched CAD/CAM System for Shipbuilding*Procedia Manufacturing* 11 pp 215-223

[15] Chen W-T, Huang K,Ardiansyah MN2018A mathematical programming model for aircraft leasing decisions*Journal of Air Transport Management*69 pp15-25

[16] LomakinVO2015Investigation of two-phase flow in axial-centrifugal impeller by hydrodynamic modeling methods*Proceedings of 2015 International Conference on Fluid Power and Mechatronics, FPM 2015* № 7337302 pp1204-1206

[17] MouJ, ChenY, GuY, ZhengS, Qian H2016Fluid-solid interaction characteristics of a centrifugal pump under different cavitation levels*ZhendongyuChongji/Journal of Vibration and Shock*35 (23) pp 203-208

[18] Lomakin V,CheremushkinV, ChaburkoP2018Investigation of vortex and hysteresis effects in the inlet device of a centrifugal pump*Global Fluid Power Society PhD Symposium, GFPS 2018* № 8472374

[19] GouskovAM, LomakinVO, Banin EP, Kuleshova MS2017Minimization of Hemolysis and Improvement of the Hydrodynamic Efficiency of a Circulatory SupportPump by Optimizing the Pump Flowpath*Biomedical Engineering* 51 (4)pp229-233

[20] LomakinVO, Chaburko PS, KuleshovaMS2017Multi-criteria Optimization of the Flow of a Centrifugal Pump on Energy and VibroacousticCharacteristics*Procedia Engineering* 176 pp476-482

[21] LomakinVO, KuleshovaMS,KraevaEA2015Fluid Flow in the Throttle Channel in the Presence of Cavitation*Procedia Engineering* 106 pp 27-35

[22] Vidmanov DA, Rumyantseva EI, Petrosyan OG2015Simulation of stochastic resonance effect in LabVIEW*International Siberian Conference on Control and Communications, SIBCON 2015 – Proceedings* № 7147018

[23] Titov I2014Using LabVIEW for building laboratory server: Pros and cons, design patterns, software architecturingand common pitfalls*IEEE Global Engineering Education Conference, EDUCON*№ 6826247

[24] UgaleRT, Gore SN, Chaudhari BN2018Web based remote and locally operated all in one electrical machine laboratory with data acquisitionfault diagnosis and protection*Proceedings - 2018 23rd International Conference on Electrical Machines, ICEM* № 8506977 pp1979-1985

[25] He C, LiJ, YangF, LiM2018Linear Motor Motion Control Experiment System Design Based on LabVIEW*MATEC Web of Conference* 160№ 02002

IOP Conf. Series: Materials Science and Engineering **492** (2019) 012008 doi:10.1088/1757-899X/492/1/012008

Application of the method of random additions for the simulation of radioactive contamination

E Moiseeva[1,2], V Tatarinov[1,3] and A Dutov[1,4]
[1]Bauman Moscow State Technical University, 5 Second Baumanskaya Street, Moscow, 105005, Russian Federation

[2]E-mail: kat-menshiko@yandex.ru
[3]E-mail: tatavictor@bmstu.ru
[4]E-mail: dutovas@bmstu.ru

Abstract. The article considers the possibility of using the method of random addition to simulate radioactive contamination of the area under conditions of significant uncertainty of the wind in the surface layer during the spread of radioactive substances. The analogy between the existing methods of construction of relief surfaces and the construction of a model of radioactive contamination is considered. The algorithm allowing to build the specified surfaces on known values of density of radioactive pollution in a small number of set points is described. The values characterizing forecast errors are estimated in advance. The article concludes a possibility of application of random additions algorithm for simulation of radiation contamination in the accident at radiation-hazardous objects (RHO). The considered modification of the algorithm makes it possible to build a model of the density of radioactive contamination by known fixed points.
Keywords: method of random addition, radioactive contamination, radiation-hazardous objects.

1. Introduction

Despite the significant improvement in the state of safety at radiation facilities in recent years, it is impossible to exclude the possibility of emergency situations, including the release of radioactive substances outside the sanitary protection zones. The reasons for such emissions can vary from natural disasters in the areas of radiation-hazardous facilities to sabotage and terrorist activities.

Currently existing approaches to radiation environment modeling are based on various solutions of the semi-empirical equation of turbulent diffusion [1-5]. Various solution methods give more or less satisfactory results for nuclear explosions and the spread of radioactive substances in a relatively stable surface air layer [6-11].

In case of accidents on radiation-hazardous objects, in conditions when the uncertainty of the space-time characteristics of the wind in the surface layer is high, other approaches to the modeling of radioactive contamination of the area are necessary. It is difficult to take into account a large number of independent causes acting on the bulk source of radioactive contamination in one or another analytical form [12-14].

IOP Conf. Series: Materials Science and Engineering **492** (2019) 012008 doi:10.1088/1757-899X/492/1/012008

Note that the appearance of the resulting level of contamination of the territory according to the results of measurements presented in figure 1a is similar to the measuring-altitude (hypsometric) map of the area (figure 1b).

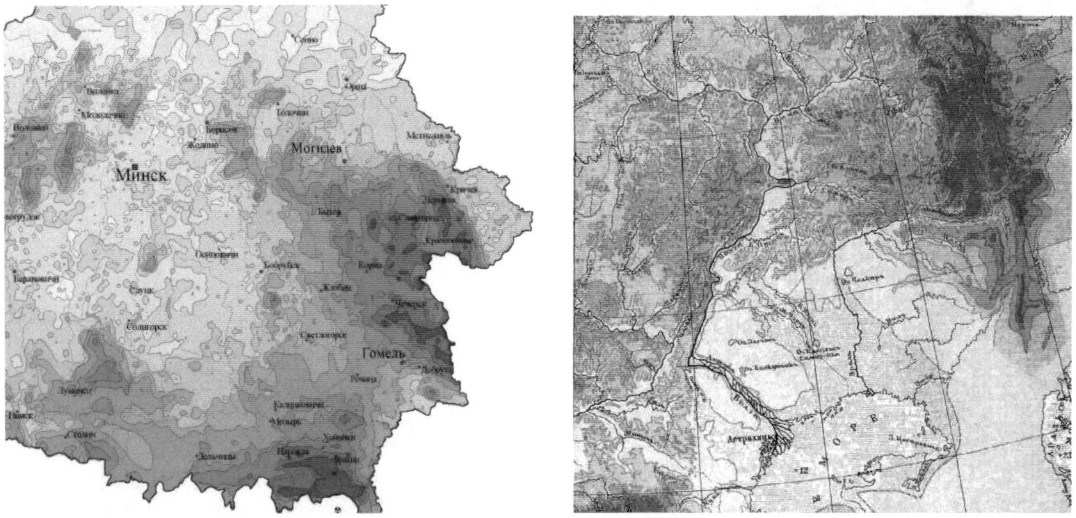

(a) (b)

Figure 1. Appearance of maps: a) radioactive contamination density of a part the territory of the Republic of Belarus ^{90}Sr, kBq/m^2, source: [4]; b) altitude, source: hypsometric map of the European part of the Russian Federation.

The reason for this similarity is clear – the formation of the relief is influenced by a large number of factors, including those associated with weathering and movement of rocks under the influence of gravity, water, etc. Accurate account of the totality of factors is impossible, but the methods of constructing surfaces that simulate the relief are known [15-17].

Their advantage is that, ignoring from the physical processes that determine the appearance of the real landscape and without giving any information about their features, they allow to describe the landscapes of a rather complex structure using a small number of variables [18-21].

2. Methods

One of these methods is the algorithm of successive random additions proposed by Foss and used in the simulation of generalized Brownian motion [22-23]. As shown by Foss, this algorithm is easy to generalize to a large number of measurements. Figure 2 shows the surfaces constructed using this algorithm, for which the program was written. Initially, the heights $Z = 0$ were fixed in four corners of the grid of 512×512 elements. Afterwards, dimensionless coordinates are used, in which the side of the area is 1. The program uses a subroutine that generates independent Gaussian random numbers $\xi_{n=0}$ with zero mean and unit variance. At the first step we get one value ξ and use it as the surface level in the center of the grid (at the $\left(\frac{1}{2},\frac{1}{2}\right)$ point). In accordance with the proposed algorithm, interpolation is performed at the second stage and elevations are found at four points with dimensionless coordinates $\left(\frac{1}{4},\frac{1}{4}\right)$, $\left(\frac{3}{4},\frac{1}{4}\right)$, $\left(\frac{1}{4},\frac{3}{4}\right)$ and $\left(\frac{3}{4},\frac{3}{4}\right)$. E.g., the elevation at the point $\left(\frac{1}{4},\frac{1}{4}\right)$ is equal to $Z\left(\frac{1}{4},\frac{1}{4}\right)=\frac{1}{4}\left(Z(0,0)+Z\left(\frac{1}{2},0\right)+Z\left(\frac{1}{2},\frac{1}{2}\right)+Z\left(0,\frac{1}{2}\right)\right)$, which would mean that the result of the interpolation is simply the arithmetic mean of the elevations at the nearest diagonal points. Elevations at the nearest two points on the border $\left(0,\frac{1}{2}\right)$ и $\left(\frac{1}{2},0\right)$ were taken equal to the arithmetic mean of elevations in the nearest corners of the area. In the developed program, changes were made to this algorithm, and the elevations at the boundary points were found as an arithmetic mean of 3 points, including $Z\left(\frac{1}{2},\frac{1}{2}\right)$. At this stage of the process, interpolated elevations were set at 13 points – five starting positions, four new interior points and four new boundary points. At the next step, 13 independent values $\xi_{n=1}$ were added to the elevations that were already calculated. In this case, Gaussian random numbers have the following dispersion at each stage:

$$\xi_n^2 = \sigma_n^2 = r^{2nH},$$

where
$r = 1/\sqrt{2}$ is the scale factor.

This procedure continues and in the next cycle points $\left(\frac{1}{2},\frac{1}{4}\right)$, $\left(\frac{1}{4},\frac{1}{2}\right)$, $\left(\frac{3}{4},\frac{1}{2}\right)$ и $\left(\frac{1}{2},\frac{3}{4}\right)$ are added. Elevations at these points are defined as the arithmetic mean of elevations at the nearest nodes, that is, at nodes that lie in directions parallel to the axes. At the points lying on the border, elevations are again determined in a special way, taking into account the central point. After each cycle, this algorithm doubles the number of points at which the elevation is set and reduces it by the factor $r = 1/\sqrt{2}$ – the distance between such points. The surfaces shown in figure 2 were obtained after 9 cycles of application of this algorithm.

The arguments describing the surface in this algorithm are the Hurst exponent H and the scale factor r. For two-dimensional surfaces, the scale factor is $1/\sqrt{2}$, and thus the only parameter describing the surface is the Hurst exponent, which will determine the “indentation” of the landscape.

3. Results

Examples of construction of three surfaces for different Hurst indicators – H are shown on the figure 2.

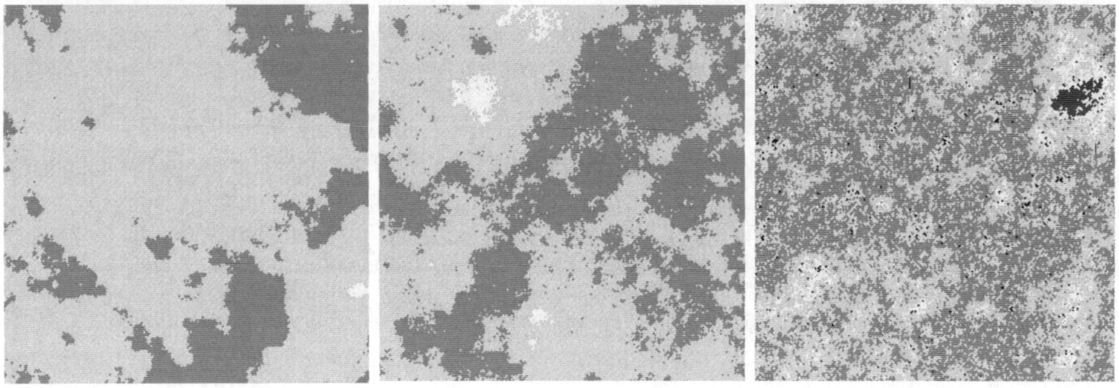

(a) (b) (c)

Figure 2. Construction of surfaces by random addition for three different values of the Hurst exponent H:a)$H = 0.9$, b)$H = 0.5$, c) $H = 0.2$.

The transition to the modeling of radioactive contamination of the area is carried out by using an obvious correlation of the density of radioactive contamination of the territory with the surface height.

As the initial data for the consideration of such a model, one could take the nature of radioactive contamination of the area, modeled in accordance with the above algorithm. If the pollution densities at some points are known, it can be assumed without loss of generality that densities at nodal points are known, for example, with relative coordinates $\left(\frac{3}{4},\frac{1}{4}\right);\left(\frac{1}{2},\frac{1}{2}\right)$ and $\left(\frac{3}{4},\frac{1}{2}\right)$, which are, respectively, 2.0; 6.0 and 2.0 relative units (if densities at other [non-nodal] points are known, this will be reduced to a change in the scale of the grid). The densities in the corners are assumed to be small, i.e. far from the density source they are almost equal to zero, which can be guaranteed by setting sufficiently large negative densities at points (0,0), (0,1), (1,0), (1,1).

Next, the construction algorithm remains the same with the following changes: for example, a point with coordinates $\left(\frac{1}{2};\frac{3}{4}\right)$ and $\left(\frac{3}{4};\frac{1}{4}\right)$ will be taken into account when finding densities in points $\left(\frac{1}{2},0\right);\left(1,\frac{1}{2}\right)$ as a weighted arithmetic mean according to the formula (1), taking into account the distance of the point from the one in which the density is calculated:

$$P(X)=\sum_{i=1}^{N}W_{(x,y)}^{i}P_i \Big/ \sum_{i=1}^{N}W_{(x,y)}^{i}, \quad (1)$$

where

$P(X)$ is the density of radioactive contamination at the design point X;

P_i are the densities at the nodal points;

$W_{(x,y)}^{i}$ are the weighting coefficients inversely proportional to the relative distance to the calculated point.

Thus, for point X the value of $X_{(0,0)}=W_{\left(\frac{1}{2},\frac{1}{2}\right)}=W_{(1,0)}=1, and\, W_{\left(\frac{3}{4},\frac{1}{2}\right)}=\dfrac{1}{\sqrt{2}/2}=\sqrt{2}$ – this factor is greater than 1 due to the point $\left(\frac{3}{4},\frac{1}{2}\right)$ being closer to point X. Similarly, it will be taken into account to find the pollution density at the point $\left(1,\frac{1}{2}\right)$ and the density value at the point $\left(\frac{1}{2},1\right)$. Then random normally distributed points (except fixed ones) are added to the found points ξ with the corresponding variance. Afterwards, the algorithm continues to find the values of the pollution density in all nodal points of the initial grid, while the specified densities remain unchanged. When at some (in this case, the 2nd) stage there will be no nodes in which densities are not determined, the dispersion decreases and the calculation is carried out according to the usual algorithm.

Due to the random nature of this model, the type of this or that implementation of the algorithm will be different from the original one that is selected. We will estimate the difference between the base model and the working one by the standard deviation σ of the power values from the base by the formula (2):

$$\sigma=\frac{\Sigma\left(P_{0i}-P_i\right)^2}{N}, \quad (2)$$

where

P_{0i} and P_i are the values of dose rates at the i-th point, respectively, the base and the working model.

IOP Conf. Series: Materials Science and Engineering **492** (2019) 012008 doi:10.1088/1757-899X/492/1/012008

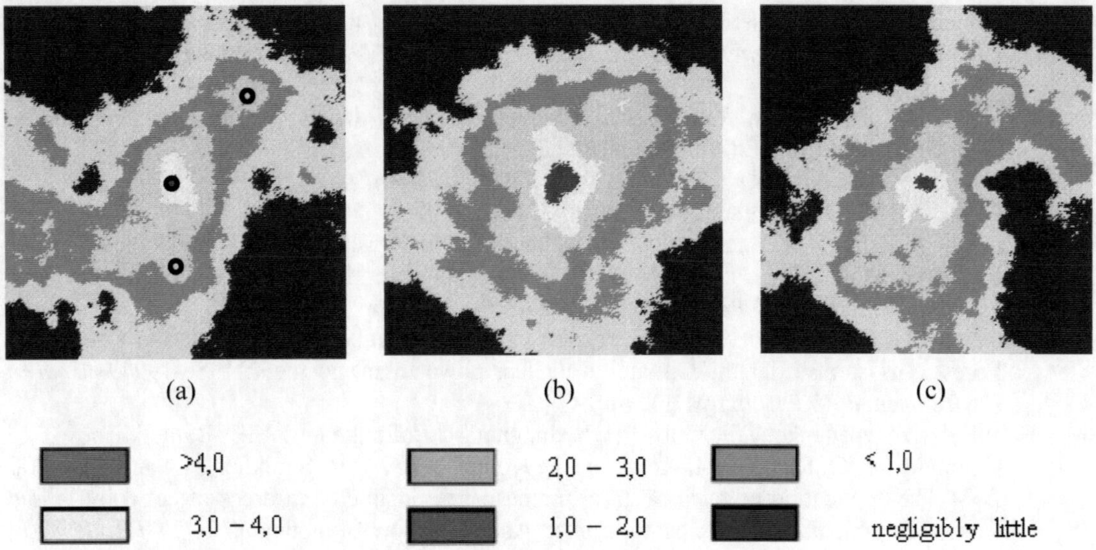

Figure 3. Results of construction of the map of density of radioactive pollution of the district, and-base model, b and c – settlement models. The number of repetitions is 9, the Hurst exponent is 0.9, a scaling factor of 1.41. Circles indicate points with a given pollution density - for point (1, 1) – 6.0 relative units, for peripheral points - 2.0 relative units.

As a result of modeling, the average deviation for 10 runs was obtained — 2.4 relative units (0.38 maximum scale) for three fixed points with Hurst exponent H = 0.9. Obviously, as the number of points with a known pollution density increases, the accuracy of the model should increase.

4. Conclusions

1. The random addition algorithm can be used to simulate radiation contamination of the area as a result of an accident at a radiation-hazardous facility (RHO).
2. The considered modification of the algorithm makes it possible to build a model of the density of radioactive contamination by known fixed points.

References

[1] Murashov M V and Panin S D 2015 Numerical modelling of contact heat transfer problem with work hardened rough surfaces *International Journal of Heat and Mass Transfer*vol 90, November, pp 72–80
[2] Orlova K N, Pietkova I R and Borovikov I F 2015 Analysis of air pollution from industrial plants by lichen indication on example of small town *IOP Conference Series: Materials Science and Engineering*vol 91 conference 1
[3] Marakhtanov M K and Okunev V S2016 mailto:okunev@power.bmstu.ruInfluence of mechanical collision macroobjects on nuclear-physical properties of components of their nuclides *Herald of the Bauman Moscow State Technical University, Series Natural Sciences*no 1 pp 61–75
[4] Pereezdchikov I V 2005 Ranking industrial plant hazards *Chemical and Petroleum Engineering*May vol 41 issue 5–6pp 340–343

[5] Mohamed A-M O and Paleologos E K2017 *Fundamentals of Geoenvironmental Engineering: Understanding Soil, Water, and Pollutant Interaction and Transport* (Book) 31 October, 688 p

[6] L'Annunziata M F 2016 *Radioactivity: Introduction and History, From the Quantum to Quarks*(Book) Second Edition 902 p

[7] Eastwood P 2008 *Particulate Emissions from Vehicles*(Book) 493 p DOI: 10.1002/9780470986516/

[8] Burtniak V, Zabulonov Y, Stokolos M, Bulavin L and Krasnoholovets V2018 The Remote Radiation Monitoring of Highly Radioactive Sports in the Chornobyl Exclusion Zone *Journal of Intelligent & Robotic Systems*June **90**Issue 3–4pp 437–442

[9] Ipatyev V, Bulavik I, Baginsky V, Goncharenko G andDvornik A1999 Forest and Chernobyl: Forest ecosystems after the Chernobyl nuclear power plant accident: 1986-1994*Journal of Environmental Radioactivity***42**(1) pp 9–38

[10] Bailiff I K, Stepanenko V F, Göksu H Y, Jungner H, Balmukhanov S.B, Balmukhanov T S, Khamidova L G, Kisilev V I, Kolyado I B, Kolizshenkov T V, Shoikhet Y N and Tsyb A F 2004 The application of retrospective luminescence dosimetry in areas affected by fallout from the semipalatinsk nuclear test site: An evaluation of potential*Health Physics***87**(6) pp 625–641

[11] Korobova E M and Romanov S L2009 A Chernobyl 137Cs contamination study as an example for the spatial structure of geochemical fields and modeling of the geochemical field structure*Chemometrics and Intelligent Laboratory Systems* 99(1) pp 1–8

[12] Masyuk V M 2003 Methods for mathematical modeling of fractal antenna arrays *4th International Conference on Antenna Theory and Techniques, ICATT 2003* 1 1239190 pp 220–223

[13] Goudie A S 2003 *Geomorphological techniques* Second edition 692 p DOI: 10.4324/9780203430590

[14] Asfahani J 2018 Estimation and mapping of radioactive heat production by aerial spectrometric gamma and fractal modeling techniques in the Syrian desert (Area-1) SyriaApplied Radiation and Isotopesvol 142 December pp 194–202

[15] Suzuki A, Fomin S, Chugunov V and Hashida T 2018 Mathematical Modeling of Non-Fickian Diffusional Mass Exchange of Radioactive Contaminants in Geological Disposal Formations *Water****10***(2) 123

[16] Huang L X, Sheng G and Li Q2006 Application of GIS and remote sensing for site characterization*Proceedings of the 11th International High Level Radioactive Waste Management Conference* IHLRWM pp 1–7

[17] Makarenko N, Karimova L and Novak M M 2003Application of fractal and morphological methods in radioecology*Health Physics***85**(3) September pp 330–342

[18] Tanaka S, Adati T, Takahashi T, Fujiwara K and Takahashi S 2018 Concentrations and biological half-life of radioactive cesium in epigeic earthworms after the Fukushima Dai-ichi Nuclear Power Plant accident *Journal of Environmental Radioactivity* 192 pp 227–232

[19] Ouyang J, Song L-J, Ma L-L, Luo M, Shao Y, Dai X-X,Yang G-S, Yang Y-G, Luo M-Y and Xu D-D 2018 Temporal variations, sources and tracer significance of Polonium-210 in the metropolitan atmosphere of Beijing, China *Atmospheric Environment*193 pp 214–223

[20] Burtniak V, Zabulonov Y, Stokolos M, Bulavin L and Krasnoholovets V 2018 Application of a territorial remote radiation monitoring system at the Chornobyl nuclear accident site *Journal of Applied Remote Sensing* **12**(4) 046007

[21] Varley A, Tyler A, Bondar Y, Hosseini A, Zabrotski V andDowdall M2018 Reconstructing the deposition environment and long-term fate of Chernobyl ^{137}Cs at the floodplain scale through mobile gamma spectrometry *Environmental Pollution*vol 240 September pp 191–199

[22] Caplin N and Willey N 2018 Ionizing radiation, higher plants, and radioprotection: From acute high doses to chronic low doses *Frontiers in Plant Science* 9 847

IOP Conf. Series: Materials Science and Engineering **492** (2019) 012008 doi:10.1088/1757-899X/492/1/012008

[23] Feder J1988 *Fractals* Springer Science & Business Media, LLC 305 p.

IOP Conf. Series: Materials Science and Engineering **492** (2019) 012009 doi:10.1088/1757-899X/492/1/012009

Jordan — Gauss successive elimination method in solution of pneumohydraulic floatation water purification problems

B Ksenofontov[1,2], K Titov[1] and A Firsova [1]

[1] Bauman Moscow State Technical University, 5 Second Baumanskaya Street, Moscow, 105005, Russian Federation

[2] E-mail:ksenofontov@bmstu.ru

Abstract. The emphasis of the article is a potential usage of Jordan – Gauss successive elimination method for reduction of a differential equation system (DES) to one differential equation and determining its solution providing all other solutions of DES. This approach is different from standard method of determining of eigen vectors and eigen values of matrix of DES, which are necessary to determine fundamental system of solutions of homogeneous differential equations of n-th order with constant coefficients denoted in norm form. In the article analytical solution is searched for in general form and can be used for other similar mathematical problems.

1. Introduction

Floatation methods are widely used for practical tasks of wastewater treatment and separation of fine suspensions in processes of mineral processing [1-10]. In this connection efficiency enhancement of floatation process may contribute to major economic benefit. Application of effective mathematical methods plays an important role at solution of many problems concerning separation processes, such as floatation process [11-30].

For example, usage of Jordan – Gauss successive elimination method for reduction of differential equation system (DES) to one differential equation and its determining its solution providing all other solutions of DES.

2. Basic concepts and suggestions

Let's consider the nature of the proposed approach in greater detail. Matrix **A** of differential equation system (DES) is given:

A:=matrix(n,n,[]):with components:

```
k1:=1/2*10^(-3):k2:=1/10*10^(-6):k3:=1/10*10^(-3):
k4:=1/10*10^(-5):
k5:=1/10*10^(-5):k6:=1/2*10^(-6):
A[1,1]:=-k1-k5:A[1,2]:=k2:A[1,3]:=k6:
A[2,1]:=k1:A[2,2]:=-k2-k3:A[2,3]:=k4:
A[3,1]:=k5:A[3,2]:=k3:A[3,3]:=-k4-k6:
```

according to which let's write down normal homogeneous DES of n-th order in vector-matrix form:

$$\frac{dX(t)}{dt} = A \cdot X \quad (1)$$

Let's define the order of DES and vector of solutions

n:=3: X:=vector(n,[seq(y[j](t),j=1..n)])

DES (1) according to [2] can be transformed to the following form

$$Z = B[s] \cdot X \quad (2)$$

Let's express vector X from (2) if inverse matrix $B^{-1}[s]$ exists

$$X = B^{-1}[s] \cdot Z \quad (3)$$

Expression (3) allows to reduce DES (1) to one differential equation (DE) of n-th order for any variable $x[s](t)$, $s = 1..n$. It can be shown that inverse matrix $B[s]$^(–1) always exists if det(A)<>0. Therefore research or solution of DES (1) should be started from determining of det(A). In case if det(A) = 0, it is necessary to find principal minor *Am in matrix A and write down new DES (4), which is equivalent to previous system (1)*

$$\frac{dX_m(t)}{dt} = Am \cdot Xm \quad (4)$$

Without loss of generality it can be assumed that principal minor *Am is placed in first* m rows and columns of matrix A, and basis solutions of (1) are first m components of vector X, which means $Xm = (x[1](t), x[2](t), ..., x[m](t))$, $m <= n$. In this case components that are left $x[m+1](t), ...,x[n](t)$ are linearly connected with basis ones. This relation is defined from linear dependence equations of A-matrix rows of DES (1):

$$\sum_{j=1}^{m} A_{j,i} \cdot u_{j,k} = A_{k+m,i}, \quad i = 1..n, \quad k = 1..n-m \quad (5)$$

$$\sum_{j=1}^{m} \frac{d}{dt} x_j(t) \cdot u_{j,k} = \frac{d}{dt} x_{k+m}, \quad k = 1..n-m$$

Where $\{u[j,k]\}$, $j = 1...m$, $k = 1...n-m$ matrix of coefficients defining linear dependence of rows of matrix A. Definition of matrix $Aa = \{A[j,i]\}$, $j = 1...m$, $i = 1...n$ allows to denote previously shown dependencies in the following vector-matrix form in Maple package

$$Aa^T \cdot col(u,k) = col(A^T, k+m), \quad (6)$$

$$\dot{X}m \cdot col(u,k) = \dot{x}_{k+m}, \quad k = 1..n-m \quad (7)$$

The last of them (7) defines relation between basis and non-basis solutions of DES. It is important to note that such transformations decrease the order of initial DES and remarkably simplify the process of its solution by reduction to one differential equation. In Maple package coefficient matrix $\{u[j,k]\}$ can be defined by means of standard tools of the package, among them is Jordan — Gauss method.

Further it is necessary to define new DES, which doesn't contain linearly dependent rows and columns and its rank is rank(Ap) = m, and then decrease its order. Decrease of order of DES, which is important for solution of differential equations [3], is carried out the following way: first m equations (5) are differentiated termwise, while other variables are excluded:

$$\dot{x}_{m+k} = \sum_{j=1}^{m} \dot{x}_j \cdot u_{j,k}, \quad k = 1..n-m \quad (8)$$

then a system of m equations with m variables is obtained, and it is necessary to introduce new variables v

$$\dot{x}_i = v_i,\ \ddot{x}_i = \dot{v}_i,\ i = 1..m \tag{9}$$

The result of termwise differentiation (5) is as follows

$$\frac{d^2}{dt^2}x_j(t) = \sum_{i=1}^{m} A_{j,i}\cdot\dot{x}_i + \sum_{i=m+1}^{n} A_{j,i}\cdot\dot{x}_i,\ j = 1..n$$

Thus, by usage of new variables, new DES is obtained

$$\dot{v}_j = \sum_{p=1}^{m}\left[A_{j,p} + \sum_{k=1}^{n-m} A_{j,m+k}\cdot u_{p,k}\right]\cdot v_p,\ j = 1..m \tag{10}$$

This DES will be equivalent to the initial one in case of fulfillment of conditions (8), defined by new variables

$$v_{m+k} = \sum_{j=1}^{m} v_j\cdot u_{j,k},\ k = 1..n-m$$

Consequently, DES coefficients (10) are defined by formulae:

$$Ap[j,p] := A[j,p] + \sum_{k=1}^{n-m} A[j,m+k]\cdot u[p,k],\ j = 1..m,\ p = 1..m \tag{11}$$

Here, as it has been mentioned above, determinant of matrix *Ap*is not equal to zero, while its rank is *m* which can be confirmed by direct verification.

Here the process of DES (10) solution can be started. Let's introduce basis vector V of new variables $v[j](t)$, $j = 1$ .. *m*and derivatives of V

dV:=map(diff,[seq(v[j](t),j=1..m)],t)

The formula defining right-hand side of DES (10) is **SD:=evalm(Ap&*V).** DES in new variables, which relation to old variables is defined by formulas (9)

sys1:=seq(dV[k]=SD[k],k=1..m); fcn1:={seq(v[j](t),j=1..m)}:

System of differential equations defining one of the new variables, for example, $v[s](t)$, $s = 1$, is

$$\begin{pmatrix} v_s{}' \\ v_s{}'' \\ \ldots \\ v_s{}^{(m)} \end{pmatrix} = \begin{pmatrix} row(Ap,s)\cdot Ap^0 \\ row(Ap,s)\cdot Ap^1 \\ \ldots \\ row(Ap,s)\cdot Ap^{m-1} \end{pmatrix}\cdot\begin{pmatrix} v_1 \\ v_2 \\ \ldots \\ v_m \end{pmatrix},\ s = 1..m \tag{12}$$

The system (12) allows to obtain a differential equation for each variable $v[s](t)$ and solve it.

When inverse matrix *Bb* is known, from (12) differential equation defining components of vector V by the following formula is obtained

$$\begin{pmatrix} v_1 \\ v_2 \\ \ldots \\ v_m \end{pmatrix} = Bb[s]\cdot\begin{pmatrix} v_s{}' \\ v_s{}'' \\ \ldots \\ v_s{}^{(m)} \end{pmatrix},\ s = 1. \tag{13}$$

or inserting expressions (9) into (13), formula defining calculations of components of vector *Xm* belonging to initial DES (5) is obtained as follows

$$\begin{pmatrix} \dot{x}_1 \\ \dot{x}_2 \\ \dots \\ \dot{x}_m \end{pmatrix} = Bb[s] \cdot \begin{pmatrix} \dot{x}_s \\ \dots \\ \ddot{x}_s \\ \dots \\ x_s^{(m+1)} \end{pmatrix}, \; s = 1..m \qquad (14)$$

Thus, for $s = 1$ differential equation of order m+1 is obtained for defining component $x[1](t)$. Let's firstly solve equation (13) using derivatives vector

dVV:=[seq(diff(v[1](t),t$j),j=1..m)]; evalm(Bb&*dVV)

and transform the equation as follows

ur:=V[1]=evalm(row(Bb,1)&*dVV);

Let's set up a task for solution of *ur*

resh:=dsolve(ur); rhs(resh);v[1](t):

On integrating solution $v[1](t)$ $xm[1](t)$ is obtained.
Solutions $v[1](t)$ and $xm[1](t)$ differ by a constant, for example, H1.
Let's define a vector of constants *h*, which is necessary for notation of DES solutions

$$xm_1 := t \to \sum_{i=1}^{m+1} h_i e^{rs_i t}$$

Other solutions are defined on the base of previously found $xm[1](t)$ with addition of a constant, occurring while integration and decreasing the order of differential equation by one.

Let's introduce a vector of solutions of the initial DES with new name *xM.* Consistency conditions (8), defining relation between non-basis variables and basis variables, are also applied to solutions of the initial DES and provide (n-m) solutions more with imperative introduction of constants $hn[i]$, $i = 1...n - m$ on the same basis

Let's calculate these solutions:

$$xM_{m+k} := \left(\sum_{j=1}^{m} xM_j U_{j,k} \right) + hn_k$$

Calculated $xM[i]$, i=1...n are valid with an accuracy to constants, but cannot be considered as solutions of the initial DES. "Excessive" constants $hb[k]$, $hn[i]$ should be expressed through $h[3]$, according to substitution of solutions $xM[i]$ in the initial DES.

As the result equations defining relations between constants are obtained, from which $(n - 1)$ constants are expressed through constant $h[3]$:

$$p6 := \left[\left[hb_1 = \frac{15020}{1003} h_3, \; hn_1 = \frac{1018025}{1003} h_3 \right] \right]$$

Defined values of constants are used in vector of solutions *xM.*

Solutions $y[i](t)$, $i = 1..n$, obtained in general form are valid for the initial DES.

Vectors *h*, *hb*stay unknown because they depend on vector of initial conditions*Cn* of the initial *DES.* That particular vector is defined and on the base of this vectorthe constants $h[i]$,$hb[j]$ are calculated, as it is shown below.

Let's introduce a vector of initial (or boundary conditions if they are determined) condition for each solution from $y[i](t)$, $i = 1...n$ for the initial DES:

Cn:=vector(n,[]):

Let's determine relation between initial conditions *Cn*[*i*], *i*=1...*n* and constants *h*[*i*], *hb*[j]. In order to do this *n*equations are defined by using $t = 0$ in the system determined by vector *xM*

s4:=seq(Cn[i]=simplify(subs(t=0,xM[i])),i=1..n):

from which we determine

s2:=solve({s4},[seq(h[i],i=1..m+1),seq(hb[j],j=1..n-m-1)]):

3. Results processing and discussion

Let's proceed to plot construction for DES. Particular values for initial conditions are defined:

Cn[1]:=100:Cn[2]:=0:Cn[3]:=0:

and constants are calculated

$$s3 := \left[\left[h_1 = 100.024,\ h_2 = -0.123,\ h_3 = 0.098\right]\right]$$

Determined constants are used in obtained solutions defined by vector *xM*.

Obtained solutions are expressed in digitalized form and plots are constructed (figures 1-4).

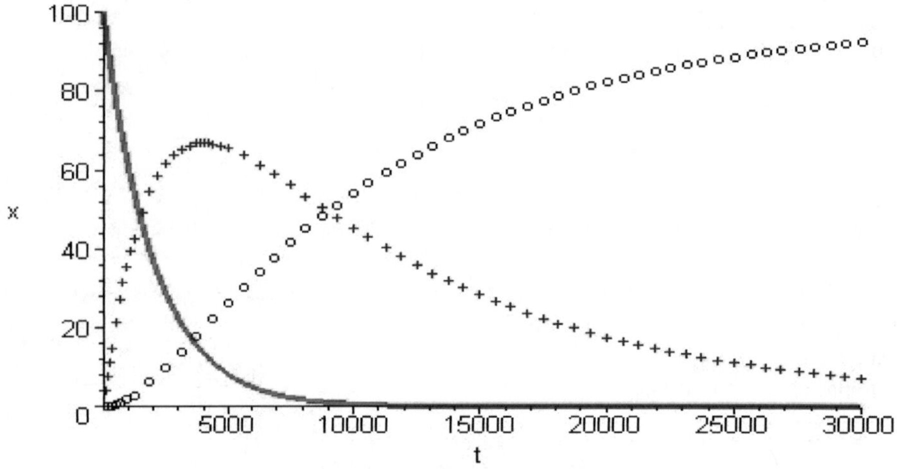

Figure 1. Solutions plot

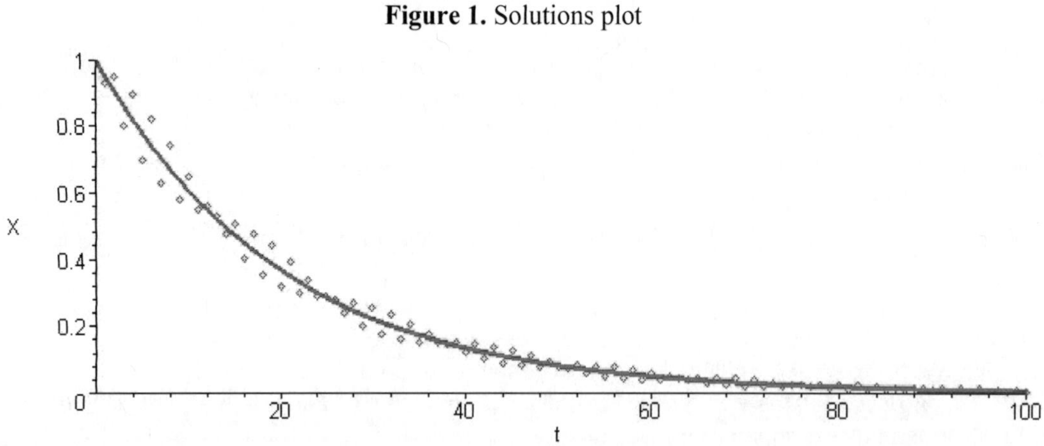

Figure 2. Analytical kinetic dependence of floatation process of fat-containing wastewaters purification: solid curve (red) — extraction of pollution particles out of water (%) to time (sec) and its experimental approve shown in discrete points.

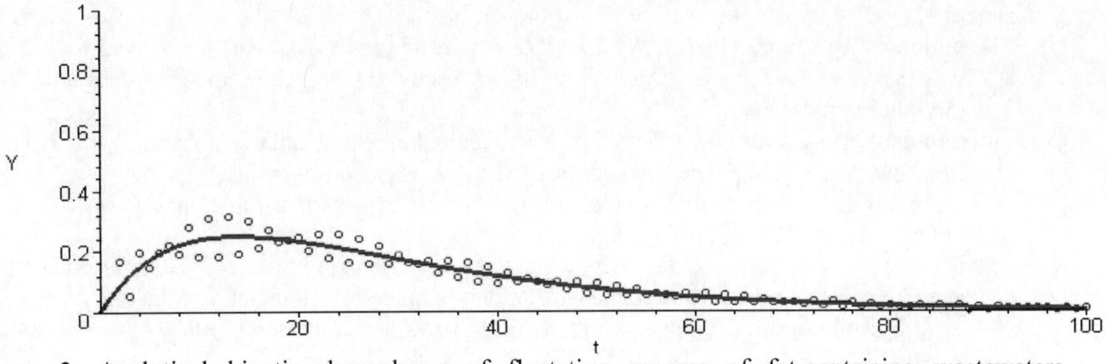

Figure 3. Analytical kinetic dependence of floatation process of fat-containing wastewaters purification: solid curve (blue) — formation of floatation complexes particle-bubble (%) to time (sec) and its experimental approve shown in discrete points.

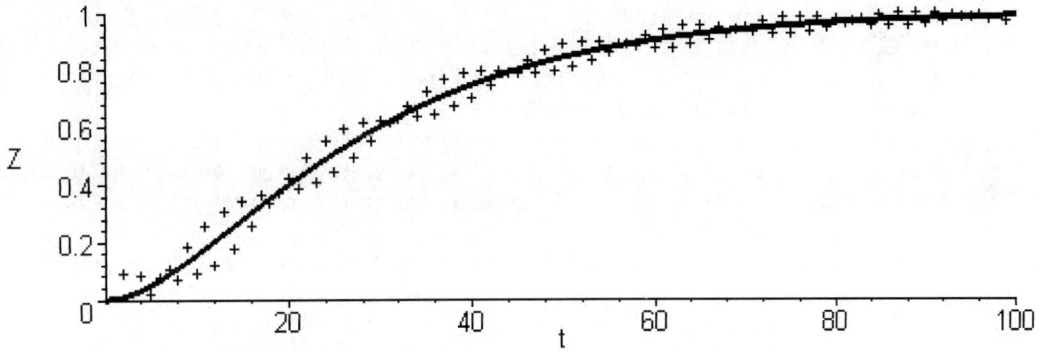

Figure 4. Analytical kinetic dependence of floatation process of fat-containing wastewaters purification: solid curve (black) — formation froth product (%) to time (sec) and its experimental approve shown in discrete points.

Each curve shown on figure 1 was approved by results of experiments conducted during floatation purification of fat-containing wastewaters. Plots of experimental calculations are shown below on figures 2-4 in reduced scale, which, however, allows to estimate accuracy of coincidence between them and theoretical curves and to draw conclusion about adequacy of mathematical model used.

Difference between experimental and theoretical data does not exceed 5-7%, which is affordable for process of wastewater floatation purification. Obtained data allows to estimate dimensions of flotators with adequate accuracy on the base of general parameter which is floatation time.

4. Conclusion

Thus, proposed approach allows to obtain basic kinetic dependencies of floatation process of separation non-homogeneous systems, in particular wastewaters and fine suspensions. Proposed approach is approved during the process of design of floatation units for purification of wastewaters of different manufactures. Upon that, the results of tests show that theoretical data are close enough to experimental data.

References

[1] Ksenofontov B S and Titov K V 2017 *Ochistka stochnyh vod: komp'yuternye tekhnologii v reshenii zadach flotacii Uchebnoe posobie* Moscow ID “FORUM”, INFRA-M Publs 240 p (Vysshee obrazovanie)

[2] Ksenofontov B S and Titov K V 2017 Reshenie kraevykh zadach ochistki vody flotaciej Moscow *Ekologiya promyshlennogo proizvodstva* issue 4 pp 31–39

[3] Ksenofontov B S 2015 *Ochistka stochnyh vod: kinetika flotacii i flotokombajny* Moscow ID “FORUM”, INFRA-M Publs 256 p

[4] Skorohodov V F, Nikitin R M, Biryukov V V and Stepannikova A S 2015 Vychislitel'nyj eksperiment nad model'yu processsa razdeleniy amineralov flotacionnym metodom. Sbornik Trudov Kol'skogo nauchnogo centra RAN 3/2015(29) *Informacionnyete khnologii* issue 6 Apatity KNC RAN Publ pp 185–211

[5] Titov K V 2015 *Komp'yuternaya matematika Ucheb. posobie* Moscow RIOR, INFRA-M Publs 282 p + Dop. materialy Available at: http://www.znanium.com (Vysshee obrazovanie) DOI: 10.12737/5954

[6] Bowen R 1976 *Theory of mixtures* Continuum Physics New York Academic Press
Doroodchi E, Galvin K P and Fletcher D F 2005 The influence of inclined plates on expansion behaviour of solid suspensions in a liquid fluidized bed — a computational fluid dynamics study *Powder Technology* **156** no 2

[7] Drew D and Lahey R 1993 *In particulate two-phase flow* Butterworth-Heinemann Boston
Hedvall P., Nordin M. 2002 Plant Designer: A crushing and screening modeling tool. *Mineral processing plant design, practice and control:Proceedings* Society for mining, metallurgy and exploration, Incvol 1

[8] Harris M, Runge K, Whiten W and Morrison R 2002 JK–SimFloat as a practical tool for flotation process design and optimization. *Mineral processing plant design, practice and control: Proceedings* Society for mining, metallurgy and exploration, Incvol 1

[9] Leeuwner M J, Eksteen J J 2008 Computational fluid dynamic modeling of two phase flow in hydrocyclone*The Journal of the Southern African Institute of Mining and Metallurgy* **108** no 4

[10] Shukla S K, Shukla P and Ghosh P 2011 Evaluation of numerical schemes for dispersed phase modeling of cyclone separators *Engineering Applications of Computational Fluid Mechanics* **5** no 2

[11] Algazin O D and Kopayev A V 2017 Filtration of liquid in a non-uniform layer with the filtration coefficient changing under the square law *Engineering Journal: Science and Innovation* issue 6 DOI: 10.18698/2308-6033-2017-6-1624

[12] Algazin O Dand Kopayev A V 2015 The solution of the mixed regional task for Laplace's equation in a multidimensional infinite layer *Herald of the Bauman Moscow State Technical University. Series Natural Sciences* no 1 (58) pp 3–13

[13] Algazin O D 2016 The exact solution of a task of Dirikhle for the elliptic equation degenerating on border like Trikomi — Keldysha in a half-space *Herald of the Bauman Moscow State Technical University. Series Natural Sciences* no 5 pp 4–17
DOI: 10.18698/1812-3368-2016-5-4-17

[14] Algazin O D 2018 Construction by method of similarity of the fundamental solution of a task Dirikhle for the equation like Keldysh in a half-space *Herald of the Bauman Moscow State Technical University. Series Natural Sciences* no 1 pp 4–15
DOI: 10.18698/1812-3368-2018-1-4-15

[15] Dimitriyenko Yu I, Gubareva E A, Markevich M N and Sborschikov S V 2016 Mathematical modeling of dielectric properties of the nanostructured composite materials by method of asymptotic averaging *Herald of the Bauman Moscow State Technical University. Series Natural Sciences* no 1 (64) pp 76–89

[16] Baskakov V D, Mark O V, Karnaukhov K A and Tarasov V A 2016 Mathematical modeling of process of impact of flat streams of ideal liquid *Herald of the Bauman Moscow State Technical University. Series Natural Sciences* no 2 (65) pp 79–90

[17] Hafizov F Sh, Alexandrov A A, Sushchev S P, Abutalipova E M and Hafizes I F 2017 Modeling and method of calculation of the cavitational and vortex office *Herald of the Bauman Moscow State Technical University. Series Natural Sciences* no 1 (70) pp 78-91

[18] Dimitriyenko Yu I and Kashkarov A I 2002 A final and element method for calculation of effective characteristics of the spatial reinforced composites *Herald of the Bauman Moscow State Technical University. Series Natural Sciences* no 2 pp 95–108

[19] Dimitriyenko Yu I, Gubareva E A and Sborschikov S V 2014 The asymptotic theory of constructive orthotropic plates with two-periodic structure *Mathematical Modeling and Computational Methods* no 1 pp 36–57

[20] Dimitriyenko Yu I, Gubareva E A and Yakovlev D O 2014 Asymptotic theory of viscoelasticity of multilayered thin composite plates *Science and Education Bauman MSTU Electron magazine* no 1 pp 359–382 DOI: 10.7463/1014.0730105

[21] Dimitriyenko Yu I, Yurin Yu V and Gubareva E A 2014 Asymptotic theory of thermocreep of multilayered thin plates *Mathematical Modeling and Computational Methods* no 4 pp 36–57

[22] Wills B A and Finch J 2015 *Wills' mineral processing technology: an introduction to the practical aspects of ore treatment and mineral recovery* Butterworth-Heinemann

[23] Grammatika M and Zimmerman W B 2001 Microhydrodynamics of flotation processes in the sea surface layer *Dynamics of Atmospheres and Oceans* **34**(2-4) 327–348

[24] Van der Westhuizen A P and Deglon D A 2007 Evaluation of solids suspension in a pilot-scale mechanical flotation cell The critical impeller speed. *Minerals Engineering* 20(3) 233–240

[25] Szpyrkowicz L 2005 Hydrodynamic effects on the performance of electro-coagulation/electro-flotation for the removal of dyes from textile wastewater *Industrial & engineering chemistry research* 44(20) 7844–7853

[26] Yianatos J B 2007 Fluid flow and kinetic modelling in flotation related processes: columns and mechanically agitated cells—a review *Chemical Engineering Research and Design* 85(12) 1591–1603

[27] Rodrigues R T and Rubio J. 2007 DAF-dissolved air flotation: Potential applications in the mining and mineral processing industry *International Journal of Mineral Processing* 82(1) 1–13

[28] Miettinen T, Ralston J and Fornasiero D 2010 The limits of fine particle flotation *Minerals Engineering* 23(5) 420–437

[29] Grattoni C, Moosai R and Dawe R A 2003 Photographic observations showing spreading and non-spreading of oil on gas bubbles of relevance to gas flotation for oily wastewater cleanup. *Colloids and Surfaces* A*: Physicochemical and Engineering Aspects* **214**(1-3) 151–155

[30] Santander M, Rodrigues R T and Rubio J 2011 Modified jet flotation in oil (petroleum) emulsion/water separations. *Colloids and Surfaces* A: *Physicochemical and Engineering Aspects*, **375**(1-3), 237–244

IOP Conf. Series: Materials Science and Engineering **492** (2019) 012010 doi:10.1088/1757-899X/492/1/012010

Methods and principles of forecasting risks of accidents on hazardous hydrotechnical objects

R Taranov[1,3] **and T Lebskaia**[1,2,4]

[1] Bauman Moscow State Technical University, 5 Second Baumanskaya Street, Moscow, 105005, Russian Federation
[2] FSBI VNII GOCHs (FC), Moscow, Russian Federation

[3] E-mail: rtaranov@bmstu.ru
[4]E-mail: tutiana95@mail.ru

Abstract. Risk analysis of accidents and emergencies on hazardous objects such as hydrotechnical objects is becoming increasingly important against the backdrop of annually increasing number of technological emergencies and significant increase of the damage they cause to the population and the environment. Methods of risk forecasting that take into consideration the object condition are highly important in this matter, but these methods have not been studied. Moreover, current forecasting methods do not take into consideration the change of object condition and its operating conditions over time. Hence creating modern approaches and methods of accident and emergency risk forecasting is an actual task aimed at increasing the level of safety for hazardous industrial objects. This paper deals with a method of forecasting accident situations and emergencies on hazardous industrial objects; the method takes into account the operating condition as well as the change of object condition, i.e. wear and tear. The presented method is a linear midterm forecast with the forecasting horizon of 1 to 5 years. The proposed method is based on the object condition statistical data. This methods can not only be used as a tool for forecasting and preventing accidents and emergencies at hazardous industrial objects, but also as a tool for optimizing expenses on developing industrial objects and ensuring its required level of safety for the population and environment.

1. Introduction

The potential of Russian technosphere for meeting the society's demands has drastically reduced over the last decade. There has been a decrease in investment on developing technosphere or even maintaining its current condition.

The percentage distribution of causes that contributed the most to the growth of the number of emergencies on hazardous industrial objects, including hydrotechnical installations, from 2009 to 2017 is shown in Figure 1 [1]:

IOP Conf. Series: Materials Science and Engineering **492** (2019) 012010 doi:10.1088/1757-899X/492/1/012010

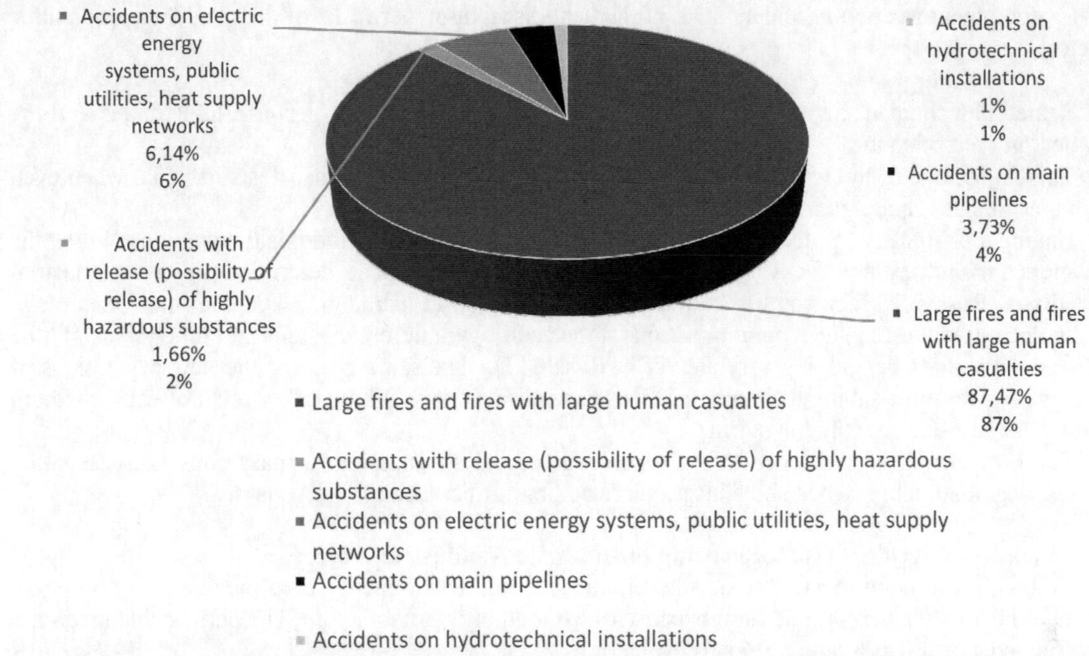

Figure 1. Percentage ratio of the accidents on hazardous industrial objects in the Russian Federation from 2009 to 2017.

The crisis phenomena in Russian technosphere and the emerging negative trends have several interrelated aspects.

There are several factors causing accidents in Russia.

Firstly, the process of replacing retired fixed assets is being disrupted. This manifests itself in the growing wear rate, especially in agriculture, on transport, in housing and utilities sector. It is caused by the sharp decrease of the expenses on replacing retired fixed assets caused by reduced funding [2 – 4].

Secondly, the infrastructure (grid lines, roads, pipelines, technological fleet in nearly all industries, utilities sector) is the accident condition. The number of dilapidated houses is increasing with each year because of late repair. This leads to deterioration of living conditions and poses a threat to people's health and safety.

Thirdly, the country's pipelines are in the similar condition, which leads to growing number of accidents caused by the spill of oil and oil products causing large economic and environmental damage.

Lastly, many of the hydrotechnical installations are high risk objects as they have been operated without maintenance for more than 50 years. Increase of their size and numbers makes vital the problems of preventing large-scale environmental disasters and ensuring safety of the people living below waterfronts and dams.

Hydrodynamic accidents are unexpected events tightly linked to significant damage to the hydraulic works (lock, dam) and uncontrollable displacement of the large masses of water leading to large area flooding and damage. The accident scale is completely determined by the conditions and parameters of the waterworks installation, the reservoir water volume, dam failure level and pattern.

There is a problem of sustaining living conditions for the people and operation of the power system objects in Russia in the autumn-winter season.

The main causes of such accidents are the following:

– unsatisfactory preparation of the power system objects for operation in winter conditions;

IOP Conf. Series: Materials Science and Engineering **492** (2019) 012010 doi:10.1088/1757-899X/492/1/012010

– equipment wear and accident state of the heat distribution networks of the residential and public sector;

– lack of energy supplies.

Sometimes "human factor" took place – local authorities could not perform their duties and the matter "got out of control".

This is just one of the ways the growing technosphere vulnerability manifests itself in, when even minor accidents require taking measures on the national level.

Having no strategy of technological development leads to a paradoxical situation. While the developed countries are focusing efforts on developing areas that determine the post-industrial paradigm, Russia is focusing on maintaining the industrial paradigm, keeping the "underdog" technological niches. IT, telecommunications, biotechnology, micromechanics and other areas of "the new economy" are developing very slowly in Russia [3]. The same goes for the new generations of energy- and resource-saving technologies. This can cause a crisis and the risk growth over the medium term and a disaster over the long term.

The growing interdependence of the technological and social spheres, mass consciousness value scales, degradation of science and education are adding further technological risks.

2. Principles of accident risk forecasting for hazardous industrial objects

From the physics point of view an accident is the situation when one or more parameters of a system changes beyond the permissible limit because of external or internal factors. The permissible levels are determined a priori, as a rule, by expert evaluation.

The violation of permissible parameter limits are determined by both subjective and objective reasons.

Objective reasons include: ignorance of the processes (natural, technical, social, etc.) by the society in general or by the individual making a decision in particular. As the society evolves this factor is going to diminish until it eventually becomes zero. This is a lengthy process and requires large expenses, mainly, by the state, on science and education [5–7].

Our present level of knowledge is not allowing us to fully evaluate the impact these factors have on the accident probability and to obtain a reliable forecast on implementing them. Therefore from the point of view of these forecasts the society will long be operating in the range of marginal estimates of the system parameters going beyond the permissible limits.

Subjective reasons include: orderliness and discipline levels of the society in general and individuals in particular, design and operation errors ("human factor"). Here the success in eliminating these reasons is achieved much faster and at lower cost.

The problem of accident prevention and control, i.e. the task of returning system parameters within permissible limits, is tightly linked to the abovementioned reasons that caused the accident. Thus understanding the nature of reasons potentially causing accidents in a given system is enabling development of adequate prevention measures aimed at preventing an accident and minimize time spent on eliminating its consequences.

The approach based on synergetics, theories of nonlinear dynamics and time series is very interesting. Often times the approach called technology transfer is used. The approach comes down to using developed and tested methods from one area (for instance, strength and fracture mechanics, automatic control theory, etc.) to describe phenomena in another area [5, 8, 9]. However all existing and emerging methods are based on computer simulation of the processes and phenomena being studied.

The possibilities of setting up a computer experiment and conducting a detailed simulation modeling have significantly expanded capabilities of researchers. It is now possible to explore scores and hundreds of simulated disasters in order to find ways of preventing the real ones. Besides, there are two interrelated factors. Firstly, computer models help create teaching programs, simulators, employee training complexes that help acting effectively in crisis situations. Secondly, they often show what the kind and volume of information is ought to be collected.

These models are closer to fundamental sciences than to engineering; however, without such models the risk of the decisions being taken would be greater because the real threats would be aggravated by ignorance. However the existing model toolbox is not adequate.

Fundamental changes in safety engineering that have taken place over the last decades are making the problem of creating new generations of models for modeling and forecasting crises and disasters very relevant. Solving any practical tasks using probability theory requires formulating it in probabilistic terms. But formulating a problem in probabilistic terms comes down to building a mathematical model of the problem. There are many ways of building mathematical models of such problems, but in each case it is required to construct a simplified object model and preserving the adequacy to the real object. Very often the model itself is so complex that even obtaining a solution causes problems of purely mathematical nature.

Sometimes the model adequacy can be verified empirically, but as a rule this kind of verification is practically unfeasible and requires involving a wide range of specialists so that the mechanism postulated in the model is in congruence with the studied phenomenon. The level of consistency between the mathematical solution and the phenomenon character depends on the model adequacy, but it may be inconsistent with the phenomenon itself, which will require further corrections of the mathematical model of the phenomenon.

Accidents can be classified by the cause (j) in the following way:

Accidents of freight and passenger trains; accidents of freight and passenger ships; air accidents; large car accidents; accidents on main and in-field oil pipelines; accidents on main gas pipelines; fires (explosions) in buildings, utilities, technological equipment of industrial objects; fires (explosions) on agricultural objects; fires (explosions) on fish industry transport and ships, on main oil and gas products main pipelines; fires (explosions) in mines and metro system; fires (explosions) in residential and public buildings; detection (loss) of unexploded ordnance and explosives; accidents with emission (possibility of emission) of chemical hazardous substances; accidents with radioactive release or possibility thereof; accidents with release of biologically hazardous substances or possibility thereof; sudden collapse of factory buildings, installations, rock fall; collapse of residential and public buildings; accidents on power electric systems; accidents on public utilities; accidents on heat supply networks in the cold time of the year; hydrodynamic accidents.

The presented list of accident causes is comprehensive enough and encompasses all spheres of technical activity; however each of the causes can include a range of accidents for different areas of the economy. A more diverse database of accidents, their causes and cause-and-effect relations is required in order to analyze accident causes, forecast them and formulate prevention measures.

The work on accident causes and accident forecasting has been conducted since the appearance of the technosphere; they are presented in depth in [1, 10 – 14].

The event of an occurring emergency (event E) is only possible if its cause is substantiated (event E_1). Probability of event superposition $E \cap E_1$ is given by (1):

$$\boldsymbol{P}\{\boldsymbol{E} \cap \boldsymbol{E}_1\} = \boldsymbol{P}\{\boldsymbol{E}_1\} \cdot \boldsymbol{P}\{\boldsymbol{E} / \boldsymbol{E}_1\} \quad (1)$$

where

$P\{E_1\}$ –probability of accident cause substantiation (hazard, fire, explosion, etc.);

$P\{E / E_1\}$ –accident situation probability if the accident cause is substantiated.

In order to simplify notation, accident situation probability for i-th hazardous industrial object within one year$P_i\{(E \cap E_1)/\text{one year}\}$will be denoted as P_{iE}.

Accident substantiation is only possible in the following circumstances:

1. Accident cause is violation of permissible limits by one or several system parameters [15] caused by initial states not accounted in design basis accidents followed by additional safety system failure compared to design basis accidents and erroneous decisions by the personnel leading to dire consequences.

2. Accident can be caused by violation of different requirements (construction, emergency prevention, environmental, safety engineering, etc.) that ensure safety of the object territory, of the population living near a hazardous technical object and the personnel or there are ignorance areas because of which the necessary requirement have not been formulated.

However numerical values of accident probability do not fully characterize the event, the most important characteristic of any accident situation is the damage (human lives, human health, economic, moral, material damage) caused by it. Therefore the term of risk is used to characterize accident situations.

2.1 The concept of risk and its cause-and-effect relationships

Right now there is no rigorous definition of such terms as: risk, risk of accident situation, forecasting risk of accident situations. As a rule, the authors use this terminology in the context that fits best this or that problem. In the papers [1, 11, 16, 17] there are several varying definitions of risk and risk of accident situation; however, the most rigorous definition from the mathematics standpoint is given in [1]. According to the definition, risk is the measure of danger that combines the probability of harmful event and the impact this event has (losses and casualties). One can evaluate the measure of danger and make decisions whether this or that measure should be taken by making combinations of these elementary events and choosing a decision rule.

We are going to adopt this definition of risk. Going further, the risk of accident situation is going to be defined as a measure of danger including the probability of the accident situation and the expected value of damage if the accident situation takes place. It is going to be denoted as follows(2):

$$\boldsymbol{R_{iE}} = \{\boldsymbol{P_{iE}}, \boldsymbol{S_{iY}}\} \quad (2)$$

where

P_{iE} –accident situation probability within one year;

S_{iY} – expected value of damage if the accident situation takes place.

The way these events are combined for risk analysis depends on the accident situation type, it's conditions and other factors.

This definition of risk of accident situation R_{iE} is formulated for a single technical object; according to [14] it is defined as a differential risk of an accident situation. As a rule, generalized indicators or integral risks of accident situations are used in statistical records and government reports.

Thus if the integral risk of accident situation is going to be used, the probability of an accident situation has to be replaced by another indicator – expected value of the number of the j-th type accident situations within one year. Hence the integral risk of the j-th type accident situations (RI_{jE}) is going to be equal to the sum as i runs from 1 to N of the i-th differential risk:

$$\boldsymbol{RI_{jE}} = \sum\{\boldsymbol{P_{ijE}}, \boldsymbol{S_{ijY}}\} = \{\boldsymbol{MP_{ijE}}, \boldsymbol{MS_{ijY}}\} \quad (3)$$

where

$i \in [1,N]$, N– number of the technical objects of the j-th type.

2.2 Methods of accident situation risk forecasting

It is necessary to be capable of forecasting accident situationrisks in order to create an efficient accident situationprevention system. Moreover, forecasts have to be as accurate as possible and have the longest possible horizons. This is where a very common contradiction appears: the longer the forecasting horizon is, the less accurate the forecast is. A forecast has to be optimized, i.e. it is necessary to find the optimal combination of parameters (forecasting horizon and accuracy) depending on a specific problem.

IOP Conf. Series: Materials Science and Engineering **492** (2019) 012010 doi:10.1088/1757-899X/492/1/012010

The analysis of the problem of accident situationrisk forecasting shows that there are three forecasting horizons and the three respective forecast types. Each type has to be approached individually:

1. Forecasting horizon within which the state change of system elements is constant, i.e. quasi-stable state.

The state of system parameter is described by the following equation:

$$\mathbf{B} = \boldsymbol{const} \quad (4)$$

where

B – column vector of parameters characterizing the state of the system elements. Quasi-stable system state and the stability of the parameters characterizing its state depends on the given system, its particularities, external environment and other factors.

Accident situationrisk forecast is done by numerical simulation of the way the processes causing the accident situationare evolving and further simulation of how the accident situationitself and its consequences are going to evolve. There are many papers[14, 15, 18–20] that outline the principles of technical system accident situationsimulation, including beyond design basis accidents. This allows one to construct a system of differential characteristics of the accident situationrisk forecast.

Let us denote the risk forecast for the accident situationin a technical system in the quasi-stable state being the cause of emergency as an operational forecast. Time horizon of an operational forecast can vary from several seconds to several tens of days.

2. Forecasting horizon within which the state change of system elements is almost linear. This is the case when the technical object does not go undergo technological shift or the production does not expand within the forecasting horizon.

The equation governing the state $\boldsymbol{Y}$ of technical system parameters is the following (5):

$$\boldsymbol{Y} = \boldsymbol{K} \cdot \boldsymbol{T} + \boldsymbol{B} \quad (5)$$

where

$\boldsymbol{Y}$ – column vector of the parameters characterizing the technical system elements state for the forecasting time T;

$\boldsymbol{B}$ – column vector of the parameters characterizing the technical system elements state at the current time instant;

$\boldsymbol{K}$ – column vector of the coefficients characterizing the change of the technical system elements state;

$\boldsymbol{T}$ – time.

Let us denote the accident situationrisk forecast for the forecasting time of linear change of technical system state parameters being the cause of accident situationas the mid-term (linear) forecast. The time horizon for such forecast can vary from one month to several (3-5) years.

It is the mid-term forecast that is of the most interest to us, because it preserves the accuracy while having a long forecasting horizon. Furthermore, there is a large database of integral and differential risk indicators that can be used to calibrate forecasting methods. Accident situationrisk forecast for the time horizon of 1 to 5 years can be very useful in forming socio-economic forecast, planning resource allocation, forming financial and material reserves for emergency prevention and recovery, for developing system of accident situationinsurance on hazardous industrial objects.

3. Forecasting horizon within which the state change of system elements is nonlinear. This is the case when the technical object starts running by new technologies, the production expansion takes place, when the personnel qualification is low or the fixed capital has not been renewed which led to irreversible degradation processes.

The equation governing the state $\boldsymbol{Y}$ of the technical system is the following (6):

$$\boldsymbol{Y} = \boldsymbol{B} + \boldsymbol{\Psi}\left(\boldsymbol{B}, \boldsymbol{T}\right) \quad (6)$$

where

IOP Conf. Series: Materials Science and Engineering **492** (2019) 012010 doi:10.1088/1757-899X/492/1/012010

$\boldsymbol{\Psi}(\boldsymbol{B},T)$ –column vector characterizing the nonlinear change of parameters characterizing the elements state of technical system for the time *T*.

Let us denote the accident situation risk forecast for the period of nonlinear technical system parameter change a long-term (nonlinear) forecast. The time horizon for this kind of forecast can vary from five years to twenty years.

Accident situation risk forecast for this horizon is rather uncertain. It can be used for national strategic planning or for planning within an individual industry. The accident situation risk forecast for the time horizon of more than five years is very uncertain and can only be useful from the standpoint of integral characteristics.

There are only a few studies in that area [1, 16, 21].

3. Methods of calculating integral and differential parameters of the accident situation risk forecast for hazardous industrial objects and analysis of their structure

3.1 Methods of calculating emergency probability and its expected value

One has to know the "history" of a problem and have full systematic information in order to make a forecast.

Right now there is a large pool of data on the accident situation causes and the factors affecting accident situation probability. This data has been collected for more than ten years and it is being published in annual government statistical reports.

Unfortunately this data is very hard to process for the forecasting purposes because it was collected according to different classifiers and the collection criteria often do not correlate, the data is often collected from different standpoints.

For each complex technical system there is a chain of cause-and-effect relationships between the state indicators $X_i(t)$ of the technical system $\mathbf{Y}=\{X_1(t),X_2(t),\ldots X_n(t)\}$ parameters, the probability P_{E1} of accident situation cause substantiation and the probability of accident situation P_E (7):

$$\mathbf{Y}=\{\boldsymbol{X}_1(\boldsymbol{t}),\boldsymbol{X}_2(\boldsymbol{t}),\ldots \boldsymbol{X}_n(\boldsymbol{t})\}\rightarrow \boldsymbol{P}_{E1}\rightarrow \boldsymbol{P}_E \quad (7)$$

where

the functional relationships $P_{E1}=F\{X_1(t),X_2(t),\ldots X_n(t)\}$;

$P_E=P_{E1}P_{E/E1}$;

$P_{E/E1}$ – probability of accident situation caused by the given cause factor.

The same can be written for the parameters characterizing the integral risk values.

For each industry there is also a chain of cause-and-effect relationship between the averaged indicator values $\mathbf{X_i}(t)$ of the system state $\boldsymbol{Z}=\{\mathbf{X}_1(t),\mathbf{X}_2(t),\ldots \mathbf{X_n}(t)\}$, the expected value of the number of accident situation causes MP_{E1} and their expected value MP_E (8):

$$\boldsymbol{Z}=\{\mathbf{X}_1(\boldsymbol{t}),\mathbf{X}_2(\boldsymbol{t}),\ldots \mathbf{X_n}(\boldsymbol{t})\}\rightarrow \boldsymbol{MP}_{E1}\rightarrow \boldsymbol{MP}_E \quad (8)$$

where

functional relationships $MP_{E1}=\Psi\{\mathbf{X}_1(t),\mathbf{X}_2(t),\ldots \mathbf{X_n}(t)\}$;

$MP_E=\vartheta(MP_{E1}, KP_{E/E1})$;

$KP_{E/E1}$ – efficiency of accident prevention measures for a given industry.

3.2 Method and algorithm for a mid-term forecast of the probabilistic characteristics of the accident situation integral risks

Let us assume that an industry has statistical information on the number of accidents *($MP_{E1}(t)$)*, emergencies *($MP_E(t)$)*, emergency damage *($MSjY(t)$)* and the generalized parameters

$Z = \{X_1(t), X_2(t), \ldots X_n(t)\}$ characterizing the change of the industry (fixed capital renewal, production levels, personnel number, etc.) for several years (m)[22].

As we have mentioned earlier, $MP_{E1} = \Psi\{X_1(t), X_2(t), \ldots X_n(t)\}$ is a linear function, thus it is possible to determine the forecasted number of emergencies in the time moment t_{m+1} by knowing the functional relationship between the number of emergencies and time and the initial conditions at the time t_m (See figure 2).

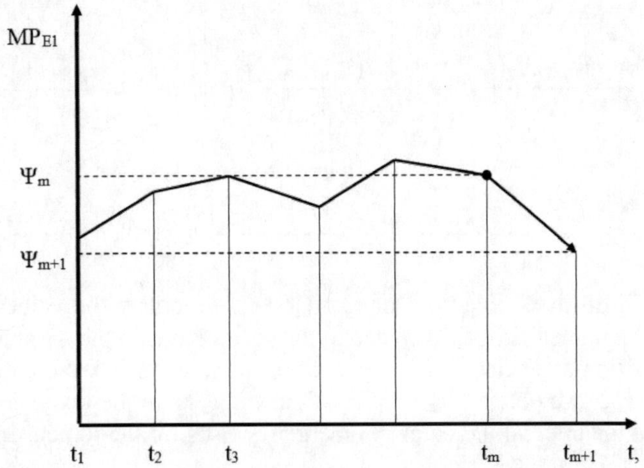

Figure 2. Schematic relationship the forecast for the number of emergencies and time history of the number of emergencies

Our task is to construct the image of function Ψ and determine its value Ψ_{m+1} at the time moment t_{m+1} by using the collected information, the function value Ψ_m at the time moment t_m and the forecasted values of its parameters $X_1(t), X_2(t), \ldots X_n(t)$

For any time interval (9), (10):

$$\frac{\Delta\Psi}{\Delta t} = \frac{\Psi_{m+1} - \Psi_m}{t_{m+1} - t_m} = \frac{d\Psi}{dt} \rightarrow \Psi_{m+1} = \Psi_m + \frac{d\Psi}{dt}\Delta t \quad (9)$$

$$\frac{d\Psi}{dt} = \frac{d\Psi\left[X_1(t), X_2(t), \ldots X_m(t)\right]}{dt} = \frac{\partial\Psi}{\partial X_1}\frac{dX_1}{dt} + \frac{\partial\Psi}{\partial X_2}\frac{dX_2}{dt} + \ldots + \frac{\partial\Psi}{\partial X_m}\frac{dX_m}{dt} \quad (10)$$

As the function Ψ is linear

$$\frac{\partial\Psi}{\partial X_1} = K_1; \frac{\partial\Psi}{\partial X_2} = K_2; \ldots \frac{\partial\Psi}{\partial X_m} = K_m \rightarrow (11)$$

$$\Psi_{m+1} = \Psi_m + \left(K_1\frac{dX_1}{dt} + K_2\frac{dX_2}{dt} + \ldots + K_m\frac{dX_m}{dt}\right)\Delta t \quad (12)$$

where

$\frac{dX_1}{dt}; \frac{dX_2}{dt}; \ldots \frac{dX_m}{dt}$ are the rates of change of the generalized parameters characterizing the condition of an economic sector. They are calculated using (13) from statistical development data:

$$\frac{dX_i}{dt} = \frac{\left(X_m(t_m) - X_m(t_{m-1})\right)}{t} \quad (13)$$

After that values of K_1, K_2,...K_m have to be determined. In order to do that one has to construct a system (14) of m linear equation for $m+1$ years using statistical information for the given economic sector. The equations are written for m time intervals

$$\Delta_1 = t_2 - t_1; \Delta_2 = t_3 - t_2; \ldots \Delta_m = t_{m+1} - t_m$$

$$\Psi_2 = \Psi_1 + K_1 \frac{X_1(t_2) - X_1(t_1)}{\Delta_1} + K_2 \frac{X_2(t_2) - X_2(t_1)}{\Delta_1} + \ldots + K_m \frac{X_m(t_2) - X_m(t_1)}{\Delta_1};$$

$$\Psi_3 = \Psi_2 + K_1 \frac{X_1(t_3) - X_1(t_2)}{\Delta_2} + K_2 \frac{X_2(t_3) - X_2(t_2)}{\Delta_2} + \ldots + K_m \frac{X_m(t_3) - X_m(t_2)}{\Delta_2};$$

....

$$\mathbf{\Psi_m} = \mathbf{\Psi_{m-1}} + \boldsymbol{K}_1 \frac{\boldsymbol{X}_1(\boldsymbol{t_m}) - \boldsymbol{X}_1(\boldsymbol{t_{m-1}})}{\Delta_m} + \boldsymbol{K}_2 \frac{\boldsymbol{X}_2(\boldsymbol{t}_3) - \boldsymbol{X}_2(\boldsymbol{t}_2)}{\Delta_m} + \ldots + \boldsymbol{K_m} \frac{\boldsymbol{X_m}(\boldsymbol{t_m}) - \boldsymbol{X_m}(\boldsymbol{t_{m-1}})}{\Delta_m} \quad (14)$$

By solving this system of linear equation for K_1, K_2,...K_m we obtain the values characterizing the influence the generalized parameters of a given economic sector have on the number of emergencies in this sector. The number of emergencies is a component of the integral risk indicator. By plugging the values of K_1, K_2,...K_m in (12) we obtain the number of emergencies in the given economic sector. It is one of the components of the integral indicator of emergency risks for the forecasting horizon Δt.

4. Conclusion

This paper deals with methods and principles of forming mid-term accident situation forecasts based on linear function of change of the parameters characterizing the industrial objects, such as hydrotechnical installations or individual economic sectors.

The proposed method can be useful for analytical and forecasting studies of the socio-economic development of the Russian Federation or its entities and also for individual economic sectors and different industrial objects taking into account the "accident factor". It can also be used in optimizing damage recovery expenses, for creating economic policies and national development strategies, in formulating requirements for the budget of the Russian Federation.

References

[1] *Government report "On the status of defending population and territories of the Russian Federation from natural and industrial emergencies in 2009–2017"* Available at: http://www.mchs.gov.ru/activities/results

[2] Prasannakumar V, Vijith H, Abinod S and Geetha N 2012 Estimation of soil erosion risk within a small mountainous sub-watershed in Kerala, India, using Revised Universal Soil Loss Equation (RUSLE) and geo-information technology *Geoscience Frontiers* **3**(2) 209–215

[3] Alpackaya I and Alpackiy D 2018 Perspectives and consequences of implementation and development digital economy *MATEC Web of Conferences* EDP Sciences vol 193 p 05087

[4] Gorlacheva E N, Gudkov A G, Omelchenko I N, Drogovoz P A and Koznov D V 2018 June Knowledge Management Capability Impact on Enterprise Performance in Russian High-Tech Sector *2018 IEEE International Conference on Engineering, Technology and Innovation (ICE/ITMC)* IEEE pp 1–9

[5] Rainer Jr R K, Snyder C A and Carr H H 1991 Risk analysis for information technology

*Journal of Management Information Systems***8**(1) 129–147

[6] Liu E M 2013 Time to change what to sow: Risk preferences and technology adoption decisions of cotton farmers in China *Review of Economics and Statistics***95**(4) 1386–1403

[7] Johnson B B and Covello V T Eds 2012 *The social and cultural construction of risk: Essays on risk selection and perception* Vol 3 Springer Science & Business Media

[8] Fedotchev A, Kruk V, Oh S J and Semikin G 2018 Eliminating pain-induced risks of operator reliability via transcutaneous electro neurostimulation controlled by Patient's breathing. *International Journal of Industrial Ergonomics***68** 256–259

[9] Carpenter M A, Pollock T G and Leary M M 2003 Testing a model of reasoned risk-taking: governance, the experience of principals and agents, and global strategy in high-technology IPO firms *Strategic Management Journal***24**(9) 803–820

[10] Biswas D, Biswas A and Das N 2006 The differential effects of celebrity and expert endorsements on consumer risk perceptions. The role of consumer knowledge, perceived congruency, and product technology orientation *Journalofadvertising***35**(2) 17–31

[11] Makhutov N A 2001 The problem of reducing risks of industrial emergencies. Institute of Machine Science RAS. *Sixth all-Russian scientific conference "Emergency risk management"* Moscow

[12] Kaliberda I V 2001 Safety regulation of the objects using atomic energy and reducing the natural and industrial disaster risks SECNRC of the Federal Environmental, Industrial and Nuclear Supervision Service of Russia. *Sixth all-Russian scientific conference "Emergency risk management"* Moscow

[13] Makhutov N A 2002 Evaluating and forecasting strategic risks in technosphere Institute of Machine Science RAS *Risk management, quarterly analytical journal*, special issue

[14] Safety of Russia. Legal, socio-economic and scientific and technical aspects Safety of the industrial complex IHF "Znanie" State Unitary Enterprise "STC "Industrial Safety", 2002

[15] Lutsenko J, Anishchenko L and Apollonova I 2018 May Bioradiolocation-based on contact fall detection method development *2018 Ural Symposium on Biomedical Engineering, Radioelectronics and Information Technology (USBEREIT)* IEEE pp 95–98

[16] Levner E and Ptuskin A 2018 Entropy-based model for the ripple effect: managing environmental risks in supply chains *International Journal of Production Research*56(7) 2539–2551

[17] Lam J 2000 Enterprise-wide risk management and the role of the chief risk officer *White paper, erisk. Com, march*, **25**

[18] Rasmussen M V 2006 *The risk society at war: terror, technology and strategy in the twenty-first century* Cambridge University Press pp 1227–1228

[19] Wynne B 2002 Risk and environment as legitimatory discourses of technology: reflexivity inside out? *Current sociology***50**(3) 459–477

[20] Lim N 2003 Consumers' perceived risk: sources versus consequences *Electronic Commerce Researchand Applications***2**(3) 216–228.

[21] Russian statistical yearbook 2010–2017 *Federal State Statistics Service* Available at: http://www.gks.ru/wps/wcm/connect/rosstat_main/rosstat/ru/

IOP Conf. Series: Materials Science and Engineering **492** (2019) 012011 doi:10.1088/1757-899X/492/1/012011

Wet part hybrid optimization method of hermetic pump

P Chaburko [1,2] **and Z Kossova** [1]

[1]Bauman Moscow State Technical University, 5 Second Baumanskaya Street, Moscow, 105005, Russian Federation

[2]E-mail:chaburko@bmstu.ru

Abstract. The article is devoted to development of the hybrid method of the wet part of hermetic pump with low specific speed optimization. As an initial approximation for the gradient descent method the results of the optimization of the pump by LP-Tau search are used. A selection of optimization parameters is mentioned. The optimization sequence is described. The mathematical model used in the numerical simulation, the boundary conditions and the parameters of the grid are shown. The results of the optimization and conclusions about the effectiveness of the method used are introduced. The experimental validation of the method is presented.

1. Introduction

Due to the constant tightening of requirements for the energy efficiency of various purpose hardware reduction of energy consumption by increasing the efficiency of industrial plants is an important and urgent task. This paper is devoted to the study of the possibilities of improving the energy efficiency of pumping equipment based on modern design methods.

The active introduction of computational fluid dynamics methods into engineering practice and the constant increase in computational power have led to the possibility of significantly improving the quality of the flow parts of pumping equipment, including in terms of its efficiency. However, computational methods themselves do not solve the problem under consideration, since they are only a tool for mathematical modeling, and not a way to optimize.

The development of methods for optimal design of pump flow parts using computational fluid dynamics methods is currently underway around the world in various scientific organizations and many results have been published [10–15].

In articles [1], a complex genetic optimization algorithm is used to increase efficiency. To apply this method, a huge number of calculation points are required, and their number increases many times with the increase in the number of optimization parameters. In this regard, genetic algorithms are used for highly specialized tasks, require highly skilled calculators and take a lot of time. In works [2, 3], optimization is carried out according to the criterion of energy efficiency, while the optimization parameters and the number of design points are determined intuitively. The result of applying this method of optimization is highly dependent on the qualifications of the researcher, since they do not use formal mathematical algorithms. In most cases, intuitive methods are applicable with one or two optimization parameters, with a greater number of them, the relationship between optimization parameters and objective functions goes beyond the scope of human imagination. In many works, for example, [4], the optimization of a specific element of the flow part was performed (in the work, a pump outlet device was selected). This method shows good results, but the lack of consideration for the mutual influence of individual elements of the flow path on each other leads to a limitation of the application of this technique. The use of the gradient method is described in [5].

This work is a continuation of the study [6], in which an attempt was made to optimize the entire pump flow section. However, in [6], the LP-Tau optimization algorithm [7] was used, which is stochastic, which means that the probability of finding the global maximum of the efficiency criterion using this method is extremely small. In this paper, this method has been modified. After the preliminary optimization of the LP-Tau method, it is proposed to use the methods of directional optimization (in this case, the gradient descent method [8]) to improve the results of the pre-optimization.

The aim of this work is to develop a method of combined optimization of the wet part of a hermetic pump based on the joint use of stochastic and directional optimization methods. According to the results of the study, it was obtained that performing pre-optimization using the stochastic method allows to obtain several models whose energy efficiency is close to the maximum. These models can be used as an initial approximation to perform a refinement optimization using the directional method. The directional method allows to further increase the energy efficiency of these models, since it is used to find the local maximum of the function in the vicinity of the point under study. Thus, the joint use of stochastic and directional methods can significantly improve the efficiency of hermetic pumps.

2. Selection of optimization parameters

In the LP-Tau search method, 7 geometric parameters of the wet part were selected. The use of such a large number of parameters in the gradient descent method leads to a strong complication of the optimization process, as well as a significant increase in optimization time.

Therefore, in this study, 3 parameters were selected that, according to the results of the analysis of the preliminary optimization results, had the greatest impact on the hydraulic efficiency (figure 1, figure 2):

1. The capacity of the channel outlet device (COD) A, mm

$$A = n \cdot b \cdot ln\frac{R_1 + a}{R_1}$$

where
n – pump shaft speed, rot/min
b – channel width of COD, mm
R_1 – radius of impeller, mm
a – channel height of COD, mm

2. The ratio of the area of output to the entrance of COD, Kd_2

$$Kd_2 = \frac{A_2}{A_1}$$

where
A_2 – channel area COD on outlet, mm^2
A_1 – channel area COD on inlet, mm^2

3. Radial clearance between the exit of the impeller and the entrance to the COD, δ, mm

IOP Conf. Series: Materials Science and Engineering **492** (2019) 012011 doi:10.1088/1757-899X/492/1/012011

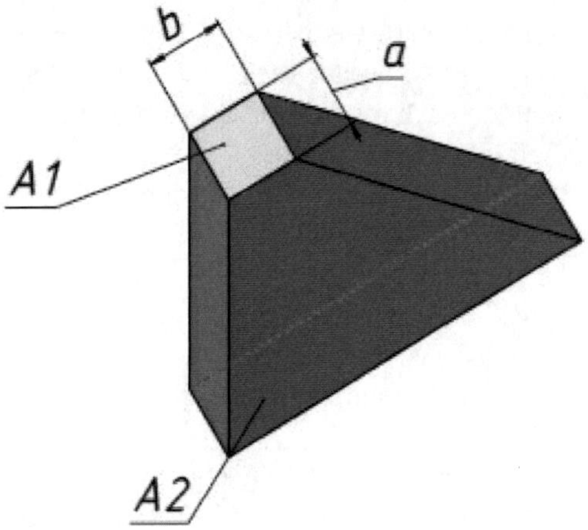

Figure 1. Channel scan of COD.

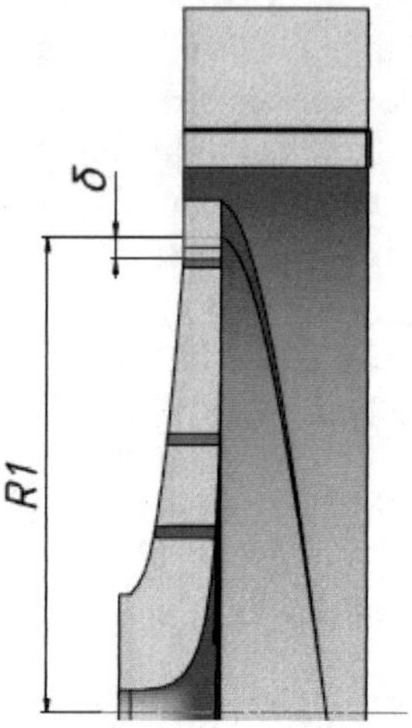

Figure 2. Meridional section of the wet part.

3. Gradient descent method

Since the functional dependence of the hydraulic efficiency value on the geometrical parameters of the pump wet part cannot be analytically derived (it is necessary to solve the basic equations of hydromechanics analytically), the algorithm for finding the global maximum of the hydraulic efficiency is carried out in the following sequence:

1. The study of the vicinity of the starting point. At this stage, all optimization parameters are reduced to a dimensionless form.

$$x_{iб} = \frac{x_i}{x_{imax} - x_{imin}} + x_{imin}$$

where

x_i – *i*-th optimization parameter,

$x_{iб}$ – *i*-th optimization parameter in dimensionless form,

x_{imin} – minimum value of *i*-th optimization parameter,

x_{imax} – maximum value of the *i*-th optimization parameter.

Then a step is made for each parameter, while the remaining parameters remain unchanged.

2. Calculation of the coordinates of the gradient vector by the equation:

$$A_i = \frac{\Delta y_i / \Delta x_i}{|A|},$$

where

Δy_i – change in the value of hydraulic efficiency when changing the *i*-th optimization parameter by Δx_i,

$|A| = \sqrt{\sum_{i=1}^{n} A_i^2}$ – gradient vector modulus,

$n = 3$ – number of optimization parameters.

3. Gradient vector step. Step size varies from 0.005 to 0.2.

4. Repeat all operations from 1item for a new point.

Schematically, the optimization sequence is presented in figure 3.

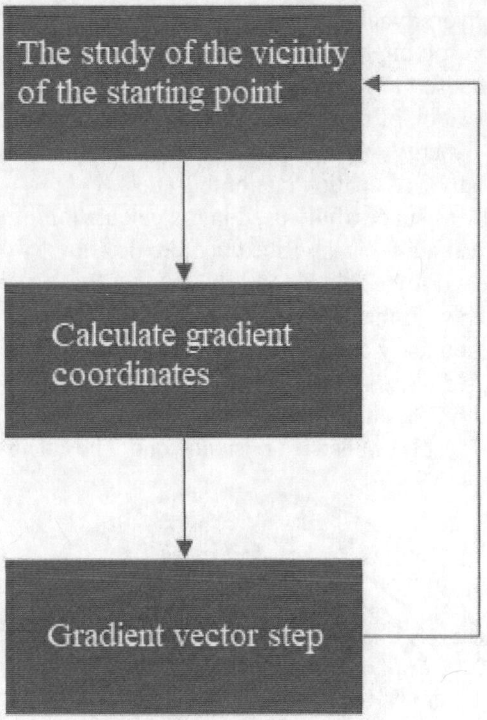

Figure 3. Optimization sequence.

Each test model obtained was calculated using computational fluid dynamics (CFD) methods. Based on the obtained geometric parameters, a 3D model is created. The model of the wet part for one of the design points is presented in figure 4.

Figure 4. Three-dimensional model of the wet part.

The model of incompressible fluid was used (ρ = const). Numerical simulation is based on solving discrete analogs of the basic hydrodynamic equations.

The calculation is based on solving the following equations:

1. The equation of mass conservation (continuity equation),
2. The equation of conservation of momentum (Reynolds averaging),
3. The equation of kinetic energy of turbulence transfer,
4. The equation of the relative dissipation rate of this energy.

This mathematical model was successfully used in the calculation of pumps of other types [4]. The application of this mathematical model is described in more detail in [6].

Also, this model was successfully used in [16–20].

All models were calculated with the same parameters of the computational grid and with the same boundary conditions. As the boundary conditions, the inlet velocity value and the outlet pressure value from the wet part were specified.

The calculated grid for all models had from 1.5 to 2 million cells. In the flow core, the mesh cells have a polyhedral shape, and for solid walls — a prismatic one. The calculated grid is shown in figure 5.

Figure 5. Calculated grid.

4. Optimization results

Optimization was carried out for two points obtained by the LP-Tau search method [6]. At the first point, the value of hydraulic efficiency was 75.4%, at the second — 74.8%.

The results of using the gradient descent method are presented in figures 6 and 7.

Thus, the value of hydraulic efficiency in the first point increased by 1.1% and amounted to 76.5%, and in the second point increased by 2.5% and amounted to 77.3%.

A comparison of the intensity of turbulent vortex formation at the optimum point obtained by the LP-Tau search method and at the optimum point obtained by the gradient descent method is presented in figures 8 and 9.

The results showed that the gradient descent method allows to improve the results obtained by the LP-Tau search method. The most optimal point obtained by the LP-Tau search method was not the most optimal starting point for the gradient method.

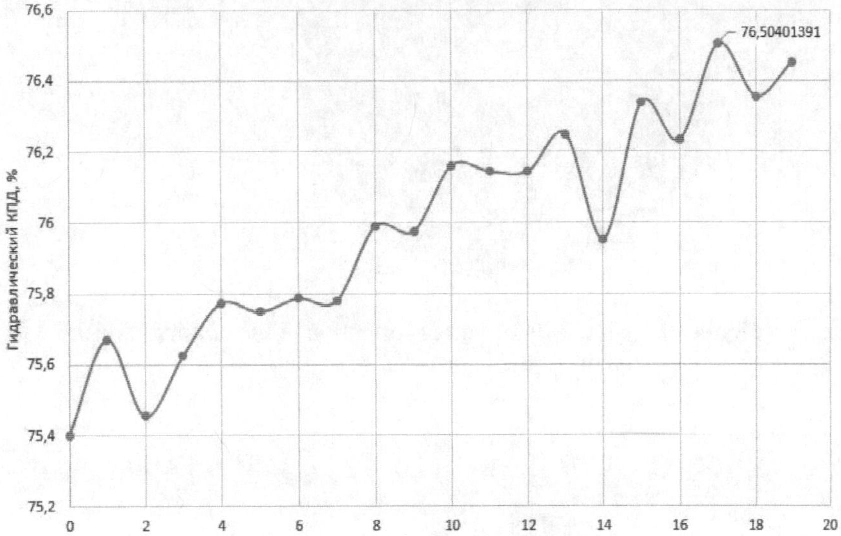

Figure 6. Optimization by the method of gradient descent at point 1.

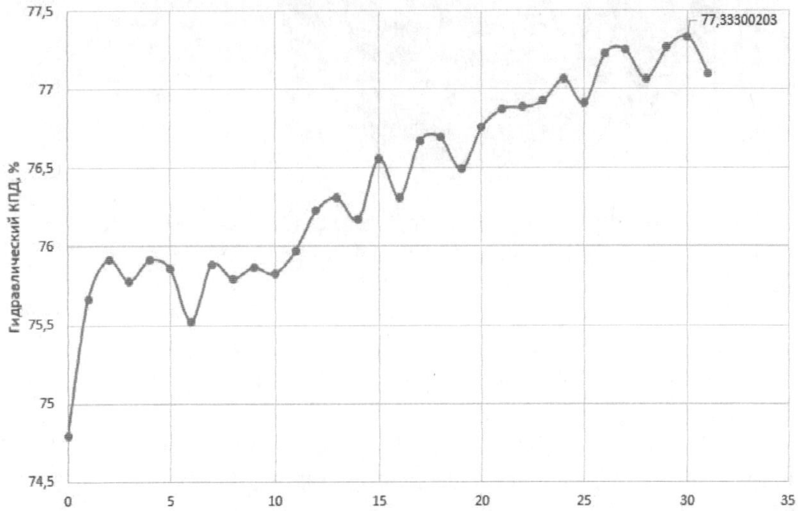

Figure 7. Optimization by the method of gradient descent at point 2.

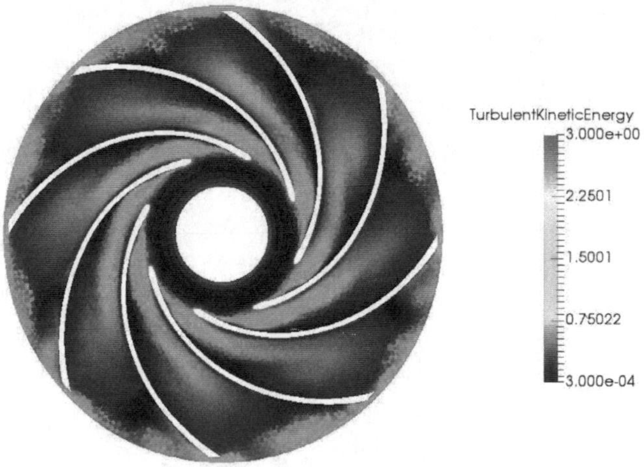

Figure 8. The intensity of vortex formation in the optimal model obtained by the LP-Tau search method.

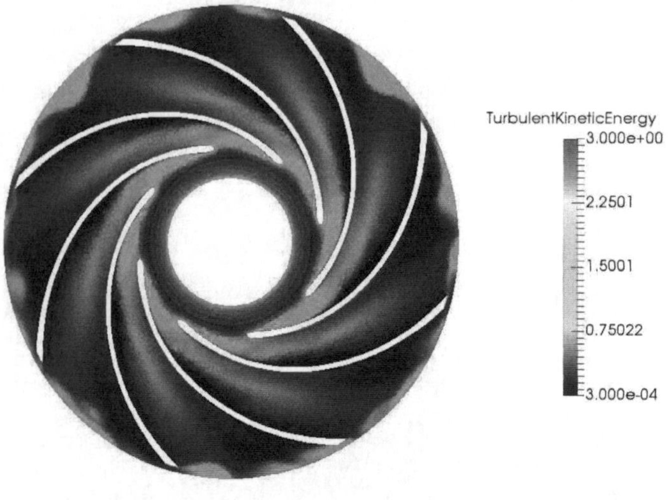

Figure 9. The intensity of the vortex in the optimal model obtained by the method of gradient descent.

5. Experimental Testing

To confirm the effectiveness of the proposed method, the final model, obtained as a result of optimization, was manufactured and tested.

Details of the elements of the wet part of the prototype were made of plastic by 3D printing. The remaining parts are made of steel. The tests were carried out in the laboratory of the Department E10 BMSTU.

The fabricated prototype is shown in figure 10.

IOP Conf. Series: Materials Science and Engineering **492** (2019) 012011 doi:10.1088/1757-899X/492/1/012011

Figure 10. Model.

According to the test results, a normal characteristic of the prototype was built. Comparison of the calculated characteristics obtained by the methods of hydrodynamic modeling (CFD) with the experimental characteristic is presented in figure 11.

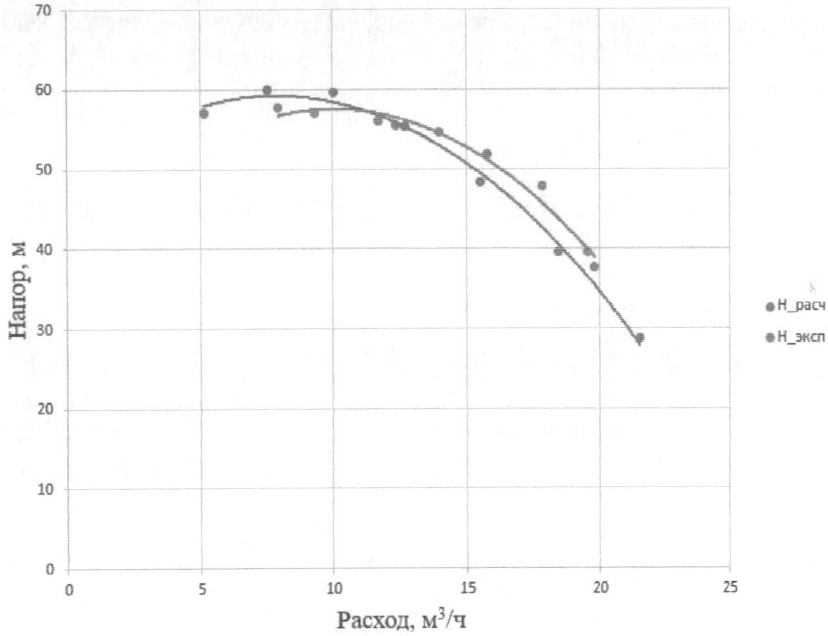

Figure 11. Comparison of the calculated and experimental characteristics of the pump.

The difference between the simulation results and the experimental data on the head near the optimal point of pump operation does not exceed 3%. Comparison of simulation results and efficiency calculations is a more difficult task, since in addition to hydraulic losses in the pump, there are other sources that reduce the efficiency of the pump. In pumps with a specific speed 34, as in this case, the share of non-hydraulic losses can reach 20-30% of power consumption.

The overall efficiency obtained experimentally was 55%, which is higher by 3-5% of the efficiency values of the analogues available on the market. In [9], a complex experimental study of pumps of various fastness was carried out. The graph of achievable efficiencies obtained in this work is given below (figure 12). For the pump with the considered parameters, the maximum efficiency according to the given schedule is 50%.

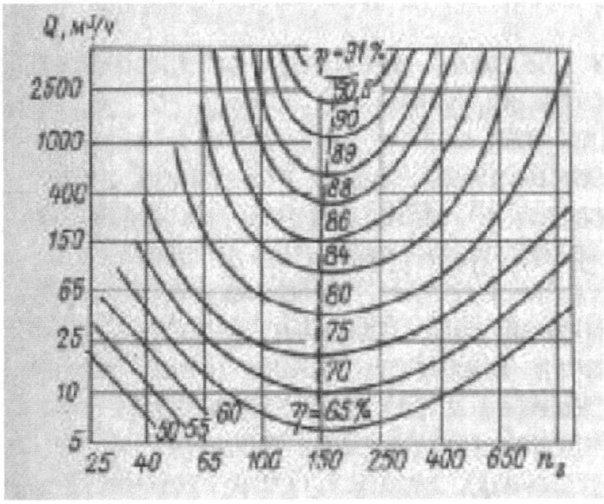

Figure 12. The effect of$n_s$on the efficiency of the pump [9].

6. Conclusion

The optimization method proposed in the paper allows profiling wet parts with high energy efficiency in a short time with a guaranteed result. Compared with the results obtained by stochastic optimization methods, the use of an integrated approach (stochastic plus directional method) makes it possible to further increase the efficiency of the wet part (in this example by 2-3%). Tests of the pump with the wet part obtained by the proposed method showed an increase in efficiency of about 5% compared with existing analogues.

Reference

[1] Hao H, Xinkai L and Bo G 2015 Hydraulic optimization of multiphase pump based on CFD and genetic algorithm *International journal of grid distribution computing***8**(6)161–170
[2] Xie S F, Wang Y, Liu Z C, Zhu Z T, Ning C and Zhao L F 2015 Optimization of centrifugal pump cavitation performance based on CFD. *IOP Conf. Series: Materials Science and Engineering* 72 doi:10.1088/1757-899X/72/3/032023
[3] Ji C, Wang Y and Yao L 2007 Numerical analysis and optimization of the volute in a centrifugal pump. *International conference on power engineering*
[4] Lomakin V O, Petrov A I and Stepanyuk A I 2012 Optimization of geometrical parameters of tap of oil main pump of NM type *Science and education: Electronic scientific publication of BMSTU*(3)Available at: http://technomag.bmstu.ru/doc/347727.html (accessed 10.11.2018)
[5] Kim J H, Oh K T, Pyun K B, Kim C K, Choi Y S and Yoon J Y 2012 Design optimization of a centrifugal pump impeller and volute using computational fluid dynamics. *IOP Conf. Series: Earth and Environmental Science***15** doi:10.1088/1755-1315/15/3/032025
[6] Chaburko P S, Lomakin V O, Kuleshova M S and Baulin M N 2016 Complex wet end part optimization of hermetic pump with lp-tau method Pumps. *Turbines. Systems***1** (18)
[7] Sobol I M 1973 Monte Carlo numerical methods Moscow NaukaPubl 311 p
[8] Nesterov Yu E 2010 Convex optimization methods Moscow ICNMO 281 p
[9] Malushenko V V and Mikhailov A K 1981 Energy pumps: Reference book. Moscow EnergoizdatPubl 200 p

[10] Zharkovsky A A and Pospelov A Yu 2014 Using 3D methods to calculate flow, predict characteristics and optimize the shape of the flow parts of hydraulic turbines *Hydraulic Engineering*(11) pp 104–109

[11] Aleksandrov A A, Devisilov V A, Sharai E Yu and Kiselyova D A 2018 Effect of geometric parameters of working channel of hydrodynamic filter with protective baffle on medium flow structure *Herald of the Bauman Moscow State Technical University, Series Natural Sciences* issue 2 pp 23–38 DOI: 10.18698/1812-3368-2018-2-23-38

[12] Ukolov A I and Rodionov V P 2018 Verification of numerical simulation results and experimental data of the cavitation influence on hydrodynamic characteristics of a jet flow *Herald of the Bauman Moscow State Technical University, Series Natural Sciences* issue 4 102–114 DOI: 10.18698/1812-3368-2018-4-102-114

[13] Kalugin M D, Korchagova V N, Kraposhin M V, Marchevsky I K and Moreva V S 2018 Using big analytics tools in performance of gas dynamics and acoustics tasks. *Herald of the Bauman Moscow State Technical University, Series Natural Sciences* issue 3 32–47 DOI: 10.18698/1812-3368-2018-3-32-47

[14] Fomichev D V and Solonin V I 2015 Structure of the turbulent flow in the BREST-OD-300 wire-spaced fuel assemblies. *5th International Youth Conference on Energy, IYCE 2015; Pisa; Italy; 27 May – 30 May 2015.* DOI: 10.1109/IYCE.2015.7180754

[15] Fedorov S V, Bolotina I A, Strukov Yu A 2018 Thermal effects during inertial stretching of high-gradient conductive rod with "frozen" magnetic field. *Herald of the Bauman Moscow State Technical University, Series Natural Sciences* issue 2 39–59 DOI: 10.18698/1812-3368-2018-2-39-59

[16] Lomakin V, Cheremushkin V and Chaburko P 2018 Investigation of vortex and hysteresis effects in the inlet device of a centrifugal pump. *Global Fluid Power Society PhD Symposium, GFPS 2018Samara Russian Federation 18 July 2018 — 20 July 2018* DOI: 10.1109/GFPS.2018.8472374

[17] Lomakin V O, Chaburko P S, Kuleshova M S 2017 Multi-criteria Optimization of the Flow of a Centrifugal Pump on Energy and Vibroacoustic Characteristics. *Procedia Engineering***176** 476–482 DOI: 10.1016/j.proeng.2017.02.347

[18] IlieM 2019 A fully-coupled CFD/CSD computational approach for aeroelastic studies of helicopter blade-vortex interaction. *Applied Mathematics and Computation***347** 15 April 122-142 DOI: 10.1016/j.amc.2018.10.069

[19] Sawant S S, Gosavi S N, Khadamkar H P, Mathpati C S, Pandit R and Lali A M 2019 Energy efficient design of high depth raceway pond using computational fluid dynamics. *Renewable Energy* April 528–537 DOI: 10.1016/j.renene.2018.10.016

[20] Shibata T, Fukushima H and SegewaK 2019 Improvement of steam turbine stage efficiency by controlling rotor shroud leakage flows Part I: Design concept and typical performance of a swirl breaker. *Journal of Engineering for Gas Turbines and Power***141** issue 4 1 April DOI: 10.1115/1.4041650

IOP Conf. Series: Materials Science and Engineering **492** (2019) 012012 doi:10.1088/1757-899X/492/1/012012

The development of the theory of calculation of the hydrodynamic coupling

V Lomakin[1], V Cheremushkin[1] and A Petrov[1,2]

[1]Bauman Moscow State Technical University, 5 Second Baumanskaya Street, Moscow, 105005, Russian Federation

[2]E-mail: alex_i_petrov@mail.ru

Abstract. This article is devoted to the study of the mechanism of operation of a hydrodynamic coupling using numerical hydrodynamic modeling, as well as through a theoretical description. Attention is paid to some aspects of theoretical calculation and it is shown how its accuracy can be increased. The article presents the basic equations of the mathematical model used. The results of flow simulation in two versions of hydrodynamic couplings are shown, as well as their comparison with theoretical calculations. Theoretically, the coupling parameters were determined both according to the dependencies proposed in the reviewed literature, and according to the additions made according to the provisions given in the article. For clarity of comparison, the results are presented in the form of graphs and fields of distribution of scalar quantities.

1. Introduction

Hydrodynamic coupling (fluid coupling) – a type of hydrodynamic transmission that provides flexible connection and transfer of power from the drive shaft to the driven shaft. The hydraulic coupling differs from the hydraulic transformer by the absence of a fixed part – the reactor; therefore, the power transfer is carried out without a change in torque and its efficiency is actually defined as its transmission ratio

$$\eta = \frac{\omega_t}{\omega_p} = i,$$

where

ω_t and ω_p – the angular velocities of the turbine and pump wheel, respectively;

i – gear ratio.

At present, quite a few works are devoted to the study of hydraulic couplings, and there are also very few works devoted to their theoretical calculation. This aspect of the development of the flow part is very important even now, with the active use of numerical calculations in engineering practice [1].

This paper presents a new method for pre-calculating the characteristics of a hydrodynamic coupling, which is the result of the development and addition of an existing algorithm. Its feature is to take into account the uneven distribution of velocities, which is usually neglected in classical methods.

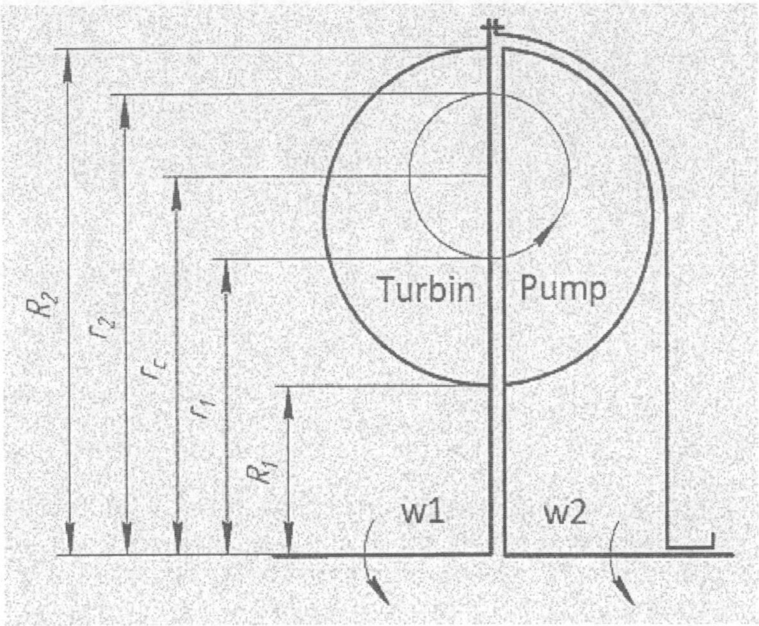

Figure1. The principal design scheme of the hydrodynamic coupling.

Figure 1 shows a schematic diagram of the hydrodynamic coupling with an indication of the main dimensions used in the calculation. It consists of pump and turbine wheels; the fluid movement in the pump wheel occurs to the periphery, in the turbine – to the axis:

R_2 – the outer radius of the coupling;

R_1 – the internal radius of the coupling;

r_1 – the radius of the fluid inlet into the pump / exit from the turbine wheel;

r_2 – the radius of the liquid exit from the pump / turbine wheel inlet;

r_c – the radius of the circle separating the flows entering and leaving the impeller (in a theoretical calculation).

2. Mathematical model and methods

In the reviewed literature, the characteristics of the fluid coupling are derived using the following basic equations [2, 3]:

Energy balance equation:

$$H_p = H_t + h_L ,(1)$$

where

H_p and H_t are the theoretical pressures of the pump and turbine wheels;

h_L is the magnitude of the losses.

The equation of change of angular momentum:

$$M = \rho Q\left(\omega_p r_{2p}^2 - U_1 r_{1p}^2\right),(2)$$

where

r_{1p} is the radius of fluid entry into the wheel;

r_{2p} – is the radius of fluid exit from the wheel;

Q –fluid flow through the impeller;

ρ – is the density of the working fluid.

The relationship between the pressure on the wheel and the moment:

$$M\omega = \rho gQH \quad (3)$$

From the above equations, a conclusion is given for the meridional velocity of the fluid and the pressure of the pump wheel:

$$c_m = \omega_p r_{2p}\sqrt{\frac{\left(1+a^2\right)\left(1-i^2\right)}{\xi}}, \quad (4)$$

$$H_p = \frac{\omega_p^2 r_{p2}^2}{g}\left(1-ia^2\right), \quad (5)$$

where

ξ is the coefficient of resistance;

a is the ratio of the entry radius to the exit radius.

The calculation is carried out on the average stream, despite the fact that he himself explicitly shows the uneven distribution of the meridional velocity by expression (4). It can be taken into account by replacing expressions (2) and (3) with the following:

$$dM = \rho\left(\omega_p r_{2p}^2 - U_1 r_{1p}^2\right)dQ, \quad (6)$$

$$\omega dM = \rho gHdQ. \quad (7)$$

Then the determination of the flow rate through the pump wheel and the moment on it is reduced to taking integrals that take into account the non-uniformity of meridional velocity:

$$Q = \int_{r_c}^{R_2} 2c_m \pi r_2 dr_2 .$$

$$M = \int_1^2 \frac{\rho gHdQ}{\omega_p} = \int_{r_c}^{R_2} 2\frac{\rho g}{\omega_p}\frac{\omega_p^2 r_{p2}^2}{g}\left(1-ia^2\right)c_m \pi r_2 dr_2 .$$

For a more adequate comparison of the two theoretical models, we also performed numerical hydrodynamic modeling of two variants of the flow parts of hydraulic couplings. The experience of using semi-empirical models of turbulence (in this case, k—ω SST) shows a fairly accurate match with the experiment for dynamic machines, like pumps [4, 5, 6, 7], and hydrotransformers [8, 9].

The basis of the turbulence model of the method used is discrete analogs of the following equations of the dynamics of a continuous viscous incompressible medium, written for solving by the finite element method [10, 11, 12, 13]:

Mass conservation equation (continuity equation)

$$\frac{\partial \tilde{u}_j}{\partial x_j} = 0,$$

where

$\tilde{u}_j$ is the averaged value of the fluid velocity in the projection onto the j-th axis (j = 1, 2, 3).

The equation for the change in the amount of motion (Reynolds averaging) in the nonstationary formulation

$$\rho\left[\frac{\partial U_i}{\partial t}+U_j\frac{\partial U_i}{\partial x_j}\right]=-\frac{\partial P}{\partial x_i}+\frac{\partial}{\partial x_i}\left[T_{ij}^{(v)}-\langle\rho u_i u_j\rangle\right],$$

where

U, P is the averaged velocity and pressure;

$\tilde{T}_{ij}^{(v)} = 2\mu\tilde{s}_{ij}$ is the viscous stress tensor for an incompressible fluid;

$\tilde{s}_{ij} = \frac{1}{2}\left[\frac{\partial \tilde{u}_i}{\partial x_j}+\frac{\partial \tilde{u}_j}{\partial x_i}\right]$ – the instantaneous strain rate tensor;

$\langle\rho u_i u_j\rangle$ – Reynolds stresses;

ρ is the fluid density;

μ is the dynamic viscosity coefficient of the fluid.

To solve the discrete analogs of the above equations, as well as the equations of the turbulence model, we use a partition of the computational domain into finite elements, i.e. construction of the computational grid. In this case, the cells have a multifaceted shape in the core of the flow and prismatic - near the walls (figure 2). The thickness of the prismatic layers is necessary to ensure the value of Y + <100 [14, 15].

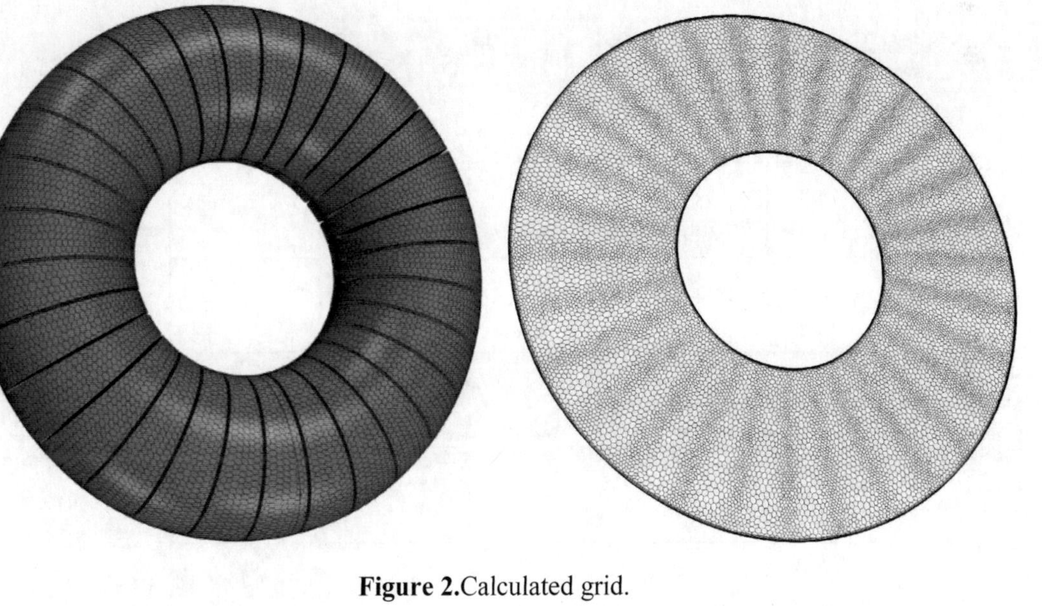

Figure 2. Calculated grid.

3. Results and analysis

Comparison of the results obtained using standard computational methods (standard) and augmented (extended), as well as by means of numerical simulation (CFD), is shown in figure 3 for two variants of the flow parts of hydraulic couplings. The results are presented in the form of dependencies on the gear ratio of the hydraulic couplings flow through the impellers and the moment of forces on them. The rotational velocity of the pump wheels is 2000 rpm, the turbine velocity varied from 2000 to 0 rpm, which corresponds to a change in the gear ratio from 1 to 0. Also for clarity of the flow process, scenes of the velocity amplitude distribution are presented (figures 4, 5).

From the data presented in the graphs, a much better similarity is observed in both the qualitative and quantitative results of the proposed extended theoretical model and numerical calculation. The

IOP Conf. Series: Materials Science and Engineering **492** (2019) 012012 doi:10.1088/1757-899X/492/1/012012

maximum errors with respect to the results of modeling the standard and augmented models for both versions of hydraulic couplings are given in table 1.

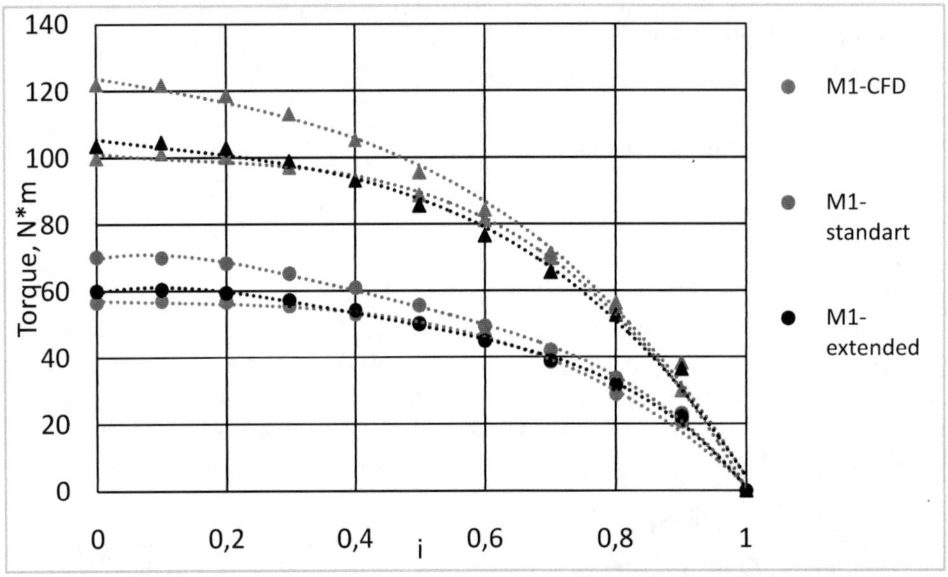

(a)

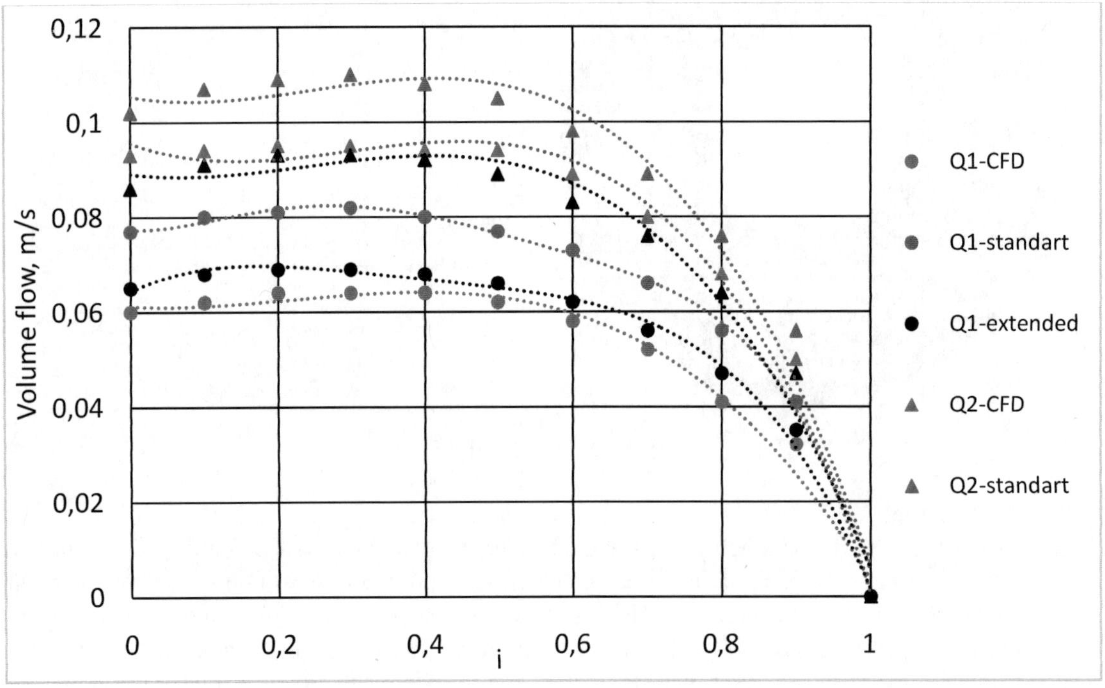

(b)

Figure 3. Graphs of dependencies for the two options for hydraulic couplings (indices 1 and 2): a) the moment on the pump wheel; b) the flow through the pump wheel.

IOP Conf. Series: Materials Science and Engineering **492** (2019) 012012 doi:10.1088/1757-899X/492/1/012012

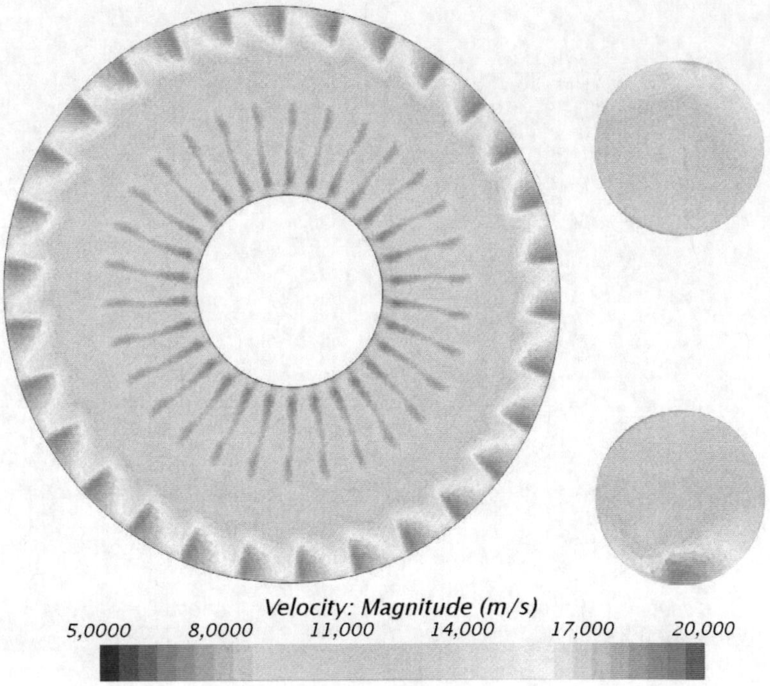

(a)

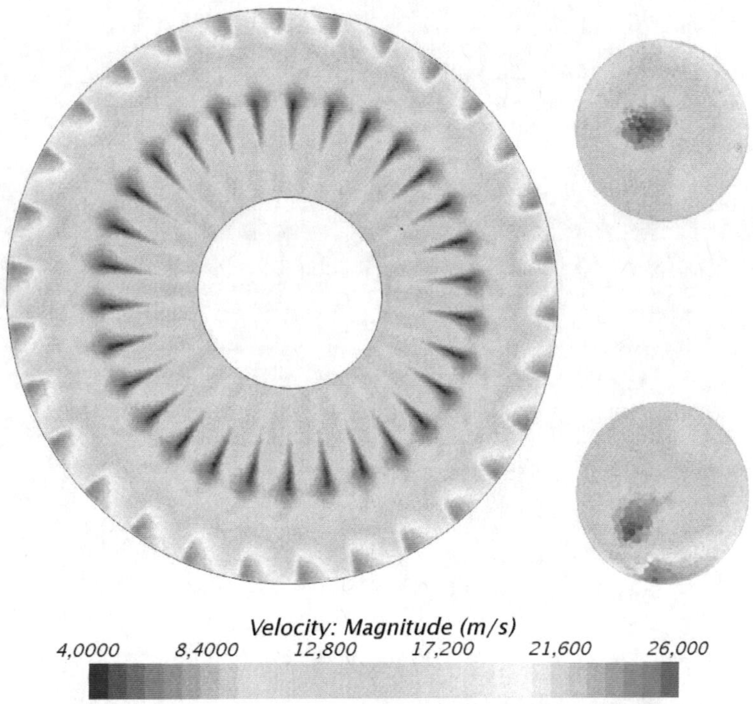

(b)

Figure4. The fields of the amplitude distribution of velocity for the 1-st version of the hydraulic coupling with the values of the gear ratio: a) 0.8; b) 0.5.

IOP Conf. Series: Materials Science and Engineering **492** (2019) 012012 doi:10.1088/1757-899X/492/1/012012

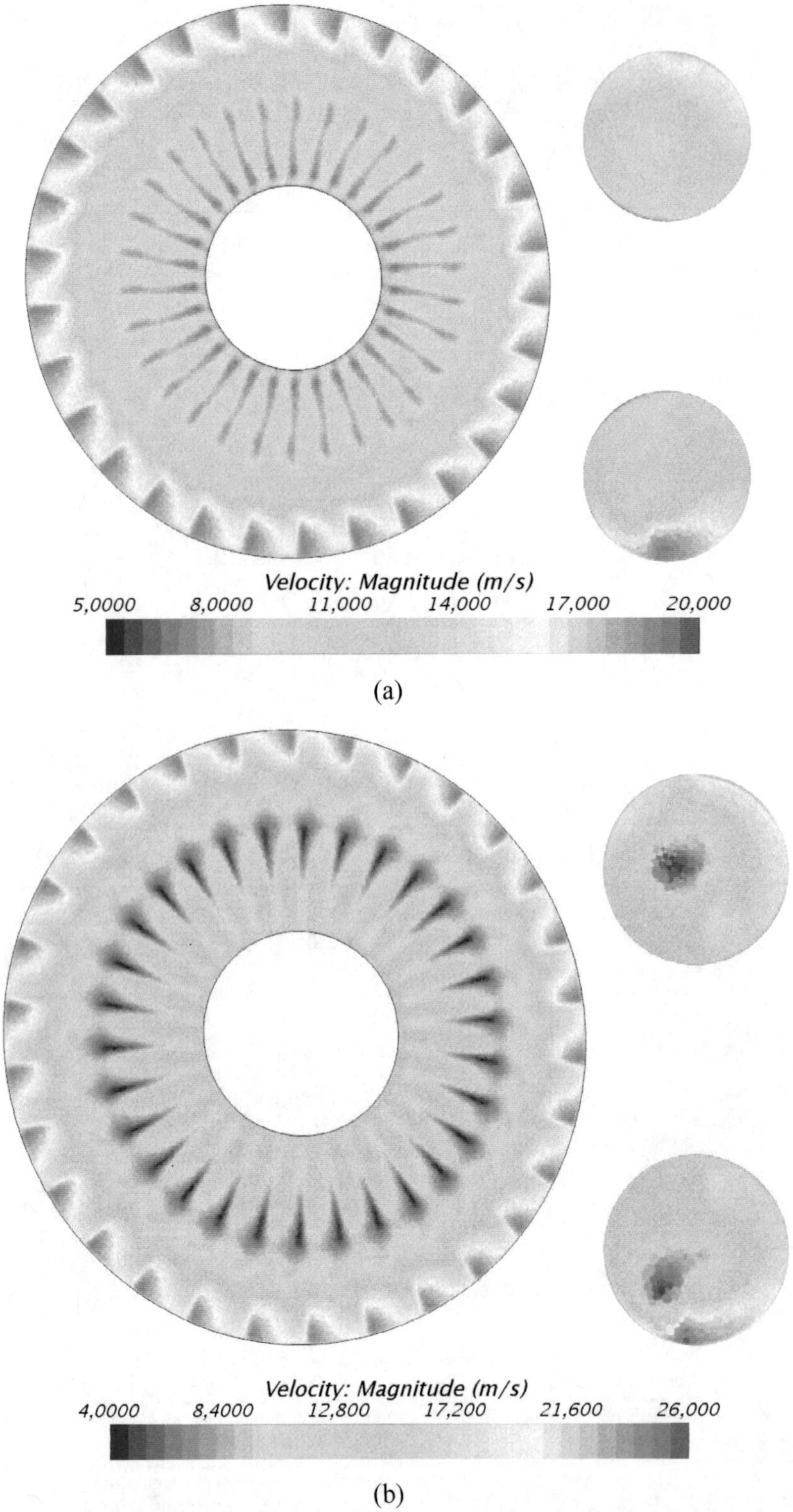

Figure 5. Fields of the amplitude distribution of velocity for the 1-st variant of the hydraulic coupling with the values of the gear ratio: a) 0.8; b) 0.5.

Table 1. Valuesoferrors.

Numberof hydraulic coupling	Calculation error, %			
	Standardmathematicalmodel		AugmentedMathematicalmodel	
	Moment	Consumption	Moment	Consumption
1	24,4	36,6	9,1	14,6
2	27,4	15,8	6,1	6,7

The larger error for the standard model has at lower transmission ratios, i.e. with large slip. This may be due to the stronger influence of the irregularity of the meridional velocity.

4. Conclusions

According to the results of the work done, the effectiveness of the augmented mathematical model was shown for theoretical calculation when compared with the results of numerical hydrodynamic modeling. This shows the importance of taking into account the phenomena associated with the uneven distribution of certain quantities characterizing the flow. Conclusions drawn from the results of work require further research to more fully describe the influence of the phenomenon under study on the characteristics of hydrodynamic couplings, as well as experimental confirmation by means of bench tests.

References

[1] Yamaguchi T and Tanaka K *Torque converter transient characteristics prediction using computational fluid dynamics* Available at: http://iopscience.iop.org/1755-1315/15/4/042020 accessed 05.05.2015

[2] Stesin S P and Yakovenko Y A 1973 *Hydrodynamic transmission (Gidrodinamicheskiye peredachi)* (Moscow: Mashinostroenie Publ.) 352 p

[3] Kartsev L V 2006 *Theory and Calculation of Hydrodynamic Transformers and Couplings* (*Teoriya i raschet gidrodinamicheskih transformatorov i muft*) (Moscow: BMSTU Publ.) 416 p

[4] Lomakin V, Cheremushkin V and Chaburko P 2018 Investigation of vortex and hysteresis effects in the inlet device of a centrifugal pump *2018 Global Fluid Power Society PhD Symposium, GFPS 2018* no 8472374 DOI: 10.1109/GFPS.2018.8472374

[5] Lomakin V O 2015 Investigation of two-phase flow in axial-centrifugal impeller by hydrodynamic modeling methods *Proceedings of 2015 International Conference on Fluid Power and Mechatronics, FPM 2015* no 7337302 pp 1204–1206 DOI: 10.1109/FPM.2015.7337302

[6] Zhu J, Zhu H, Zhang J and Zhang H-Q 2019 A numerical study on flow patterns inside an electrical submersible pump (ESP) and comparison with visualization experiments *Journal of Petroleum Science and Engineering* **173** pp 339–350 DOI: 10.1016/j.petrol.2018.10.038

[7] Wu Y,Zhu L and Luo Y 2018 Computational Fluid Dynamics Analysis of an Injection Suspension Blood Pump on the Hydraulic, Suspension and Hemolysis Property *Jixie Gongcheng Xuebao/Journal of Mechanical Engineering* **54** (20) pp 52–58 DOI: 10.3901/JME.2018.20.052

IOP Conf. Series: Materials Science and Engineering **492** (2019) 012012 doi:10.1088/1757-899X/492/1/012012

[8] Chunbao Liu, Changsuo Liu and Wenxing Ma 2015 *Rans, detached Eddy simulation and large Eddy simulation of internal Torque converters flows: A comparative study* Available at: http://www.tandfonline.com/doi/pdf/10.1080/19942060.2015.1004814 accessed 10.05.2015

[9] Sun Z, Chew J, Fomison N and Edwards D *Analysis of fluid flow and heat transfer in industrial fluid couplings* Available at: http://dx.doi.org./10.1243/09544062JMES1478 accessed 02.05.2015

[10] Lomakin V O, Chaburko P S and Kuleshova M S 2017 Multi-criteria Optimization of the Flow of a Centrifugal Pump on Energy and Vibroacoustic Characteristics *Procedia Engineering* **176** pp 476–482 DOI: 10.1016/j.proeng 2017.02.347

[11] Gouskov A M, Lomakin V O, Banin E P and Kuleshova M S 2017 Minimization of Hemolysis and Improvement of the Hydrodynamic Efficiency of a Circulatory Support Pump by Optimizing the Pump Flowpath *Biomedical Engineering* **51** pp 229–233 DOI: 10.1007/s10527-017-9720-9

[12] Timushev S F,Knyazev V A,Panaiotti S S,Soldatov V A and Rohatgi U S 2009 Numerical procedure for assessment of centrifugal pump cavitation erosion *ASME International Mechanical Engineering Congress and Exposition, Proceedings* **14** pp 73-78 DOI: 10.1115/IMECE2008-67177

[13] Kochevsky A N, Kozlov S N, Aye K M, Schelyaev A Y and Konshin V N 2005 Simulation of flow inside an axial-flow pomp with adjustable guide vanes *Proceedings of the American Society of Mechanical Engineers Fluids Engineering Division Summer Conference* **1B** pp 1397–1404

[14] Parveen Banu J and Mani A 2019 Numerical studies on ejector with swirl generator *International Journal of Thermal Sciences* 137 pp 589–600 DOI: 10.1016/j.ijthermalsci.2018.11.033

[15] Śmierciew K, Gagan J and Butrymowicz D 2019 Application of numerical modelling for design and improvement of performance of gas ejector *Applied Thermal Engineering* pp 85–93 DOI: 10.1016/j.applthermaleng.2018.12.030

IOP Conf. Series: Materials Science and Engineering **492** (2019) 012013 doi:10.1088/1757-899X/492/1/012013

Investigation of the influence of centrifugal pump wet part geometry on hemolysis index

A Gouskov[1,3,4], V Lomakin[1], E Banin[1,2], M Kuleshova[1] and P Chaburko[1]

[1] Bauman Moscow State Technical University, 5 Second Baumanskaya Street, Moscow, 105005, Russian Federation
[2] National Research Center "Kurchatov Institute", Moscow, Russian Federation
[3] Mechanical Engineering Research Institute of the Russian Academy of Sciences, Moscow, Russian Federation

[4] E-mail: gouskov_am@mail.ru

Abstract: The present study focuses on the influence of the circulatory support pump wet part geometry on the level of damage experienced by blood cells in it. The object of the study is a centrifugal pump: axial input, impeller and volute. The values of density and viscosity are constant and are close to the physiological parameters of blood of a healthy human. In order to estimate hemolysis blood cells are considered as spherical particles, which do not interact with each other. At each point of the trajectory of the particle equivalent shear stress is calculated according to Huber-Mises criterion. Hemolysis is then calculated in accordance with Lagrangian approach to particle movement in a flow. As an experimental hemolysis curve a power law of equivalent shear stress depending on the time the particle has experienced this stress was chosen. For a wet part optimization criterion of maximum efficiency and minimum hemolysis are considered.

1. Introduction

Interest in wearable circulatory systems has increased significantly over the last ten years. Every year the number of publications and patent applications in the world on this subject is steadily increasing (according to lens.org[1] the number of patents and patent applications on this subject in 2014 reached 5 784). Based on the large amount of works, it can be concluded that in the medium term, miniature assist devices will be the main means to save patients' lives at the terminal stages of heart failure, in case the heart transplantation is impossible. However, during the period of circulatory support, wearable assist pumps have complications [1–5]: bleeding, thrombosis, stroke, etc.

Auxiliary circulation pumps are divided into two types: rotary and volumetric. Currently, according to statistics from [1] rotor pumps of two types prevail in medical practice: axial and radial. The presented schemes differ structurally from each other and suggest different approaches to implantation, but in general there is now little data that suggests the absolute advantages of one type

[1] Request: "(Blood pump) AND implantable"

IOP Conf. Series: Materials Science and Engineering **492** (2019) 012013 doi:10.1088/1757-899X/492/1/012013

over the other. However, in a number of works similar comparisons are made. The article [6] presents data on the implantation of 105 axial pumps HeartMate 2 (Thoratec Inc.) and 34 centrifugal HVAD (HeartWare Inc.). Groups of patients were selected to ensure comparability of obtained results. As a result of the study, no statistically significant difference in medical indications was found for the case of implantation of an axial or radial pump. The study [7] (the sample consists of 194 HeartMate 2 devices and 209 HVAD) also notes that the final patient parameters are nearly the same, but pump replacement (due to breakage, unit wear, scheduled replacement) is more often required for axial HeartMate 2. We note here the study [8], in which 68 patients participated. The patients had radial pumps Duraheart (n = 15), Evaheart (n = 23) and axial pumps HeartMate 2 (n = 22), Jarvik 2000 (n = 8) installed. As a result of the study, the survival rates in two groups of patients (with an installed radial or axial pump) had similar survival rates and the frequency of infection of the supply cable, the researchers also noted that patients with radial pumps more often have a violation of the cerebral circulation. Some additional data on the results of clinical practice can be found in [9–11].

The interest of this work is focused on the analysis of design flaws in the radial HVAD. Based on the analysis, a modification of the centrifugal pump is proposed and calculations of hydraulic parameters are given, theoretical hemolysis in the wet part of the pump is also calculated.

2. Geometric model

The simulation object is a centrifugal pump: axial input, impeller and volute. A pump HVAD [12] (HeartWare Inc., USA) is considered as a prototype. Figure 1 shows that this pump has an unusual for centrifugal pumps geometry – there are no curved blades in impeller, interblade channel is narrowed.

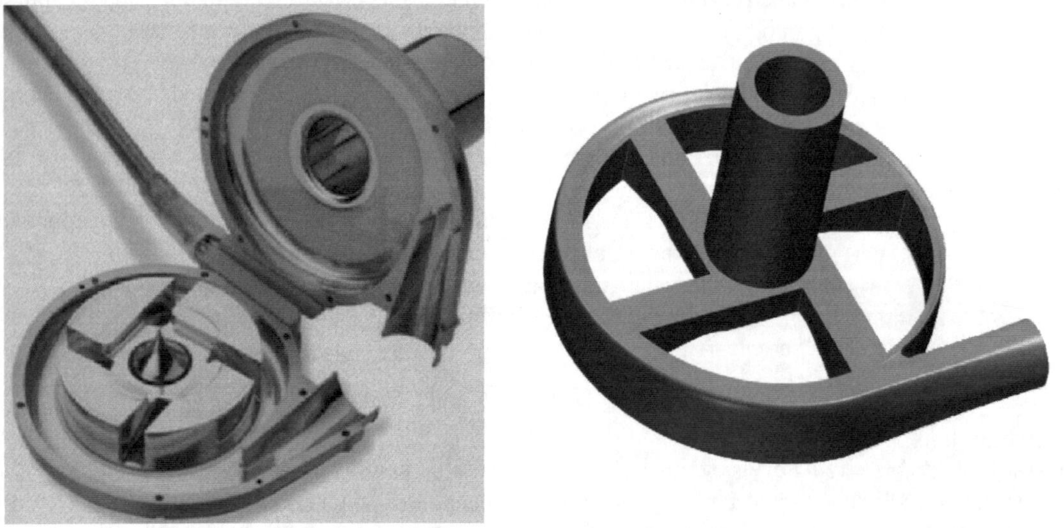

Figure 1. Geometry of HVAD wet part.

3. Mathematical simulation methods

Hydrodynamics is simulated using STAR-CCM+ software with additional C++ macro providing assessment of the accumulated damage. In order to solve differential equations a finite volume method is used. The flow region is discretized into small cells, in this case there are 1.5 million of them, in each cell such parameters as pressure, velocities, etc. are calculated. Reynolds number inside the pump can reach the value of 72 000 (at the interblade channels), at the same time there are areas with laminar flow there.

Shear Stress Transform Method (SST k–ω[13–16]) is used to describe the flow. This model has already showed its feasibility during other simulations [13]. Since the flow cannot be considered as fully turbulent more common k–ε and k–ω models can lead to computational errors [17–19]. Hybrid SST k–ω model is a combination of the models mentioned above. It uses k–ω model at the areas by the

IOP Conf. Series: Materials Science and Engineering **492** (2019) 012013 doi:10.1088/1757-899X/492/1/012013

wall (where laminar flows are concentrated) and $k - \varepsilon$ model for turbulent areas; transient zones are simulated with special transition algorithms.

In order to save machine time as a first approximation all models are calculated at steady state, then a few models with best parameters are simulated at unsteady state. Mass flow of 5 l/min is assigned as an inlet boundary condition, 0 Pa pressure is assigned to the outlet boundary, the impeller rotational speed is 2 900 rpm. Characteristics of the fluid under consideration are close to those of blood: Newtonian fluid with dynamic viscosity of 0.0035 kg/m·s, which corresponds to a hematocrit level of 33% (the most probable class of patients to have such a device installed), density ρ = 1050 kg/m^3. Control particles are non-interacting, their trajectories do not affect the flow.

4. Optimization parameters

Simulation of flow in the base model (original HVAD geometry) resulted in an unsatisfying flow pattern – a pressure drop can be seen in the channels and intensive vorticity is obtained (figure 2). This configuration of the wet part also has relatively low values of hydrodynamic parameters (head – 64 mm Hg, efficiency – 51.8%, HI – 0.0285).

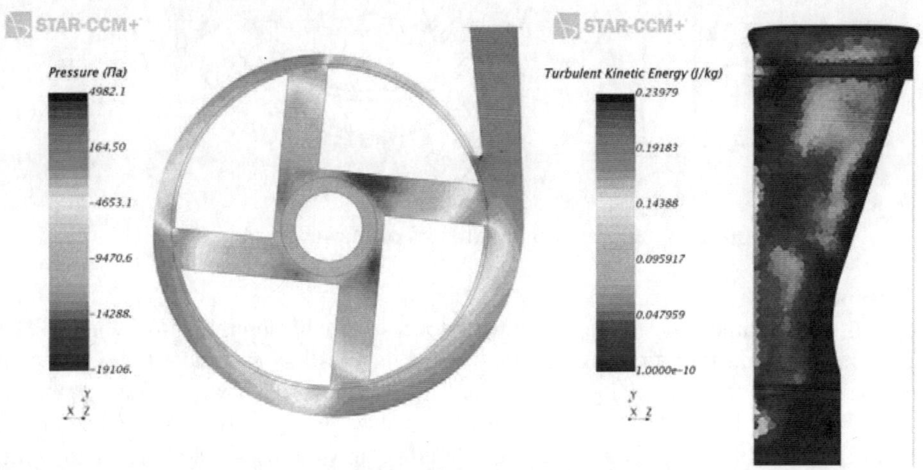

Figure 2. Pressure (left) and turbulent kinetic energy (right) distribution at the pump cross section.

For further analysis geometry with curved blades was chosen (figure 3). It helps to obtain more uniform flow with better hydraulic characteristics, which in its case leads to decreased blood cells damage.

Figure 3. Geometry of the suggested pump wet part.

The following parameters were changing during optimization process (figure 4): number of impeller blades – z (varying from 3 to 6), impeller outlet width – b_2 (varying from 3.5 mm to 6 mm), impeller outer diameter – d_2 (varying from 24 mm to 46 mm), blade angle at impeller outlet – β_2 (varying from 20° to 40°), blade wrap angle – φ (varying from 90° to 170°).

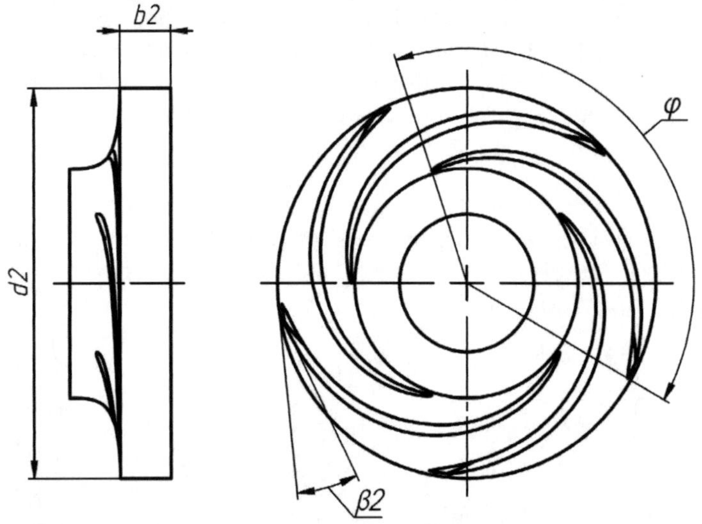

Figure 4. Optimization parameters of the suggested pump.

5. Optimization algorithm

A huge amount of machine time required for the process of pump optimization with CFD methods imposes constraints on optimization algorithms, number of designs considered and parameters, that can be chosen. In this case optimization is based on LPτ sequences [20]' which finds global minimum of the function relatively fast and is flexible in terms of chosen number of designs to be considered.

During the optimization 2 criteria were considered – maximum of efficiency and minimum of hemolysis.

Efficiency is calculated according to the following formula:

$$\eta = \frac{Q(p_2 - p_1)}{\omega M}, \tag{1}$$

where

Q – volumetric flow, m^3/s;

p_2– total pressure at the outlet of the pump, Pa;

p_1 – total pressure at the inlet of the pump, Pa;

M – torque, N·m;

ω – impeller rotational speed, rad/s.

*Hemolysis.*The value of shear stress experienced by blood cell during a period of time is usually considered as a source of cell damage. At the moment in order to obtain the level of hemolysis the following power law is used:

$$HI(\%) = \frac{\Delta Hb}{HB} \cdot 100 = C\tau^{\beta} t^{\alpha}, \tag{2}$$

where

HI – hemolysis index for one particle, %;

ΔHb – change of extracellular hemoglobin concentration in plasma;

IOP Conf. Series: Materials Science and Engineering **492** (2019) 012013 doi:10.1088/1757-899X/492/1/012013

HB – total hemoglobin concentration;
C– empirical constant;
τ – equivalent shear stress;
t – time interval of the cell experiencing the stress.

Formula (2) is one of the most common one (excluding more complex models, where there is still a power law with some additional features to obtain special blood cells behavior [21]). To apply formula (2) it is required to describe stress tensor in a scalar way. The following calculations use Huber—Mises criterion

$$\tau = \sqrt{\frac{1}{2}\left[(\sigma_1-\sigma_2)^2+(\sigma_2-\sigma_3)^2+(\sigma_1-\sigma_3)^2\right]}, \quad (3)$$

where
σ_i– primary stress components.

Values of constants for (2) are based on [22].

To estimate the level of cells damage the index of hemolysis is averaged over all control particles [23,24]:

$$HI_{total} = \frac{1}{N}\sum_{p=1}^{N} HI_p \quad (4)$$

where
HI_p shows the amount of cell *p* damage accumulated on its way through the pump;
N – number of control particles.

Simulation was carried for 128 design variants. In order to decrease the amount of numerical error each model was calculated using the same mesh parameters and boundary conditions.

6. Optimization results

The results of steady state simulations for the most representative designs are shown in table 1, the distributions of physical values are shown on figures 5–9.

Table 1. Some of simulation results.

No.	z	b_2	d_2	β_2	φ	Efficiency, %	H,m	H,mm Hg	HI
13	5	4.8	33.7	22.1	135	64.5	1.41	108.94	0.0451
14	4	6.3	44.2	31.1	96	57.7	2.73	210.92	0.249
20	4	4.5	38.3	21.6	142	67.0	1.73	133.66	0.0617
31	5	5.5	24.8	38.2	153	22.2	0.18	13.91	0.0263
32	4	4.0	35.3	29.2	114	68.4	1.43	110.48	0.0297
40	4	5.0	36.6	34.8	148	66.2	1.85	142.93	0.0495
48	4	3.7	36.0	23.0	151	67.4	1.39	107.39	0.0257
93	5	5.5	45.0	29.9	106	47.1	2.74	211.69	0.444
109	4	4.5	44.0	25.1	95	52.3	2.69	207.83	0.351

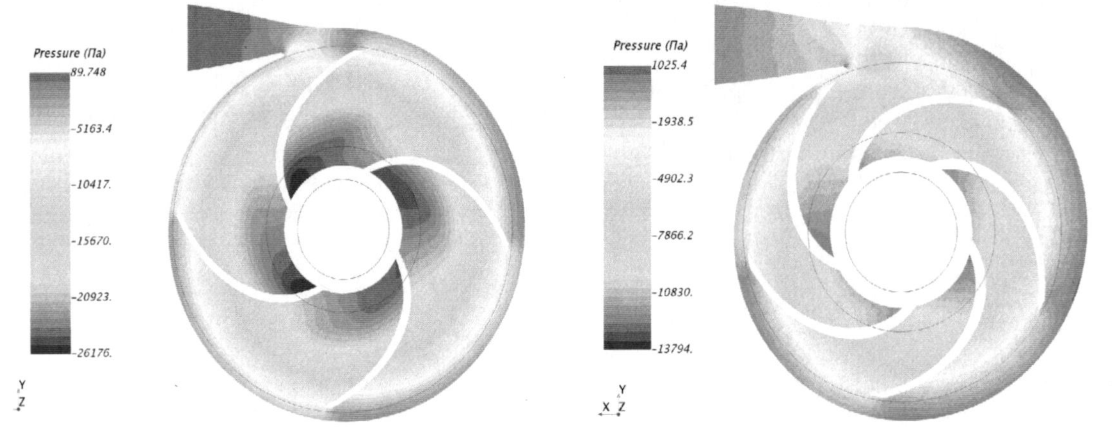

Figure 5. Pressure distribution at the cross sections of models no. 14 (left) and no. 13 (right).

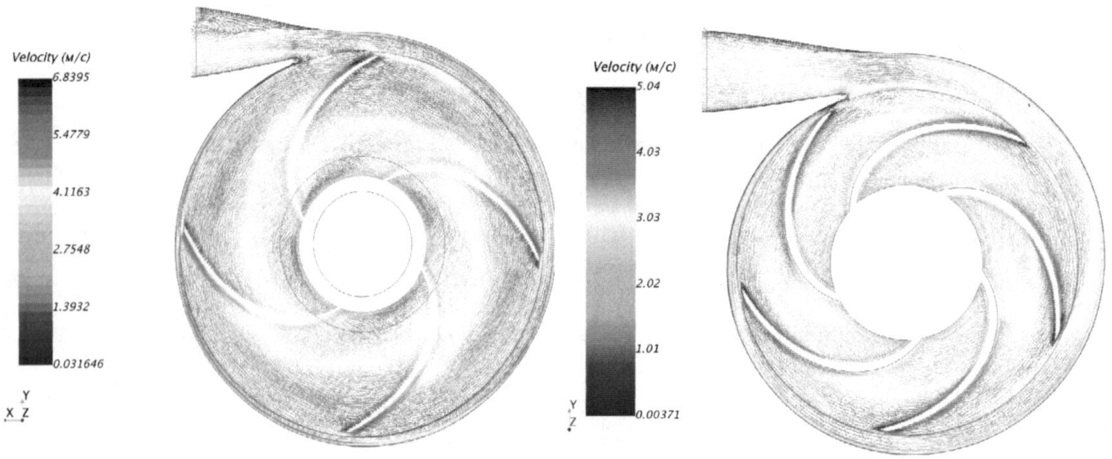

Figure 6. Velocity distribution at the cross sections of models no. 14 (left) and no. 13 (right).

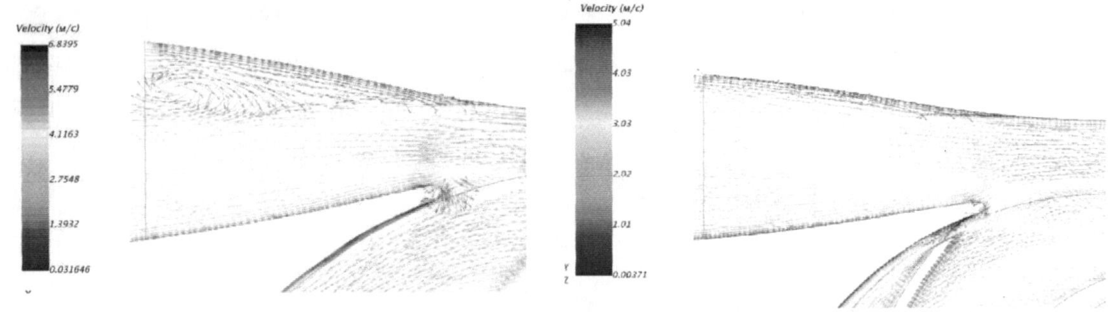

Figure 7. Velocity distribution at the cross sections of models no. 14 (left) and no. 13 (right) diffusors.

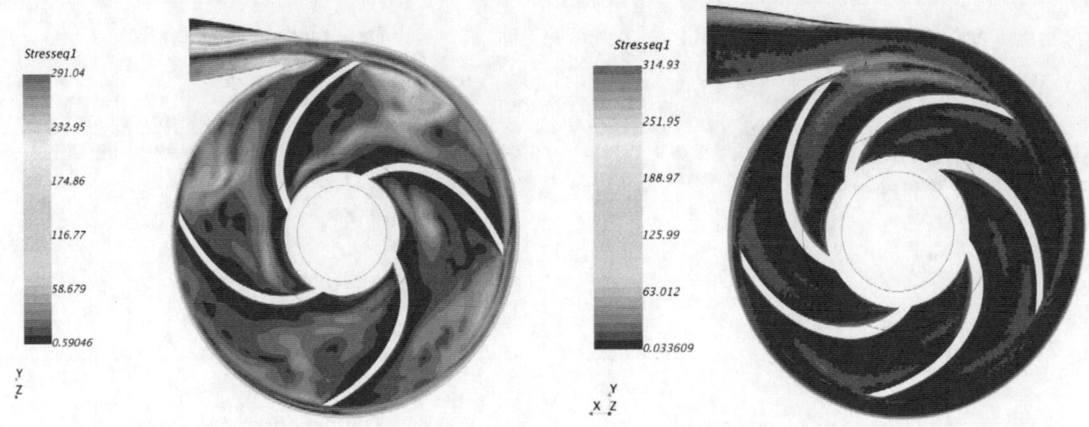

Figure 8. Equivalent shear stress distribution at the cross sections of models no. 14 (left) and no. 13 (right).

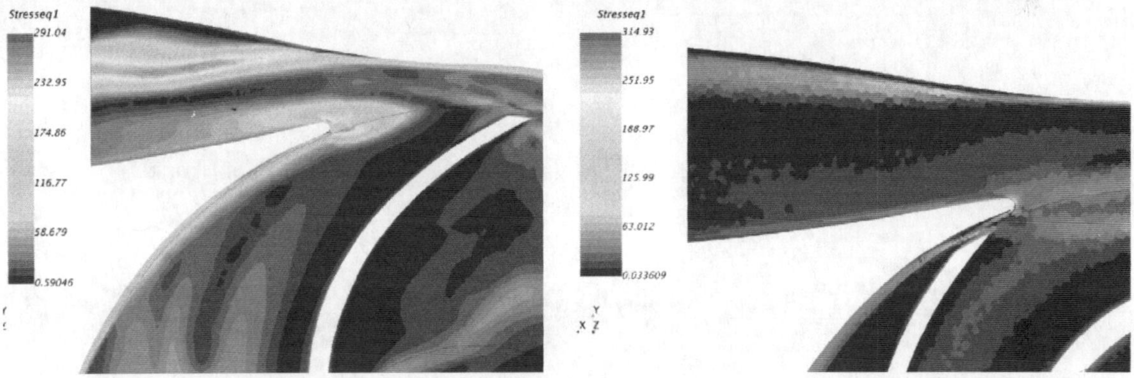

Figure 9. Equivalent shear stress distribution at the cross sections of models no. 14 (left) and no. 13 (right) diffusors.

In accordance with the data obtained the following dependences of the criteria under consideration on the parameters described above were pictured (figure 10).

A few of the designs with the best criteria values were calculated then at an unsteady state. The results of these simulations are represented in table 2.

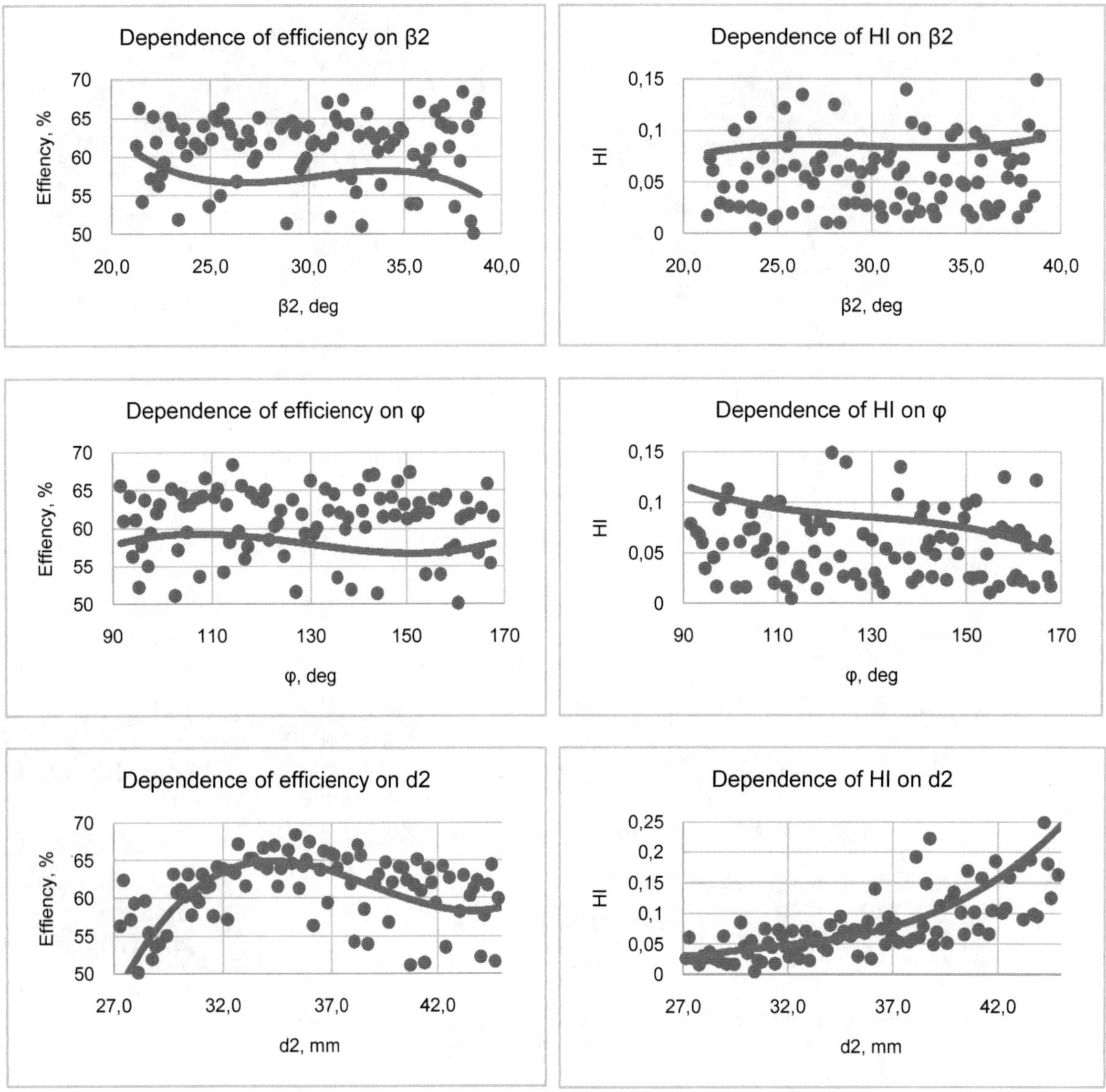

Figure 10. The effect of optimization parameters on efficiency (left) and hemolysis index (right), approximation line (red) — polynomial of the third degree.

Table 2. Results of unsteady simulation.

No.	z	b_2	d_2	β_2	φ	Efficiency, %	H,m	H,mm Hg	HI
133	5	4.8	33.7	4.8	135	65.2	1.44	111.25	0.0252
20	4	4.5	38.3	21.6	142	67.7	1.84	142.16	0.0345
32	4	4.0	35.3	29.2	114	69.1	1.5	115.89	0.0166
40	4	5.0	36.6	34.8	148	68.7	1.88	145.25	0.0277
48	4	3.7	36.0	23.0	151	68.2	1.47	113.57	0.0143

IOP Conf. Series: Materials Science and Engineering **492** (2019) 012013 doi:10.1088/1757-899X/492/1/012013

In accordance with optimization results the wet part geometry with the best criteria combination was chosen (design no. 32, table 2) and is shown on figure 11.

Figure 11. Optimized wet part geometry of the pump under consideration.

7. Comparison of wet part geometry configurations in a wide range of mass flow values

Requirements imposed on such pumps can change during the day, depending on patient's kinesis and the medicine he is taking [25]. In particular, it will affect the mass flow the pump should work on. Therefore, in order to compare two wet part configurations described above, it is necessary to compare them in a wide range of mass flow values.

Both variants were calculated in the same conditions: unsteady flow, mesh and solver parameters were kept the same; volumetric flow was varying from 20% to 120% from BEP (from 1 l/min to 6 l/min). Some of the results to compare are shown on figures 12–14.

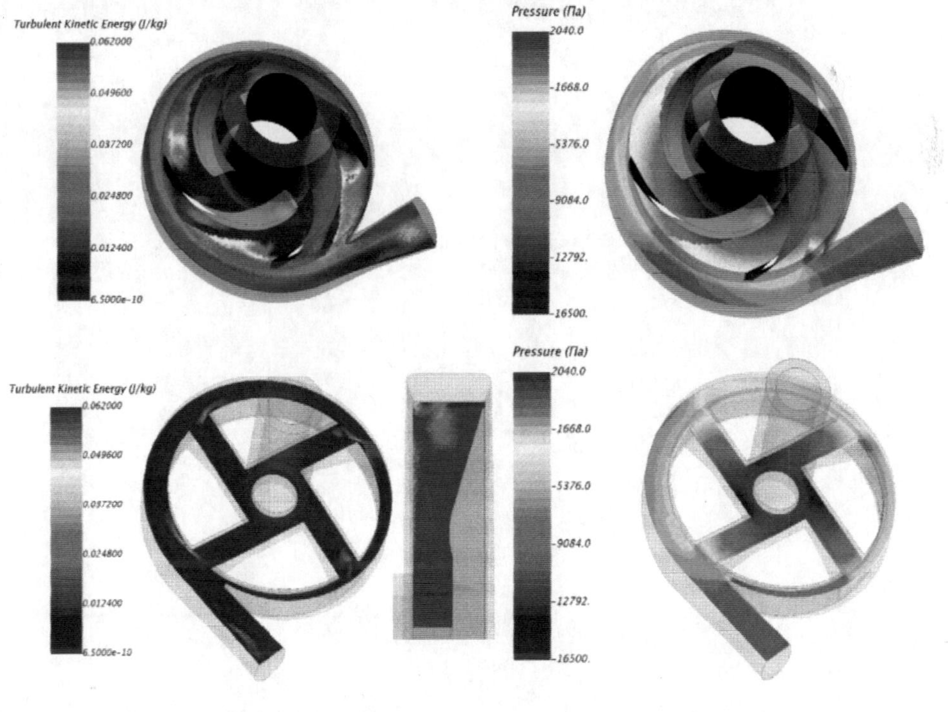

Figure 12. Simulation results at $Q = 0{,}4Q_{BEP}$.

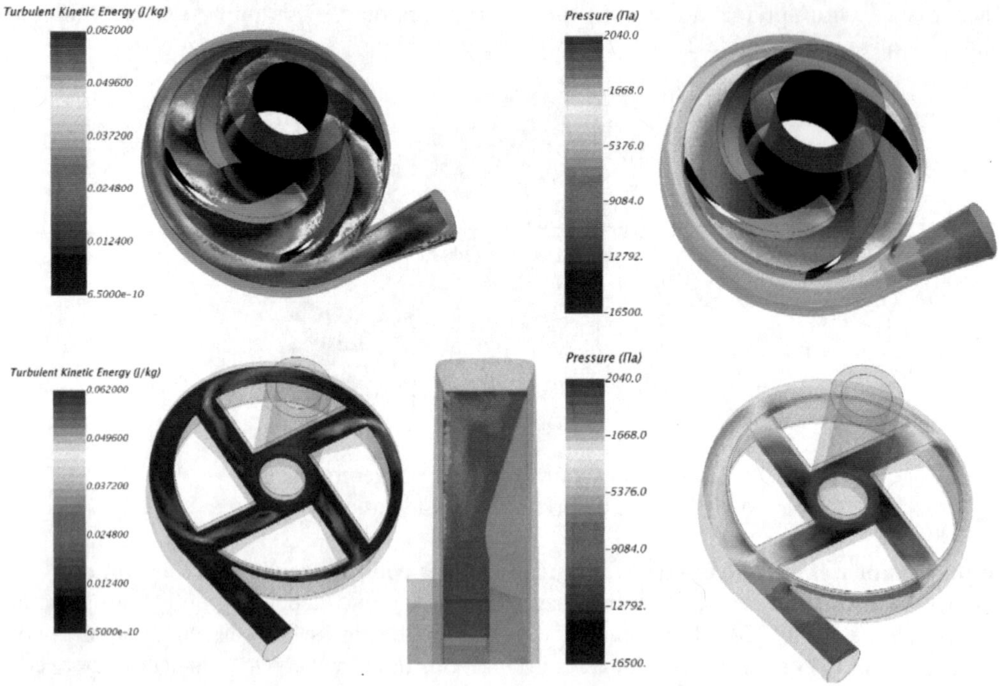

Figure 13. Simulation results at $Q = 0{,}8Q_{\mathrm{BEP}}$.

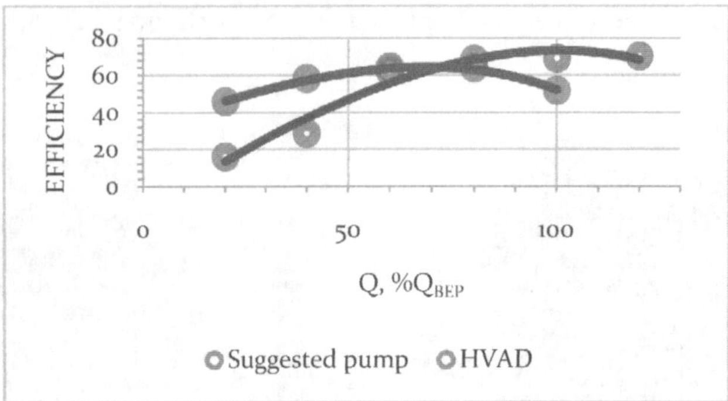

(a)

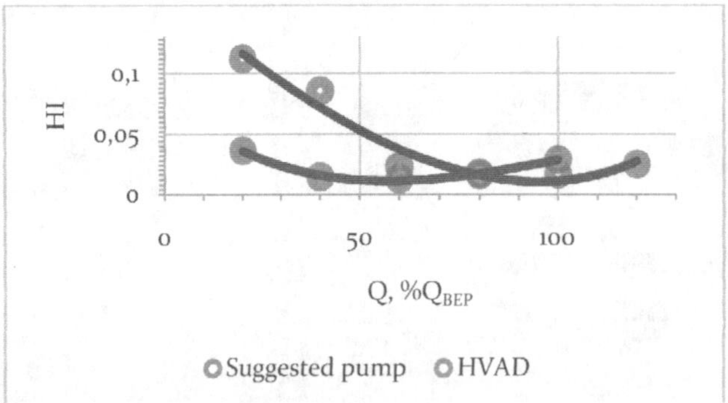

(b)

Figure 14. Pump characteristics in a wide range of mass flow value: a) comparison of efficiency curves; b) comparison of hemolysis index curves

IOP Conf. Series: Materials Science and Engineering **492** (2019) 012013 doi:10.1088/1757-899X/492/1/012013

8. Discussion and Conclusions

Conducted calculations allowed to conclude, that the suggested pump configuration with curved blades demonstrates better behavior than HVAD geometry at BEP and near it (at volumetric flow from 3 l/min to 6 l/min). High vorticity is obtained in the geometry with narrow channels. Under given conditions the suggested geometry efficiency is 14% higher and hemolysis index is nearly twice lower at BEP. Moreover, the mass of the device is also decreased in this case, what make it more comfortable to use. Nevertheless, in the zone of relatively low mass flows (from 1 l/min to 3 l/min) configuration with curved blades behaves significantly worse than the one with narrow channels (for volumetric flow 1 l/min its efficiency is 30% lower and hemolysis index is nearly 3-times higher). Such details should be taken into account when choosing frequency control system.

Obtained criteria dependences on parameters allow to conclude on the recommended intervals of the values to choose from when designing such devices: impeller width at the outlet (b_2) is recommended to vary between 3.5…4 mm, impeller diameter (d_2)between 32…37 mm, blade angle at the outlet (β_2) between 220°…270°, blade wrap angle (φ) between 120°…140°. The data calculated is also a good base for the following analysis and additional optimization to be carried.

References

[1] Kirklin J K et al 2017 Eighth annual INTERMACS report: Special focus on framing the impact of adverse events *J. Hear. Lung Transplant.* **36**(10) pp 1080–86

[2] Gouskov A M, Sorokin F D and Banin E P 2016 Simulation of an Inlet Structure of an Implantable Axial Blood Pump *Biomed. Eng.* (NY)

[3] Su B, Chua L P and Zhong L 2013 Numerical Studies of an Axial Flow Blood Pump With Different Diffuser Designs *J. Mech. Med. Biol.* **13** (3) pp 1350029

[4] Wu J et al 2012 Computational Fluid Dynamics-Based Design Optimization for an Implantable Miniature Maglev Pediatric Ventricular Assist Device *J. Fluids Eng.* **134** (4) pp 041101

[5] Karimov J H et al 2017 The axial continuous-flow blood pump: Bench evaluation of changes in flow associated with changes of inflow cannula angle *J. Hear. Lung Transplant.* **36** (1) pp 106–112

[6] Gaffey A C et al 2018 Is there a difference in bleeding after left ventricular assist device implant: centrifugal versus axial? *J. Cardiothorac. Surg.* **13**(1) pp 22

[7] F.H. S. et al 2017 Characteristics and outcomes of LVAD patients who undergo pump exchange: Axial flow vs. centrifugal flow therapy J. *Hear. Lung Transplant.* **36** (4) pp S99–S100

[8] Kimura M et al 2017 Cerebrovascular Accident Rate Is Different Between Centrifugal and Axial-Flow Pumps, but Survival and Driveline Infection Rates Are Similar *Transplant. Proc.* **49** (1) pp 121–124

[9] Kimura M et al 2015 Midterm outcome of implantable left ventricular assist devices as a bridge to transplantation: Single-center experience in Japan *J. Cardiol.* **65** (5) pp 383–389

[10] Lalonde S D et al 2013 Clinical Differences Between Continuous Flow Ventricular Assist Devices: A Comparison Between HeartMate II and HeartWare HVAD *J. Card. Surg.* **28** (5) pp 604–610

[11] Coffin S T et al 2015 Adverse neurologic events in patients bridged with long-term mechanical circulatory support: A device-specific comparative analysis *J. Hear. Lung Transplant.* **34** (12) pp 1578–1585

[12] Bell P E 1992 *United States Patent.* **2** (12)

[13] Versteg H and Malalasekera W 2007 *An introduction to Computational Fluid Dynamics: The*

IOP Conf. Series: Materials Science and Engineering **492** (2019) 012013 doi:10.1088/1757-899X/492/1/012013

Finite Volume Method. 2nd ed. (Pearson Education Limited, Harlow)

[14] Bui V et al 2017 Numerical study of the influence of flow blockage on the aerodynamic coefficients of models in low-speed wind tunnels (Springer) **24** (6) pp 857–866

[15] Isaev S A et al 2010 Influence of the Reynolds number and the spherical dimple depth on turbulent heat transfer and hydraulic loss in a narrow channel *Int. J. Heat Mass Transf.* **53** (1–3) pp 178–197

[16] Isaev S A et al 2016 Numerical simulation of the turbulent air flow in the narrow channel with a heated wall and a spherical dimple placed on it for vortex heat transfer enhancement depending on the dimple depth *Int. J. Heat Mass Transf.* **94** pp 426–448

[17] Fraser K H et al 2011 The use of computational fluid dynamics in the development of ventricular assist devices *Medical Engineering and Physics* **33** (3) pp 263–280

[18] Devisilov V A, Sharai E Y 2016 Numerical study of the flow structure in a hydrodynamic filter *Theor. Found. Chem. Eng.* **50** (2) pp 209–216

[19] Kulik V V, Parkin AN and Navasardyan E S 2016 Numerical Modeling Procedure for Micromachined Cryogenic Cooler Elements Using ANSYS Fluent Software and Viscous Flow in a Small-Diameter Channel with Heat Transfer as an Example *Chem. Pet. Eng.* **52** (7–8) pp 531–538

[20] Chaburko P S, Lomakin V O and Kuleshova M S B.M.N. 2016 Comprehensive optimization of the hermetic pump's flow part using LP-TAU search method *Pumps. Turbines. System* **1** (18)

[21] Yu H et al 2017 A Review of Hemolysis Prediction Models for Computational Fluid Dynamics *Artif. Organs.* **41** (7) pp 603–621

[22] Giersiepen M et al 1990 Estimation of Shear Stress-Related Blood Trauma in Heart Valve Prostheses in vivo Comparison of 25 Aortic Valves *Artif. Organs.* **12** (5) pp 130–136

[23] Gouskov A M et al 2017 Minimization of Hemolysis and Improvement of the Hydrodynamic Efficiency of a Circulatory Support Pump by Optimizing the Pump Flowpath *Biomed. Eng.* (NY)

[24] Gouskov A M et al 2016 Assessment of Hemolysis in a Ventricular Assist Axial Flow Blood Pump *Biomed. Eng.* (NY)

[25] Demidova M and Tikhonenko V 2001 Circadian rhythm of heart rate variability in healthy subjects *Journal of Arrhythmology* **23** pp 52–58

IOP Conf. Series: Materials Science and Engineering **492** (2019) 012014 doi:10.1088/1757-899X/492/1/012014

Independent component analysis of hydrodynamic bubbles' properties

S Gavrilev[1], O Ivanova[1,3] and T Decaen[2]

[1]Bauman Moscow State Technical University, 5 Second Baumanskaya Street, Moscow, 105005, Russian Federation
[2]Oliver Wyman, 1 Rue Euler, 76008 Paris, France

[3]E-mail: olgaivanova@bmstu.ru

Abstract. The paper considers the existing acoustic methods for determining the dispersion composition of the gaseous phase in liquid medium used in hydromechanical systems. The shortcomings of these methods were identified. A new method based on the analysis of independent components is proposed. The specialties of the use of the method based on the analysis of independent components are described. An experimental plant was assembled and a number of measurements were carried out.

1. Introduction

With an increase in the volume of hydromechanical systems including the substances in different phase states, there is a need of study the hydrodynamic processes occurring in them.

Special attention is paid to determining the dispersion composition of the gaseous phase in liquid medium, in particular to air bubbles in water [1–3].

At present, there are three types of methods for determining the dispersion composition of the gaseous phase in liquid medium: optical, photometric and acoustic. Optical and photometric methods are the most common [4–7], but they have one significant weakness — they are applicable only in optically transparent media.

Acoustic methods are based on the hydroacoustic properties of bubbles, such as radiation, absorption and reflection of sound waves by bubbles. The noise emitted by bubbles has been the subject of research for many decades. This question was paid attention by scientists such as: Leyton (1994), Shtrassberg (1956), Longet—Khiggins (1990), Din iStoks (2008) and many others. However, their works are mainly dedicated to the study of the acoustic of a single bubble or a small group of bubbles.

The most modern acoustic methods for determining the distribution of bubbles by size are based on measurements of the acoustic wave's velocity attenuation waves passing through a layer of a bubble cloud. These methods are called "active". Examples of the realizations of these methods are described in [8, 9]. However, we should take into account that the acoustic wave generated for the bubbles irradiation can influence on the bubbles. This phenomenon can be noted in the results of [10]. The authors of this work carried out a series of experiments, changing the emitted acoustic wave intensity and in each of the experiments obtained different results. Thus, in a number of hydrodynamic systems, where the maintenance of working conditions with a certain bubble size is especially important, the use of active methods is inadmissible.

Another type of methods are passive methods. They all originate from the research of Minaert. He became the first to describe the sound waves generated by a bubble in 1933. According to his research, under adiabatic conditions, the frequency of the sound wave emitted by a bubble relates with its size as:

$$f = \frac{1}{R}\left(\frac{3\gamma P}{(2\pi)^2 \rho}\right)^{\frac{1}{2}}, (1)$$

where
f – is the frequency of the sound wave emitted by the bubble;
P – is the absolute liquid pressure;
γ – is the coefficient of gas specific heat,
ρ – is the density of the liquid;
R – is the bubble radius.

The principle of passive methods is that the noise emitted by bubbles is measured, and the size of the bubbles is calculated from equation 1. Examples of the implementation of these methods are described in [11–14]. The disadvantage of these works is that under this approach, it is possible only to understand the size of bubbles, but impossible to get the size distribution.

In [12], the author associated the noise spectrum of bubbles, measured by a single hydrophone, with their size distribution. It is worth noting that the author did not take into account the properties of the acoustic field, where the measurements were carried out. Also, the geometry of the collocations of the hydrophone and the bubbles was not taken into account. In other words if we move the hydrophone or slightly change the acoustic field conditions, we will get a completely different spectrum, and, therefore, a different distribution.

At present, the growth of computing power and a variety of new algorithms of data processing allows to liquidate the disadvantages of the methods described above. In this article, to improve the passive methods for determining the size distribution of bubbles, it is proposed to use the algorithms of analysis of independent components. These algorithms are used in virtually all areas of contemporary science [15–19].

2. Formulation of the problem

There are n signals being recorded from hydrophones. Each of the signals is a mixture of acoustic waves emanating from a variety of sources (bubbles). In a first approximation, assume that the sources have a certain number n, i.e. there are only n different sizes of bubbles. The goal of the task is to identify our initial signals from the incoming mixtures of signals. In matrix form, this can be written as:

$$\mathrm{X} = \mathrm{A} \times \mathrm{S}, \ (2)$$

where
X – is the matrix of recorded signals;
A – is the confusion matrix;
S – is the matrix of the original signals.

In other word, the goal is to find the confusion matrix A to count initial signals:

$$\mathrm{S} = \mathrm{A}^{-1} \times \mathrm{X}. \ (3)$$

In this paper, the FastICA algorithm [20,21] was chosen as the basic algorithm for the analysis of independent components. It refers to standard algorithms and does not take into account the temporal structure of the input signals. But as a test of the feasibility of further works, FastICA algorithm is the optimal choice.

IOP Conf. Series: Materials Science and Engineering **492** (2019) 012014 doi:10.1088/1757-899X/492/1/012014

3. Experimental part

It was supposed that surfacing bubbles execute a harmonic vibration with frequencies determined by formula 1. The measurements were carried out in a hydroacoustic anechoic chamber at the academic chair «Ecology and Industrial Safety», Bauman Moscow State Technical University. Installation diagram is shown in figure 1.

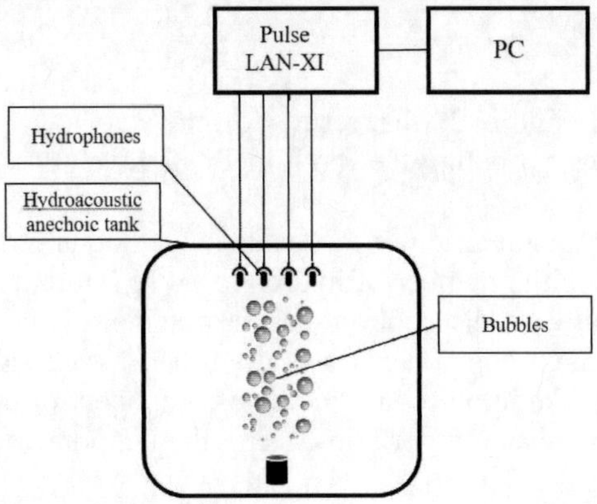

Figure 1. Schematic diagram of experimental plant.

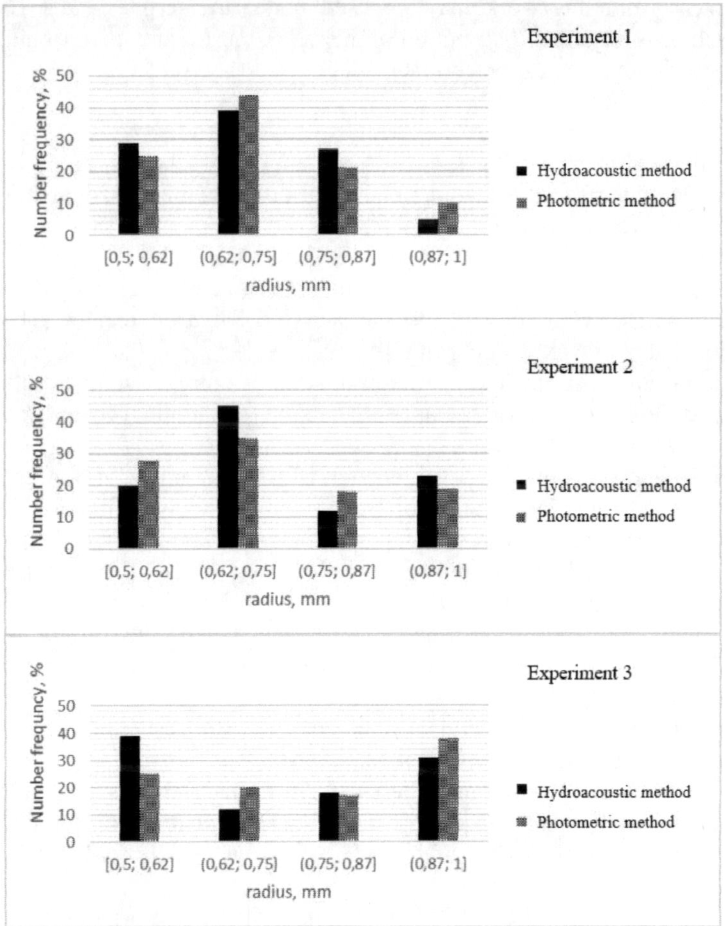

Figure 2. Diagrams of bubble size distribution.

Bubble noise was recorded using four Bruel&Kjaer type 8103 hydrophones with the Pulse LAN-XI multichannel analyzer. Next, the recorded signals were processed by an algorithm of analysis independent components and the bubble size distribution was calculated. At the same time, bubble cloud images were taken, and these images were processed in the Shadow sizing software of Dynamic studio by Dantec Dynamics [22]. Thus three experiments were carried out. The results are presented in figure 2.

4. Results

As it can be seen from the figure 2, the results of the acoustic and photometric method are quantitatively different. It is worth noting that both methods give approximately the same picture of the distribution.

One of the factors limiting the precision of the acoustic method is that only four hydrophones were used. In other words, increasing the number of hydrophones in future, and thus the matrix size of the recorded signals X, a more precise distribution can be obtained. It is also worth noting that as an algorithm of analysis independent components was used the FastICA algorithm, which refers to the basic algorithms and does not take into account the time component of the input signals. In the future, it is expected to use another family of algorithms, using the temporal structure of input signals, such as: AMUSE, TDSEP, SOBI [23–25].

References

[1] Bakhtin B I, Ivashov A I, Kuznetsov A V and Skorokhodov A S 2014 Experimental investigation of the specific features of formation of cavitation zones in intense ultrasound fields *J. Eng. Phys. Thermophys.* **87** 672–85

[2] Kichatov B V, Polyaev V M and Runkovskii S V2001 Gas bubble dynamics with instability in the operation of a system of transpiration cooling *Heat Transf. Res.* **32** 273–80

[3] Ganchev B G and Peresad'ko V G 1985 Processes of hydrodynamics and heat exchange in descending bubble flows *J. Eng. Phys.* **49** 879–85

[4] Mazahernasab R and Ahmadi R 2016 Determination of bubble size distribution in a laboratory mechanical flotation cell by a laser diffraction technique *Physicochem. Probl. Miner. Process.* **52** 690–702

[5] Baroni D B, Filho J S C, Lamy C A, Bittencourt M S Q, Pereira C M N A and Motta M S 2011 Determination of Size Distribution of Bubbles in a Bubbly Column Two-Phase Flows By Ultrasound and Neural Networks

[6] Zhang Y and Sun H 2012 Measurement of bubble size distribution in liquids by optical and acoustical methods *Proc. Int. Conf. Commun. Syst. Netw. Technol. CSNT 2012* 671–4

[7] Gaillard T, Honorez C, Jumeau M, Elias F and Drenckhan W 2015 A simple technique for the automation of bubble size measurements *Colloids Surfaces A Physicochem. Eng. Asp.* **473** 68–74

[8] Wu X J and Chahine G L 2010 Development of an acoustic instrument for bubble size distribution measurement *J. Hydrodyn.* **22** 325–31

[9] Xue J 2004 Bubble Velocity , Size and Interfacial Area Measurements in Bubble Columns 1–210

[10] Al-Masry W A, Ali E M and Aqeel Y M 2005 Determination of Bubble Characteristics in Bubble Columns Using Statistical Analysis of Acoustic Sound Measurements *Chem. Eng. Res. Des.* **83** 1196–207

[11] Greene C A and Wilson P S 2012 Laboratory investigation of a passive acoustic method for measurement of underwater gas seep ebullition *J. Acoust. Soc. Am.* **131** EL61-EL66

IOP Conf. Series: Materials Science and Engineering **492** (2019) 012014 doi:10.1088/1757-899X/492/1/012014

[12] Manasseh R, LaFontaine R F, Davy J, Shepherd I and Zhu Y-G 2001 Passive acoustic bubble sizing in sparged systems *Exp. Fluids***30** 672–82

[13] Chen L, Wood S, Moore S and Nguyen B 2012 Acoustic Emission of Bubbly Flow and Its Size Distribution Spectrum *Proc. Acoust.* 1–6

[14] Husin S and Mba D 2010 Correlation between Acoustic Emission (AE) and Bubble Dynamics **II** 0–5

[15] Anishchenko L N 2016 Independent component analysis in bioradar data processing *2016 Prog. Electromagn. Res. Symp. PIERS 2016 - Proc.* 2206–10

[16] Anishchenko L, Razevig V and Chizh M 2018 Blind separation of several biological objects respiration patterns by means of a step-frequency continuous-wave bioradar *2017 IEEE Int. Conf. Microwaves, Antennas, Commun. Electron. Syst. COMCAS 2017***2017–Novem** 1–4

[17] Hyvärinen A, Karhunen J and Oja E 2001 Independent Component Analysis *Appl. Comput. Harmon. Anal.***21** 135–44

[18] Knyazev B, Barth E and Martinetz T 2017 Recursive autoconvolution for unsupervised learning of convolutional neural networks *Proc. Int. Jt. Conf. Neural Networks***2017–May** 2486–93

[19] Chernomyrdin N V., Zaytsev K I, Lesnichaya A D, Kudrin K G, Cherkasova O P, Kurlov V N, Shikunova I A, Perchik A V., Yurchenko S O and Reshetov I V. 2016 Principle component analysis and linear discriminant analysis of multi-spectral autofluorescence imaging data for differentiating basal cell carcinoma and healthy skin 99760B

[20] Heaton C and Ripley B D 2017 Package ‘ fastICA ’

[21] Langlois D, Chartier S and Gosselin D 2010 An Introduction to Independent Component Analysis: InfoMax and FastICA algorithms *Tutor. Quant. Methods Psychol.*

[22] S. R, H. H, W.Y. S, E. S-P, R.C. C, W. B, M. D, E. D and S.L. T 2010 Optimizing outcomes of IVM treatment cycles-results of a new and improved protocol *Hum. Reprod.*

[23] Liu W and Chi C-Y Research In AMUSE : A Blind Identification Algorithm *Electr. Eng.***0**

[24] Frølich L and Dowding I 2018 Removal of muscular artifacts in EEG signals: a comparison of linear decomposition methods *Brain Informatics*

[25] De Lathauwer L and Castaing J 2008 Blind identification of underdetermined mixtures by simultaneous matrix diagonalization *IEEE Trans. Signal Process.*

IOP Conf. Series: Materials Science and Engineering **492** (2019) 012015 doi:10.1088/1757-899X/492/1/012015

The method of constructing the cavitation characteristics of a screw centrifugal pump using the methods of hydrodynamic modeling

N Egorkina[1, 2] **and A Petrov**[1]

[1] Bauman Moscow State Technical University, 5 Second Baumanskaya Street, Moscow, 105005, Russian Federation

[2] E-mail: negorkin@bk.ru

Abstract. By the example of a centrifugal pump with an upstream screw, the possibilities of using hydrodynamic modeling for constructing a cavitation characteristic of a screw centrifugal pump are shown. An example is given of how the cavitation characteristic of a screw centrifugal pump was obtained on the basis of a simplified model of the Rayleigh–Plesset model. It illustrates and shows how the pump characteristic changes with varying empirical coefficients. According to the results of the study, conclusions were made that in the future will allow to improve the calculation of such pumps and to obtain more accurate calculation results.

1. Introduction

Cavitation refers to the process of disrupting the continuity of flow as a result of the formation of vapor or gas bubbles in a fluid flow in a region of reduced pressure and their subsequent condensation in a region of elevated pressure.

Cavitation is a complex set of the following phenomena:

- vapor and dissolved gases from the liquid in areas where the pressure of the liquid is equal to or less than the pressure of saturated vapors;
- local increase in the velocity of the fluid in the place where the vaporization occurred, and random fluid motion;
- condensation of vapor bubbles entrained by the flow of fluid to the area of high pressure. Condensation of each of the bubbles leads to a sharp decrease in volume and hydraulic shock in microscopic zones; however, the collapse in large quantities on the cavitating surface leads to large areas of destruction. Repeated mechanical effects during condensation of bubbles cause the mechanical process of destruction of the wheel material, which is the most dangerous consequence of cavitation (figure 1).

Cavitation is accompanied by a decrease in the flow, head, power and efficiency of the pump; sound phenomena (noise, crackling, shock) and vibration of the pumping unit; erosion of the material of the channel walls; reduced reliability of the pump.

In centrifugal pumps, cavitation occurs on the back side of the impeller blade near its inlet edge, since here the pressure is much lower than the pressure in the suction inlet pipe due to hydraulic losses in the pump inlet device and local increase in the relative velocity of the fluid.

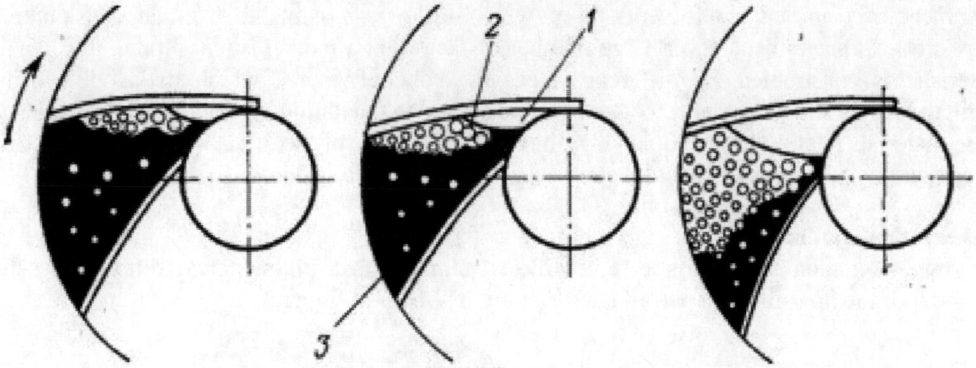

Figure 1. Cavitation in the centrifugal wheel: 1- cavity, 2- hydrodynamic trace, 3 - fluid flow.

The excess of the specific energy of the liquid at the entrance to the pump over the specific energy of the vapor of this liquid, reduced to the center of gravity of the inlet section, is called the cavitation margin Δh, which is determined by

$$\Delta h = \frac{p_{pump}}{\gamma} + \frac{v_0^2}{2g} - \frac{p_{satursted\ stream}}{\gamma}, \quad (1)$$

The evaluation of the cavitation qualities of the pumps is made on the basis of the cavitation characteristics obtained as a result of tests on special stands. Cavitation characteristic — the dependence of the allowable cavitation stock on the pump feed (figure 2).

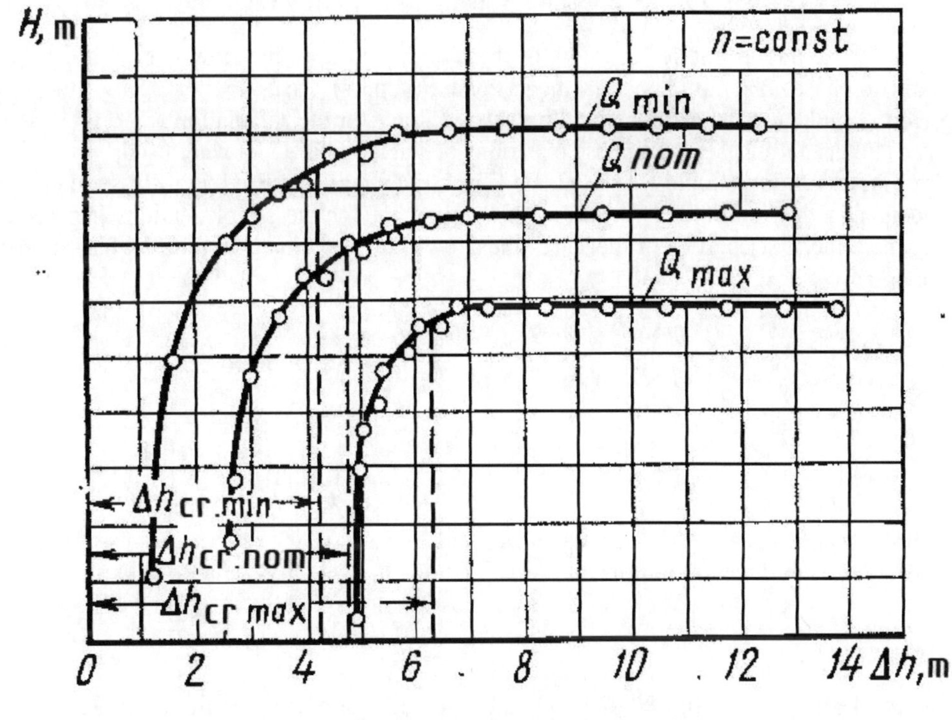

Figure 2. Cavitation characteristics.

In this paper, the calculation and construction of the cavitation characteristic will be performed using hydrodynamic modeling. For this, a simplified Rayleigh–Plesset equation is used, with the establishment of empirical coefficients in it, which make it possible to exclude consideration of complex physical phenomena and a large number of factors in a mathematical model. The cavitation characteristic is constructed for different values of the coefficients, which makes it possible to compare and reveal the dependence of their influence on the cavitation characteristic. In many papers there are materials relating to this topic [1-4], however, the study of the effect of empirical coefficients on the cavitation characteristic of screw centrifugal pumps was not made.

2. Mathematical model

To build the cavitation characteristic, a centrifugal pump with a pre-switched screw was selected. A 3D-model of the flow-through part of such a pump is shown in figure 3.

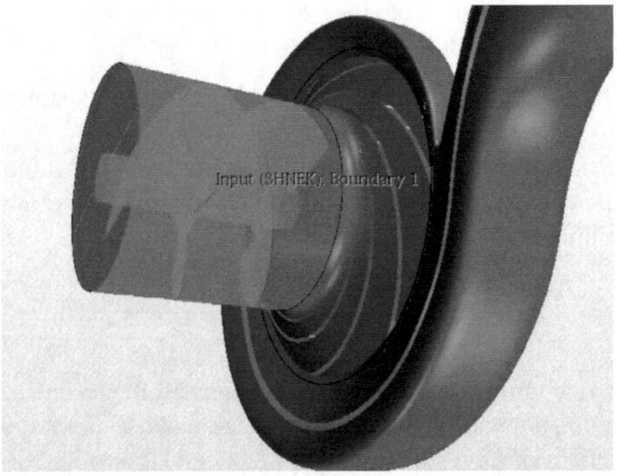

Figure 3.3D-model of the flow-through part of the centrifugal pump.

The calculation for cavitation implies the flow of the working fluid in the flow part of the pump, which consists of two phases: the liquid phase + the gas phase (vapors) [5,6].

This fact greatly complicates the mathematical model necessary for the calculation.

Such models as Euler multiphase, multiphase equations of state, VOF-model were used.

The growth of the cavitation bubble was calculated from the simplified Rayleigh-Plesset equation with the introduction of the coefficients k_1 and k_2 to compensate for the simplifications introduced into the model, where the first coefficient measures the growth rate of the cavitation bubble, and the second – the collapse rate:

$$\left(\frac{dR}{dt}\right)=k_1\sqrt{\frac{2}{3}\left(\frac{p_{pump}-p}{\rho}\right)} \quad if\ p_{pump}>p, \tag{2}$$

$$\left(\frac{dR}{dt}\right)=-k_2\sqrt{\frac{2}{3}\left(\frac{p_{pump}-p}{\rho}\right)} \quad if\ p_{pump}<p, \tag{3}$$

Construction of cavitation characteristics was carried out by varying the reference pressure (Total Pressure). The Total Pressure function returns the static pressure and velocity pressure values:

$$T_p=p+\frac{\rho V^2}{2}. \tag{4}$$

The computational grid consists of 480,000 computational cells (figure 4), it is structured at solid walls (prismatic boundary layer) and is not structured in the flow core (polyhedra) [7, 8].

Figure 4. Calculated grid of screw centrifugal pump.

3. Simulation results

Calculation of the cavitation characteristic was made by gradually lowering the reference pressure, because this method allows to reduce the pressure level in the entire flow section simultaneously, the working fluid is water at a temperature of 25 °C.

The construction of the cavitation characteristic and the calculation of the values of pressure were made for different values of empirical coefficients: 1) 1 and 1; 2) 2 and 0.5; 3) 5 and 0.2; 4) 10 and 0.1.

Based on the obtained numerical values, the cavitation characteristic of the pump was obtained at coefficients 1 and 1 (figure 5). As can be seen from this characteristic, the pump does not have a pressure drop even at values of the cavitation margin close to zero, which cannot be [9-12].

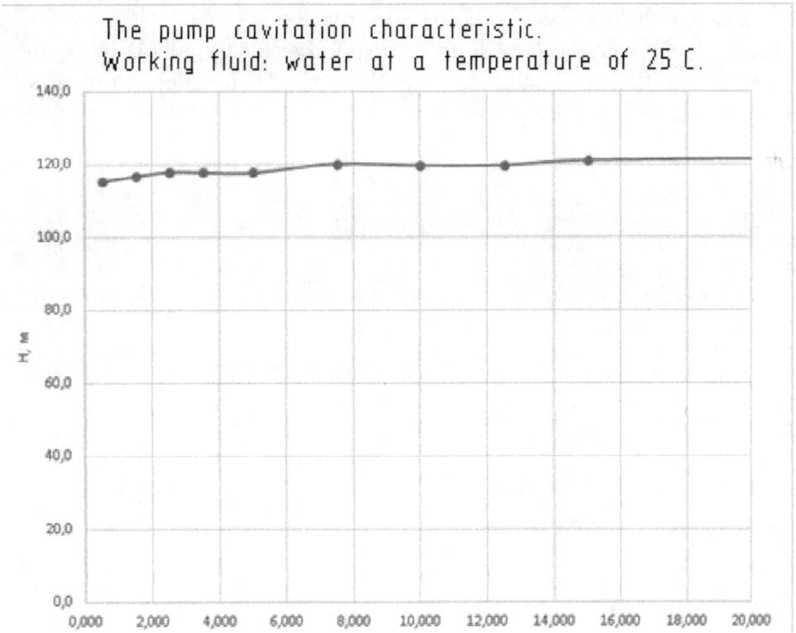

Figure 5. Cavitation characteristic of a centrifugal pump with empirical coefficients 1 and 1.

Then the cavitation characteristic of the pump was obtained with coefficients 2 and 0.5 (figure 6). As can be seen from this characteristic, the pump also does not have a pressure drop even at values of the cavitation margin close to zero, as well as at values 1 and 1, which cannot correspond to reality [13-15].

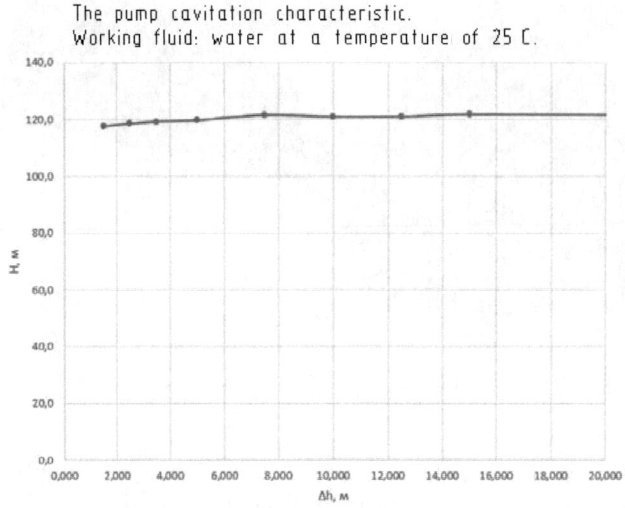

Figure 6. Cavitation characteristics of a screw centrifugal pump with empirical coefficients of 2 and 0.5.

Further calculations were performed for the coefficients 5 and 0.2, as well as 10 and 0.1, respectively. Figure 7 shows a comparison of the cavitation characteristics of the pump at different coefficients. As we see from the graphs, the disruption of the pump begins when the values 5 and 0.2 are chosen. At coefficients of 10 and 0.1, the disruption of work occurs at a higher pressure, that is, at a higher value of the safety factor.

According to the calculations, the cavitation reserve in the pump calculated by the formula (4) is equal to 2 m. Thus, the coefficients 5 and 0.2, as well as 10 and 0.1, reflect more closely the reality of the cavitation process occurring in the hydromachine [16, 17].

$$\Delta h_c = \frac{v_0^2}{2g} + \lambda_{cr}\frac{W_0^2}{2g}.$$

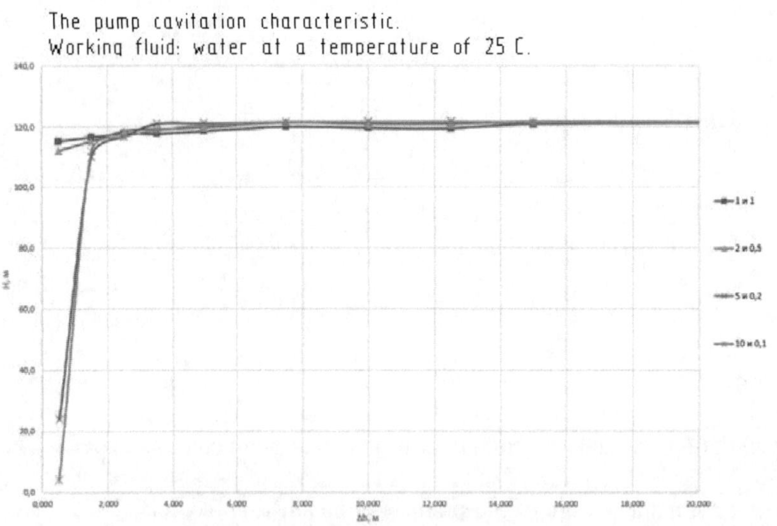

Figure 7. Comparison of the cavitation characteristics of the pump with different empirical coefficients.

IOP Conf. Series: Materials Science and Engineering **492** (2019) 012015 doi:10.1088/1757-899X/492/1/012015

Figure 8 illustrates the process of cavitation development when the coefficients in the Rayleigh–Plesset equation change [18-20].

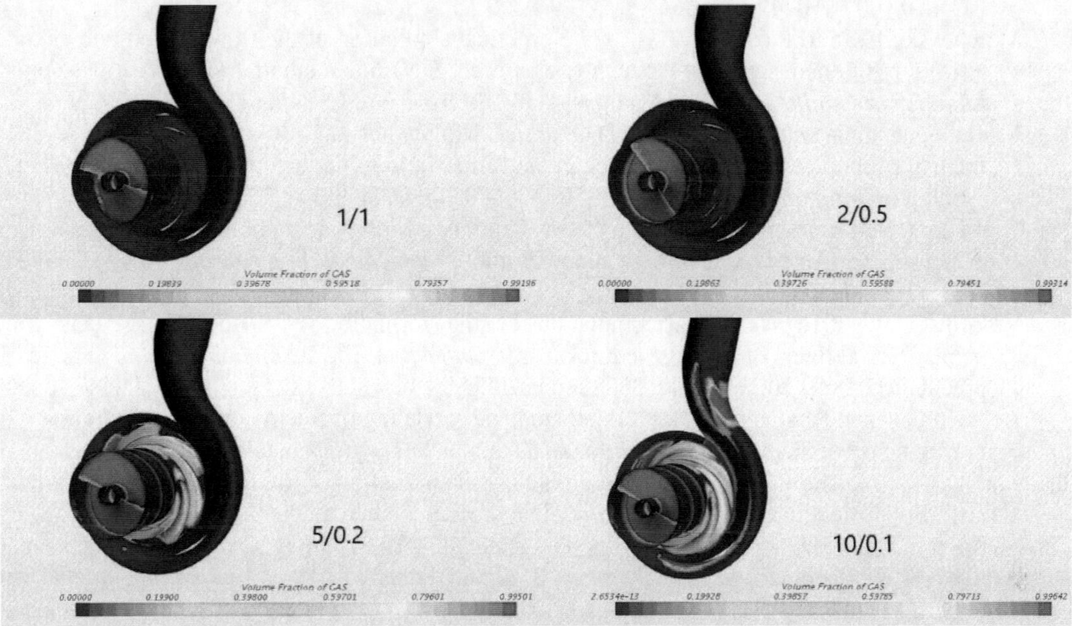

Figure 8. Volume fraction of gas with a cavitation reserve of 2.5 m with various empirical coefficients.

4. Findings

The dependence between the coefficients in the Rayleigh-Plesset equation and the cavitation characteristic of a centrifugal pump is revealed. With the mutual increase of the first coefficient and decrease of the second, a change in the cavitation characteristic and the cavitation reserve of the screw-centrifugal pump is observed.

It is necessary to take into account the coefficients in the Rayleigh-Plesset equation, since simplifying the calculation model negatively affects the final result of the study.

Due to mathematical modeling, it is possible to predict the operation of the pump without preliminary full-scale testing, which will allow pre-identification of possible shortcomings in the work. Thus, the percentage of tests that meet the requirements of developers will be increased, which means that the cost of the finished product can be reduced.

References

[1] Petrov A I and Isaev N Yu 2017 Gidrodinamicheskoe modelirovanie raboty centrobezhnogo nasosa v zone otricatel'nyh podach *Gidravlika* (3)

[2] Petrov A I 2016 Metodika nepreryvnogo poluchenija harakteristik lopastnogo nasosa dlja peremennoj temperatury i vjazkosti rabochej zhidkosti pri ispytanijah v termobarokamere *Inzhenernyj vestnik. Elektronnyj zhurnal* (10)

[3] Lomakin V O and Bibik O Yu 2017 Vlijaniej empiricheskihkoj efficientov v modeli Releja-Plesetta na raschetnye kavitacionnye harakteristiki centrobezhnogo nasosa *Gidravlika* (3)

[4] Lomakin V O, Kuleshovav M S and Bozh'eva S M 2016 Numerical Modeling of Liquid Flow in a Pump Station *Power Technology and Engineering***49** (5) pp 324–327 DOI: 10.1007/s10749-016-0623-9

[5] Marchevskii I K and Puzikova V V 2017 Numerical simulation of the flow around two circular airfoils positioned across the stream using the LS-STAG method *Journal of Machinery Manufacture andReliability***46** (2) pp 114–119 DOI: 10.3103/S105261881702011X

[6] Yudin A, Vozhdaev A, Sukhotskiy D, Salmina M, Sukhotskaya T and Sukhotskiy V 2017 Intensive robotics education approach in the form of a summer camp *Advances in Intelligent Systems and Computing***560** pp 246–250 DOI: 10.1007/978-3-319-55553-9_23

[7] Gouskov A M, Lomakin V O, Banin E P and Kuleshova M S 2016 Assessment of Hemolysis in a Ventricular Assist Axial Flow Blood Pump *Biomedical Engineering* **50** (4) 233–236 DOI: 10.1007/s10527-016-9627-x

[8] Fedorov S V 2016 Numerical simulation of the formation of shaped-charge jets from hemispherical liners of degressive thickness *Combustion, Explosion and Shock Waves* 52 (5) pp 600–612 DOI: 10.1134/S0010508216050117

[9] Borovin G K and Kostyuk A V 2002 Mathematical modeling of the hydraulic control system of a walking robot *Izvestiya Akademii Nauk. Teoriya I Sistemy Upravleniya*

[10] Marchevskii I K and Puzikova V V 2017 Numerical simulation of the flow around two circular airfoils positioned across the stream using the LS-STAG method *Journal of Machinery Manufacture and Reliability* **46** (2) pp 114–119 DOI: 10.3103/S105261881702011X

[11] Arefyev K Y, Voronetsky A V, Suchkov S A and Ilchenko M A 2017 Computational and experimental study of the two-phase mixing in gas-dynamic ignition system *Thermophysics and Aeromechanics* **24** (2) pp 225–237 DOI: 10.1134/S086986431702007X

[12] Lomakin V O, Kuleshova M S, Kraeva E A 2015 Fluid Flow in the Throttle Channel in the Presence of Cavitation *Procedia Engineering* **106** p 27

[13] Popov Ye P 1995 New method for composing the equations of a spline from specified points *Journal of Computer and Systems Sciences International***33** (4) pp 135–146

[14] Guskov A M, Lomakin V O, Banin E P and Kuleshova M S 2017 Minimization of Hemolysis and Improvement of the Hydrodynamic Efficiency of a Circulatory Support Pump by Optimizing the Pump Flow path *Biomedical Engineering* **4** pp 229–233

[15] Lomakin V O 2015 *Proceedings of 2015 International Conference on Fluid Power and Mechatronics*

[16] Arefyev K Y, Prokhorov A N and Saveliev A S 2018 Study of the breakup of liquid droplets in the vortex wake behind pylon at high airspeeds *Thermophysics and Aeromechanics***25** (1) pp 55–66DOI: 10.1134/S0869864318010055

[17] Serdyukov V I, Serdyukova N A and Shishkina S I 2018 Improving Operational Reliability by Means of Artificial Intelligence *Russian Engineering Research***38** (1) pp 15–18 DOI: 10.3103/S1068798X1801015X

[18] Gouskov A, Nikolaev S, Kuts V, Nizametdinov F, Korovaitseva E and Yuan S 2018 Analysis of displacement fields of particle shaping surface during nanoscale ductile mode cutting of brittle materials *International Journal of Advanced Manufacturing Technology***95** (5-8) pp 1911–1918DOI: 10.1007/s00170-017-1233-x

[19] Shen J, Wei X, Kraposhin V S and Vekshin B S 2012 High cold plastic deformation of die steel 4Kh5VF1S subjected to hardening and tempering *Metal Science and Heat Treatment* **53** (9-10) pp 503–504 DOI: 10.1007/s11041-012-9423-5

[20] Blinov D S and Shatilov A A 2013 Stress-strain state of slotted grips *Russian Engineering Research***33** (4) pp 188–193 DOI: 10.3103/S1068798X13040047

IOP Conf. Series: Materials Science and Engineering **492** (2019) 012016 doi:10.1088/1757-899X/492/1/012016

Methodology of dynamic pumps testing on high-viscosity liquids

K Dobrokhodov[1] and A Petrov[1, 2]

[1]Bauman Moscow State Technical University, 5 Second Baumanskaya Street, Moscow, 105005, Russian Federation

[2]E-mail: alexeypetrov@bmstu.ru

Abstract. The article describes the experience of testing various types of dynamic pumps with high-viscosity liquids, both at low and at high temperatures of the operating environment. A systematization of possible methods for such tests operations and a comparative analysis of their advantages and disadvantages are given. Simplified schemes of test benches are given. Approximate limits of applicability of various methods are determined. The ways of solving the problems of hydrodynamic similarity adherence with these types of tests and such test benches development prospects are briefly proven.

1. Introduction

The problem of securing operation of dynamic pumps (rotary type, axial, disk pumps) with high-viscosity liquids (up to 3000cSt and more) is widely spread nowadays. Such pump operating conditions are typical for petroleum and chemical industries, food industry, in medical equipment and for thermostabilisation systems of different applications. In general, dramatically increasing working fluid viscosity causes dramatic drop in energy performance of dynamic pumps. Quick development of computational fluid dynamics allows to design, research and optimize pump design directly for high-viscosity liquids. These problems are described in papers [2-12]. However, high-viscosity testing of such pumps is a complex engineering and scientific problem. Solving it would require selecting a working liquid and designing the test bench capable of maintaining constant properties on the working medium. This paper is dedicated to analysis of possible test methods and means of dynamic pumps testing with high-viscosity liquid, based on previous testing experience of BMSTU and Central Research Institute of Automation and Hydraulics NPO Gidromash OJSC scientists.

2. Testing methods

The main similarity parameter on rotary pump tests is Reynolds number which for rotary pump is written for example as [13]

$$\mathrm{Re} = \frac{\omega R_2^2}{\nu},$$

where
ω – rotor rotational speed, rad/s;
R_2 – wheel output radiusby stream tube, m;
ν – kinematic coefficient of viscosity, m^2/s.

During simulative pumps testing for abidance of similarity consideration it is important that Reynolds numbers for simulative and real pumps are be either equal (it's usually unfeasible because of the dramatic difference between simulative and real pumps dimensions) either be within the frames of self-similarity [14].

For simulative and real pumps Reynolds numbers must be much greater than 2×10^6which is corresponds to pump operation within the frames of self-similarity [15-18]. More specifically, Reynolds numbers for vane pumps corresponding to start of the self-similarity frames, can be dramatically lower, because of the turbulizing factor in terms of rotating vane grid slightly change boundary-layer flow. However liquid viscosity increasing, for example from 200 to 2000 c St lead to Reynolds number changing inversely to liquid viscosity (in this case an order of magnitude). The similar problem appears in real pump testing in a wide range of viscosities, but in this case it should be noted that the self-similarity condition is violated in the same way as real pump self-similarity trouble similar with real operation condition troubles which is acceptable.

Otherwise, impact of scale effecton testing pumps models with high-viscosity working liquid is practically unexplored. The significant factor is the change of the flow pattern in real and simulative pumps flow section which is dramatically depending on the flow sections dimensions. Particularly, during testing pumps with low flow rate in altitude-temperature chamber (Central Research Institute of Automation and Hydraulics) for Reynolds numbers calculated for wheel rotating vane grid on the order of $1\cdot10^3$and for Reynolds numbers for flow in channels on the order of10^2 was found out laminar flow regime of the liquid in the guide vanes channels.

A typical relationship between working liquid characteristics (kinematic coefficient of viscosity and density) and its temperature is presented in the figure 1. The example of dynamic pump characteristic change cause by increasing working liquid viscosity is presented in figure 2.

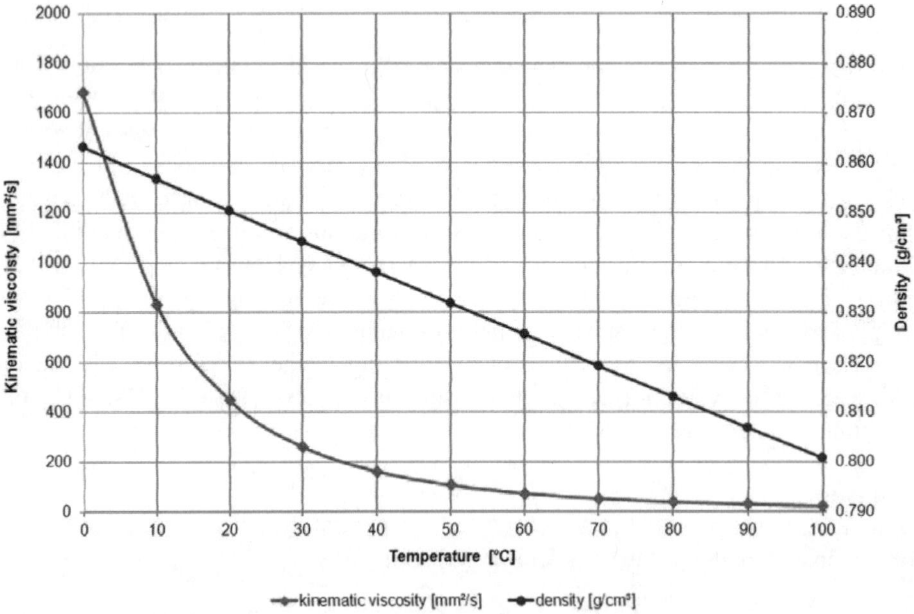

Figure 1. The relationship between working fluid characteristic and temperature.

IOP Conf. Series: Materials Science and Engineering **492** (2019) 012016 doi:10.1088/1757-899X/492/1/012016

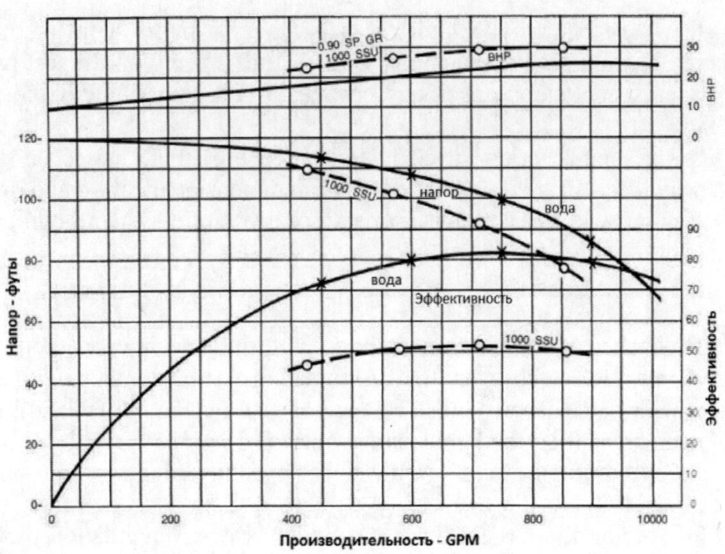

Figure 2. Change of Rotary vane pump characteristics if the working liquid viscosity increases.

Having analyzed both these charts it is now possible to formulate the following problems occurring during high-viscosity liquid dynamic pump testing.

1. According to pump operating mode in test bench (operating mode changing many foldduring testing) its capacity and efficiency change. All the pump powerinput, except external mechanical losses (which are usually relatively insignificant) heat the working liquid in test bench. Hence, the pump in test bench is an unstable source of heat generation. In this case heating rate in test bench can be evaluated by the formula:

$$\tau = \frac{\Delta T \cdot V_l \cdot \rho \cdot C_v}{N_{heat}},$$

$$\tau = \frac{\Delta T \cdot V_l \cdot \rho \cdot C_v}{N_{heat}},$$

where

τ – liquid heating rate in test bench, s;

ΔT – temperature drop from the original to final condition, K;

V_l – total volume of the working liquid in the test bench, m^3;

ρ – liquid density, kg/m^3;

C_v – specific heat capacity of the working liquid, J/kg·K;

N_{heat} – heat generation capacity, which can be written as power input times the external mechanical efficiency of the pump, W.

1. Working liquid viscosity dramatically depends on the temperature, especially in the left part of diagram. Even negligible heating of the operating environment leads to decrease of the viscosity and in turn to change of the pressure headand efficiency changing.

2. Operating environment density is weakly dependent on temperature and it can be easily taken into account.

3. Gas content in the liquid will dramatically influence its viscosity. Deaeration of the operating environment in test bench is necessary. Based on testing experience, it's a significant factor for high-viscosity liquids because it is difficult to deaerate all volume of liquid in test bench in this case. Usually lengthy operation of the test benchin recirculation mode is required.

Thus, in the system “test bench with high-viscosity liquid – pump with intermittent operation” complicated oscillating heating processes appear, which is influencing on the test process. Operating mode is changed – heat generation capacity is changed – working liquid viscosity is changed – operating mode is changed.

The ultimate aim is getting a variety of characteristic each of them has to be read in with fixed working liquid viscosity with fur there laboration and nondimensionalization and/or design equation. During operation the pump will be working in open-loop hydraulic system or in circulating system with rather huge amount of liquid.

3. Results

Testing pumps in test benches with high-viscosity working liquids were carried out in Russia in such organizations as the Central Research Institute of Automation and Hydraulics (with antifreeze agentin cooling chambersand with high-viscosity oils. Test are carried out by the authors), in BMSTU (with liquid mixtures contain emulsifiers and thickening agents. Test are carried out with the authors' participation), in NPO Gidromash OJSC (with hydroglyceric solutions and oils) and in other organizations.

What solutions for solving the heat balance and liquid viscosity regulation problem in the “test bench – pump” system can be applied? Firstly, it is necessary to determine what working liquid viscosity characteristic area the test bench will be operated in. Two of applicable working temperature limits for extremely high-viscosity liquid L_1 at room temperature and for relatively low-viscosity liquid L_2 illustration is presented in the figure 3 (for example, L_2 is an oil, L_1 variable concentration guar gum aqueous solution).

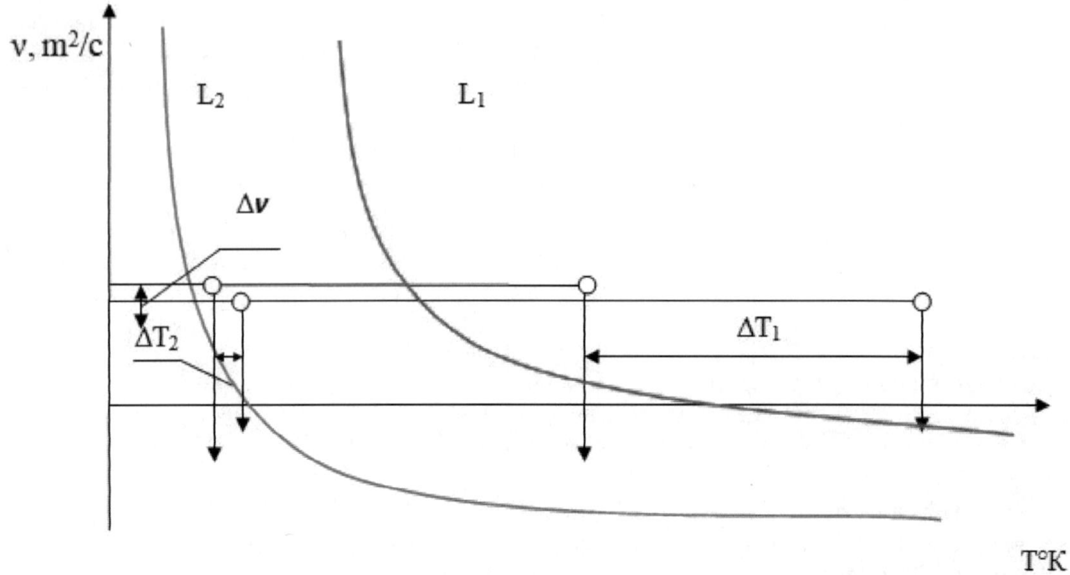

Figure 3. Example of different viscosity working liquid application ranges.

Based on the example it can be concluded thatthe range of accepted temperatures $\Delta T1$ for high-viscosity liquid in demanded range of viscosity $\Delta \boldsymbol{v}$ is rather big, which allows to operate the test bench without high-accuracytemperature stabilization system.

The reverse side of this solution is impossibility of viscosity increasing or decreasing during operation. Such increasing can be achieved only by changingemulsifying agentconcentration (in this case guar gum) by carefully stirring the liquid during test bench operation and consequent viscosity measurement of the obtained water solution.

IOP Conf. Series: Materials Science and Engineering **492** (2019) 012016 doi:10.1088/1757-899X/492/1/012016

On the other hand if we are working with a liquid that has a type L_2 relationship between viscosity and temperature in the temperature range ΔT_2, even negligible heat generation increase would lead to sudden change of the working liquid viscosity.

In this case an appropriate temperature range in very narrow, it requires extremely high-accuracy temperature stabilization system with high operation speed (to in time react for heat leak changing in pump operation changing).

This method allows changing the liquid viscosity in wide ranges by simply changing the temperature (by both cooling and heating) without changing its composition. It is difficult to calculate heating oscillatory processes in such test bench because of the effect of the variable pump operation mode, change of test bench heat generation to environment and nonlinearchange of the liquid flow pattern through the test bench parts.

In light of this, there are three of appropriate test benches designs for dynamic pumps with high-viscosity liquids and with different testing methodologies:

1. Test bench with thermostabilisation servosystem (cooling machine and heat-exchange unit are embedded, an additional heating system for L2 working liquid, with working liquid temperature feedback and PID controller). Such scheme is presented in figure 4. For big pump testing thermostabilisation system is done as a separate unit with external circulation liquid circuitconnected to the tank. Vacuum system for working liquid deaeration is required too.

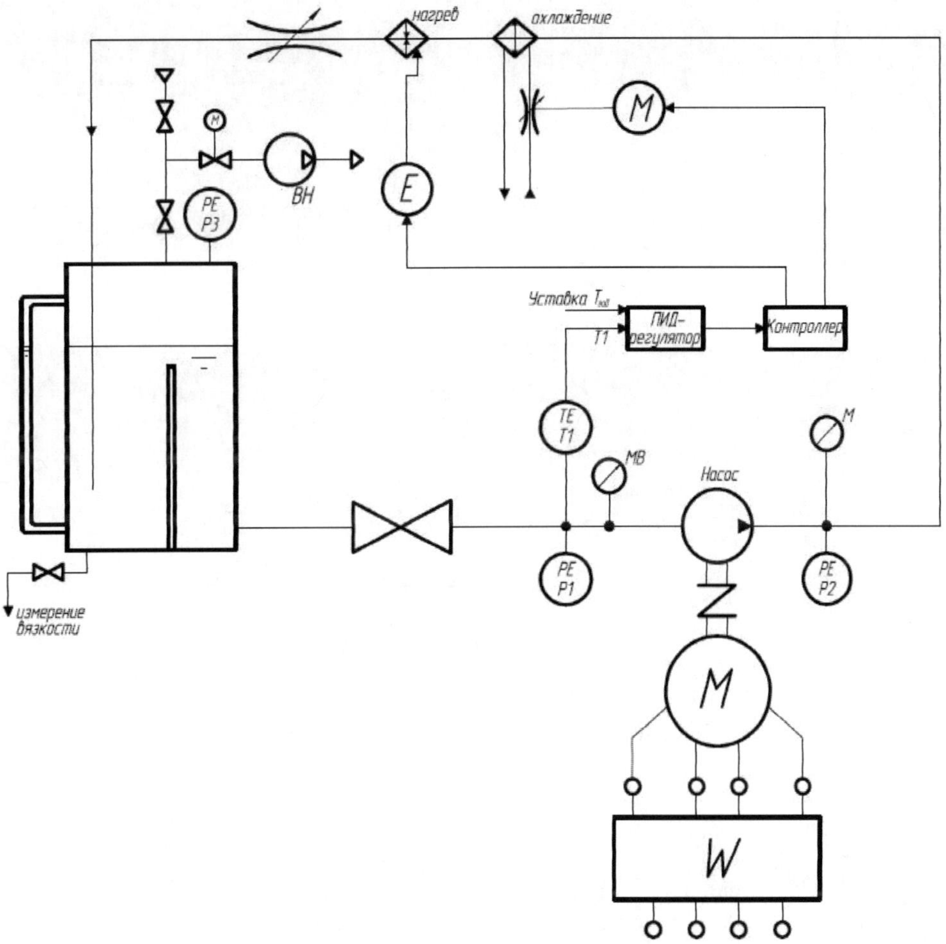

Figure 4. Scheme of the test bench with working liquid thermostabilisation servo-mechanism.

2. Test bench with huge amount of high-viscosity working liquid (like L1) and relatively low pump output what provides thermostabilisation by large temperature inertia (low heating rate) of the test bench, but at the same time is makes operation of the test bench more complicated because of the necessity to change supplemental additive concentration in a huge amount of the liquid. Working liquid amount in the test bench has to allow full testing without significant increasing the viscosity. It can be calculated based on the previous formulas and viscosity change diagram for the liquid.

3. Test bench operating on extremely high-viscosity liquid (type L1) with powerful unregulated heat-exchange unit (viscosity stabilisation provided by operation on the first stage of the viscosity characteristic) and with stepped variation of the emulsifying supplemental additive concentration. In addition, viscosity analyzer in test bench is required (figure 5).

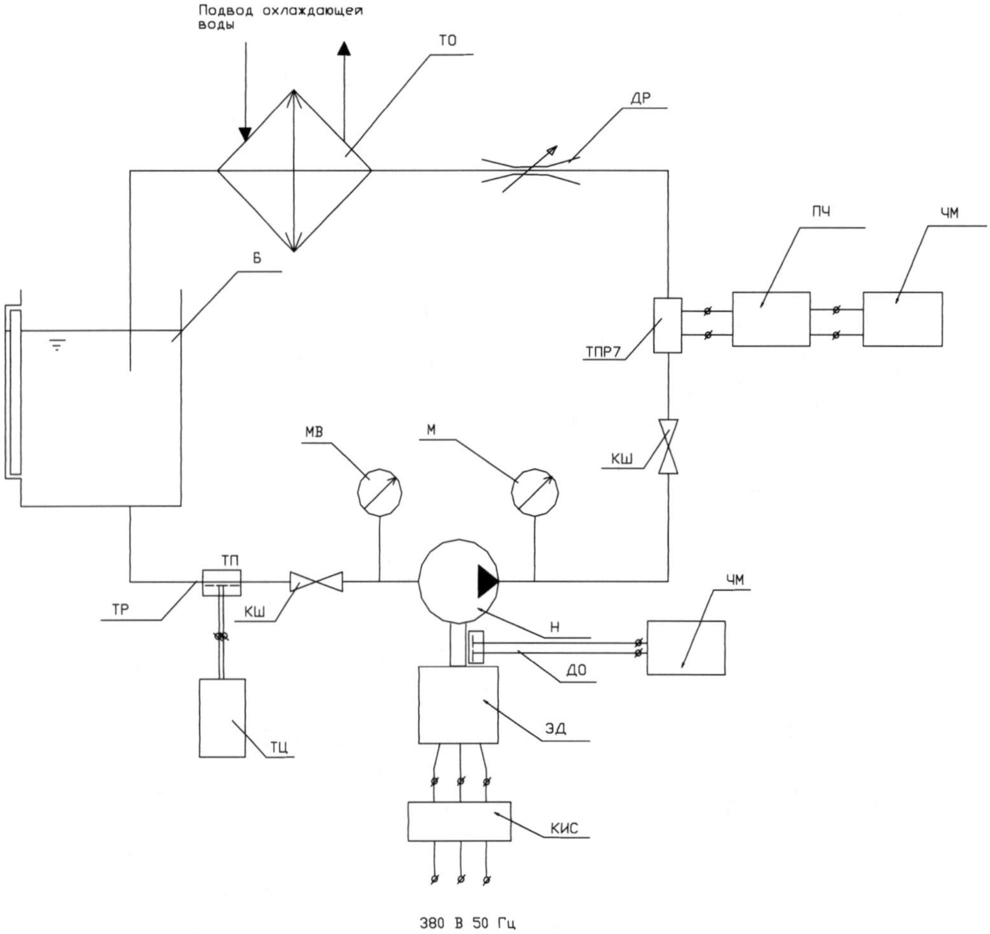

Figure 5. Scheme of the test bench with extremely high-viscosity working liquid.

In case there is no possibility to implement thermostabilisation servo-mechanism to the test bench with L_2working liquid it is possible to test pump with continuous rating [20], but it is very labor-intensive and lengthy method.

The example of the final dimensionless characteristic for the low specific speed rotary pump characteristic obtained using the proposed method is presented on figure 6.

Figure 6. Relationship between the dimensionless pressure head and the Reynolds number.

4. Discussion

1. Possible testing methods of dynamic pumps with working liquids in huge range of viscosities presented in the paper are tested through practiceand can be used the wide range of similar problems.

2. It is required to do additional research for studying similarity and scale effect during testing such pumps with high-viscosity liquids. Right now this research is carried out at Hydraulics department E10 of BMSTU.

References

[1] Guskov A M, Lomakin V O, Banin E P and Kuleshova M S 2017 Minimization of Hemolysis and Improvement of the Hydrodynamic Efficiency of a Circulatory Support Pump by Optimizing the Pump Flowpath*Biomedical Engineering* 4 pp 229–233

[2] Lomakin V O, Chaburko P S and Kuleshova M S 2017 Multi-criteria Optimization of the Flow of a Centrifugal Pump on Energy and Vibroacoustic Characteristics *Procedia Engineering* 176 pp 476–482

[3] Gouskov A M, Lomakin V O, Banin E P and Kuleshova M S 2016 Assessment of Hemolysis in a Ventricular Assist Axial Flow Blood Pump *Biomedical Engineering* 4 pp 12–15

[4] Lomakin V O, Kuleshova M S and Bozh'eva S M 2016 Numerical Modeling of Liquid Flow in a Pump Station Power *Technology and Engineering* 5 pp 324–327

[5] Lomakin V O 2015 *Proceedings of 2015 International Conference on Fluid Power and Mechatronics*

[6] Lomakin V O, Kuleshova M S, Kraeva E A 2015 Fluid Flow in the Throttle Channel in the Presence of Cavitation *Procedia Engineering* 106 pp 27–35

[7] Kraposhin M V, Banholzer M, Pfitzner M and Marchevsky I K 2018 A hybrid pressure-based solver for non-ideal single-phase fluid flows at all speeds *International Journal for Numerical Methods in Fluids***88** (2) pp 79–99

[8] Pelevin F V, Avraamov N I, Ir'yanov N Yu, Orlin S A, Lozovetskii V V and Ponomarev A V 2018 Intensification of Heat Exchange in the Regenerative Cooling System of a Liquid-Propellant Rocket Engine *Journal of Engineering Physics and Thermophysics*, **91** (3) pp 601–610
[9] Shevchenko S Yu, Melnik Yu A, Smirnov A E and Htet W Y M 2017 Comparative evaluation of methods for the determination of heat transfer coefficients of liquid and gaseous quenching media *Mechanics and Industry*, **18**(7) no 703
[10] Polyanin A D and Zhurov A I 2016 Non-isothermal flows of liquid metals and melts: Qualitative features, asymptotic models, problems, exact and approximate solutions *International Journal of Non-Linear Mechanics* 82 pp 104–113
[11] Kulik V V, Parkin A N and Navasardyan E S 2016 Numerical Modeling Procedure for Micromachined Cryogenic Cooler Elements Using ANSYS Fluent Software and Viscous Flow in a Small-Diameter Channel with Heat Transfer as an Example *Chemical and Petroleum Engineering***52** (7-8) pp 531–538
[12] Volkov V Yu, Belova O V, Krutikov AA and Skibin A P 2013 A coolant flow simulation in fast reactor wire-wrapped assembly *Thermal Engineering* (English translation of Teploenergetika) 60 (6) pp 429–433
[13] Devisilov V A and Sharai E Yu 2012 Hydrodynamics of a rheologically complicated liquid in a self-cleaning filter *Theoretical Foundations of Chemical Engineering***46** (6) pp 594–600
[14] Min'kov L L, Krokhina A V and Dueckc J 2011 Hydrodynamic mechanisms of the influence of injection on the classification characteristics of a hydrocyclone*Journal of Engineering Physics and Thermophysics***84** (4) pp 807–819
[15] Isaev S A, Kornev N V, Leontiev A I and Hassel E 2010 Influence of the Reynolds number and the spherical dimple depth on turbulent heat transfer and hydraulic loss in a narrow channel *International Journal of Heat and Mass Transfer***53** (1-3) pp 178–197
[16] Ionaitis R R, Chain P L 2009 Principles for engineering active-passive hydrodynamic control-and-safety systems for nuclear reactors *Atomic Energy* **106** (3) pp 175–184
[17] Pelevin F V, Avraamov N I, Ir'yanov N Yu, Orlin S A, Lozovetskii V V and Ponomarev A V 2018 Intensification of Heat Exchange in the Regenerative Cooling System of a Liquid-Propellant Rocket Engine *Journal of Engineering Physics and Thermophysics***91** (3) pp 601–610
[18] Aref'ev K Yu, Voronetskii A V, Il'Chenko M A 2013 Dynamic characteristics of a resonant gas-dynamic system for ignition of a fuel mixture *Combustion, Explosion and Shock Waves* 49 (6) pp 657–661
[19] Krapivtsev V G 2017 Model Studies of Coolant Flow Hydrodynamics in VVER-1000 In-Reactor Pressure Channel *Atomic Energy***122** (5) pp 304–310
[20] Marchevskii I K and Puzikova V V 2016 Numerical simulation of the flow around two fixed circular airfoils positioned in tandem using the LS-STAG method *Journal of Machinery Manufacture and Reliability***45** (2) pp 130–136

IOP Conf. Series: Materials Science and Engineering **492** (2019) 012017 doi:10.1088/1757-899X/492/1/012017

Enhancement of monitoring systems for the transport of dangerous goods by road

V Tatarinov[1,2] and A Kirsanov[1]

[1]Bauman Moscow State Technical University, 5 Second Baumanskaya Street, Moscow, 105005, Russian Federation

[2]E-mail: tatavictor@bmstu.ru

Abstract. Methods to improve the safety level of the transport of dangerous goods by roads through the introduction of advanced information and satellite technologies are considered. The common architecture and components of the proposed system are described. The proposed system allows to reduce the response time to car accidents with dangerous goods by automatically providing a real time information about the accident.

1. Introduction

Road transport in Russian Federation plays a leading role in terms of goods traffic. Leadership of road transport in the amounts of goods transportation is attributed to its high maneuverability and the capability of "door-to-door" goods delivery as urgent as necessary [1].

According to the global experience, motorization, along with its undoubted positive impact on the social and economic development of the state, also poses negative consequences associated with a large number of road traffic accidents, dead and wounded, enormous material damage, and negative impact on the ecological state of the environment [2].

As for environmental safety, of particular concern are accidents with dangerous goods [3, 4], especially with hazardous chemicals, occurring within the boundaries of big towns and settlements. The damage caused by such emergency situations is usually of great size, since, in addition to eliminating the aftermaths of the accident itself, we have to deal with the secondary damaging effects. The accidents distinctive feature associated with the release of hazardous chemicals requires to take urgent measures aimed at people protection and localization of the pollution source [5].

2. Theoretical analysis

Classically, the management cycle of emergency response and the administrative and technical activities include:

- transferring (receiving) of information about the vehicle accident;
- clarification of the problem and assessment of the situation;
- making decisions on emergency response, assignment of the forces and facilities involved;
- giving orders to perform emergency response tasks;
- march of forces and facilities to the emergency area;
- emergency response activities.

The presented list of activities clearly demonstrates the importance of timely informing the management body in charge of emergency response. The sooner the accident is reported, the earlier the emergency response activities will be started, since the duration of the management cycle (the making of administrative and technical decisions) is relatively constant for the corresponding levels of the EMERCOM hierarchy. Thus, the response promptness of the forces and facilities aimed at localization of the pollution source depends primarily on the time for the information about the emergency situation (about the accident with the vehicle of increased danger) to pass from the accident site to the management body.

Hence, the rapid transferring of information about an emergency situation should be carried out automatically [6-8] and in real time directly from the vehicle. Information coming from the sensor should include:

- information on hazardous chemicals (name and amount of the substance);
- the time of the accident;
- geographical coordinates of the accident site;
- consignor and consignee;
- state registration number of the vehicle;
- vehicle owner (transport company delivering goods)

3. Practical implementation

Information sensors monitor the state of a vehicle of increased danger during its movement and parking and provide a signal of an emergency event occurrence when a vehicle overturns or a tank containing hazardous chemicals is depressurized (in case of a pressure drop inside the tank to atmospheric levels).

At the same time, information sensors located on a vehicle must be customized to be able to work in advanced information technology systems, including satellite technologies. Transfer of a signal from the accident site to the EMERCOM Control Center through special sensors of GPS or GLONASS satellite systems [9, 10] will allow managers and specialists to quickly assess the situation and determine the plan of operations for the response of an emergency situation with dangerous goods.

The receipt of a signal of an emergency situation (accident) with the vehicle of increased danger by the management body is the basis for making calculations to forecast the chemical situation in the accident area.

The calculations are carried out in accordance with existing guidance documents and methodologies [10-13]. The name of the substance, its quantity in tons, air temperature, wind speed in the surface layer and the degree of vertical stability of air are necessary to perform the calculations. To display the area of possible pollution on the map (screen) of the operations duty officer, the topographic coordinates of the accident are necessary [14, 15].

The system includes several modules (figure 1).

1. Module “Database” collects data from sensors and perform data storage and processing. Data collection is carried out in real time [16]. The database is stored on the local MySQL server. The advantage of this server is provided by its operational stability and cross-platform (multi-OS) support [17].

2. The module of the cartographical service provides representation of the terrain map, the paths of vehicle traffic and the scale of pollution in case of an accident. Google Maps and its API libraries are used as a map service [16, 17]. The service is accessed via an Internet connection.

3. The module for automatic forecast of the chemical pollution scale calculates the forecast of the pollution spread or the environmental damage by means of existing methods.

4. The module for acquisition of the weather conditions performs the automatic collection of information on weather conditions in real time (air temperature, wind direction, wind speed, etc.) in the area where the accident occurred, accessing available weather Internet services such as GISMETEO.RU. In addition, the module determines the degree of vertical stability of the air at the site of an accident, which is necessary to forecast the scale of chemical pollution.

IOP Conf. Series: Materials Science and Engineering **492** (2019) 012017 doi:10.1088/1757-899X/492/1/012017

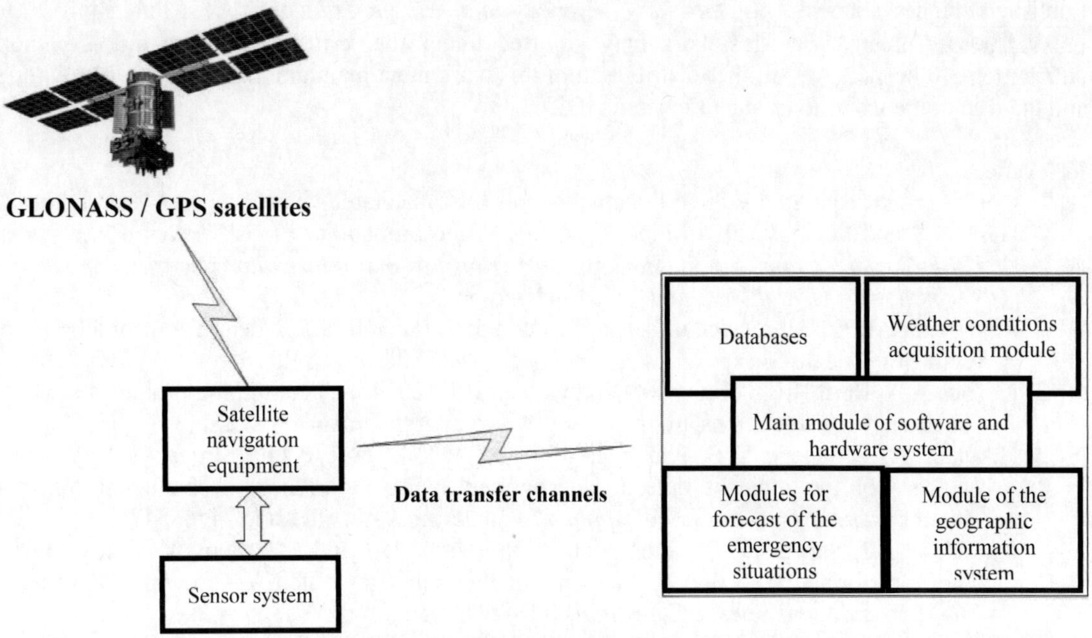

Figure 1. System configuration.

The interaction of all modules is carried out by the main module of the system [20]. It monitors the data flow from sensors to the database; displays the map of the tracking object movement using Google Maps; visually provides all the necessary information about the object [21]. As soon as the signal of a vehicle accident is received, the module automatically requests the weather conditions of the area where the accident occurred. These data are transferred to the module aimed to forecast the scale of chemical pollution. Then the results of the forecast are displayed on the map as an area of possible chemical pollution. The module performs all operations instantly, which allows the operator to correct activities aimed at elimination of the accident aftermaths as quickly as possible and promptly inform the regional EMERCOM management bodies of the necessity to evacuate the population of those settlements that are situated the chemical pollution area.

4. Conclusions

This low-cost system reunites all the latest available GPS / GLONASS, Google Maps and GISMETEO.RU technologies. The system effectively solves the problem of responsiveness to the aftermaths of accidents with the transport vehicles carrying hazardous chemical goods.

The need for a permanent stable Internet connection should be noted as a disadvantage of the proposed system, but the modern infrastructure of the telecommunications network and the Internet communications of the Russian Federation has sufficient reliability and stability.

The proposed implementation of the monitoring enhancement will expand the capabilities of the existing information and control systems of the duty dispatcher services of the Russian Unified Emergency Rescue Service of the EMERCOM of Russia. This system takes into account the distinctive features of the emergency situations that occurs with the road transport delivering the dangerous goods, providing the decision-makers with the necessary real-time information for making management decisions on emergency response. This leads to increased responsiveness of emergency services and effective emergency response.

References

[1] Gørtz I L and Nagarajan V 2016 Locating depots for capacitated vehicle routing *Networks* **68**(2)

[2] Guskov S and Levin V 2016 Model estimates of the probability of risk events in the system *IEEE Conference on Quality Management, Transport and Information Security, Information Technologies, IT and MQ and IS* 7751902 61–64.

[3] Imkhovik N A 2010 Influence of dense inert additives (W and Pb) on detonation conditions and regime of condensed explosives *Journal of Energetic Materials* 28 (SUPPL. 1) 216–230

[4] Kovacs A A, Hartl R F, Parragh S N and Golden B L 2014 Vehicle routing problems in which consistency considerations are important: A survey *Networks* **64**(3) 192–213

[5] Nikolaev A B, Sapego Y S, Ivakhnenko A M, Mel'nikova T E and Stroganov V Y 2017 Analysis of the incident detection technologies and algorithms in intelligent transport systems *International Journal of Applied Engineering Research* **12**(15) 4765–4774

[6] Nikolaev A B, Sapego Y S, Jakubovich A N, Berner L I and Stroganov V Y 2016 Fuzzy algorithm for the detection of incidents in the transport system *International Journal of Environmental and Science Education* **11**(16) 9039–59

[7] Andreev A, Berezkin D and Kozlov I 2017 Method for forecasting of situations development based on event detection in text stream *CEUR Workshop Proceedings* **2022** 367–374

[8] Li T, van Heck E, Vervest P, Kauffman R J, Dellaert B G C 2014 Consumer informedness and firm information strategy *Information Systems Research* **25**(2) 345–363

[9] Shaytura S V, Kozhaev Y P, Ordov K V, Minitaeva A M and Feoktistova V M 2018 Geoinformation services in a spatial economy *International Journal of Civil Engineering and Technology* **9**(2) 829–841

[10] Samimi P, Mukhtar M and Teimouri Y 2016 A combinatorial double auction resource allocation model in cloud computing *Information Sciences* **357** 201–216 94–103

[11] Chen W, Delgrossi L, Kosch T and Saito T 2016 Automotive Networking and Applications *IEEE Communications Magazine* **54**(12) 158

[12] Deitch M J, van Docto M and Feirer S T 2016 A spatially explicit framework for assessing the effects of weather and water rights on streamflow *Applied Geography* **67** 14–26

[13] Korotkov S, Makuev V, Lopatnikov M, Sirotov A and Stonozhenko L 2016 Forest-use issues in Moscow Region at the beginning of the 21st century *Bulletin of the Transilvania University of Brasov, Series II: Forestry, Wood Industry, Agricultural Food Engineering* **9**(2) 17–24.

[14] Kanhabua N, Blanco R and Nørvåg K 2015 Temporal information retrieval *Foundations and Trends in Information Retrieval* **9**(2) 91–208

[15] Chen X, Zhou Y and Li M 2016 A conceptual framework for the design of geo-dynamics visualization *International Conference on Geoinformatics*

[16] Gupta M and Bendersky M 2015 Information retrieval with verbose queries *Foundations and Trends in Information Retrieval* **9**(3-4) 209–354

[17] Choudhury G S 2016 Data management at scale *Information Services and Use* **36**(1-2) 27–33

[18] Pike A, MacKinnon D, Dawley S, Cumbers A and McMaster R 2016 *Doing evolution in economic geography*

[19] Buscombe D 2016 Spatially explicit spectral analysis of point clouds and geospatial data *Computers and Geosciences* **86** 92–108

[20] Karanasios S, Allen D and Finnegan P 2015 Information systems journal special issue on: Activity theory in information systems research *Information Systems Journal* **25**(3) 309–313

IOP Conf. Series: Materials Science and Engineering **492** (2019) 012017 doi:10.1088/1757-899X/492/1/012017

[21] Sakulin S, Alfimtsev A, Solovyev D and Sokolov D 2018 Web page interface optimization based on nature-inspired algorithms *International Journal of Swarm Intelligence Research* **9**(2) 28–46

IOP Conf. Series: Materials Science and Engineering **492** (2019) 012018 doi:10.1088/1757-899X/492/1/012018

Simulation of hydrological regimes of a river with a specified roughness coefficient

O Ivanova[1,2], Zh Makhambetova[1] and M Ivanov[1]

[1]Bauman Moscow State Technical University, 5 Second Baumanskaya Street, Moscow, 105005, Russian Federation

[2]E-mail: olgaivanova@bmstu.ru

Abstract. The result of simulating the hydrological regimes of the river in the Mike11 software and predicted the flush of water to the reservoir dam due to a theoretical study of existing methods for calculating the roughness coefficient on a river section, verification of the coefficient using mathematical modeling of the river outflow, clarifying the method of calculating the roughness coefficient are presented in the paper. It is shown that water level in the river has positively associated connection to the roughness coefficient. The highest water level in the lower reaches of the river will be achieved in the case of calculating the roughness coefficient using the Manning formula

1. Introduction

Problem of forecasting and simulating natural and man-made disasters is one of the most important and relevant nowadays. In particular, the risk of emergency situations associated with the flooding of territories remains very high, the number of victims and the economic damage caused by floods continue to increase [1]. At the same time, it is no longer possible to achieve high quality and speed of work on simulating the prediction of natural processes and phenomena without automation aids, as well as the associated emergency situations. Geo-information systems ideally cover this requirement [2].

The following methods are possible for simulating a breakthrough wave: physical experiment, laboratory experiment, numerical computer simulation. Numerical simulation of breakthrough waves can be divided into one-dimensional, two-dimensional and three-dimensional modeling. So one-dimensional modeling does not give accurate results and does not allow to derive two-dimensional results, but it is the fastest. One-dimensional modeling is carried out, for example, in the software product MIKE11 [3]. Two-dimensional and three-dimensional modeling in existing software systems, such as, for example, MIKE21, Flow Vision, requires high-class specialists and high-performance computers [4].

Estimation of the quality and accuracy of simulation the progress of a breakthrough wave along a river course is a complicated issue. Using any of the methods, only approximate data about the flooded areas resulting from the passage of a breakthrough wave is obtained. The accuracy of the obtained results is difficult to estimate. One of these reasons is the difference in the methods of models calibration [5]. To calibrate such models, we use the selection of roughness coefficients, which allows us to bring the model closer to actual conditions.

2. The Roughness coefficient of the rivers and canals

In the canals of the large cross section channels, a large number of factors that cause changes in the values of roughness coefficient n and hydraulic friction coefficients can be identified.

The factors affecting the value of the roughness coefficient of the earthen canals of the channels will be listed.

1) Surface roughness [6] — characterized by the shape and size of the grains of the material composing the channel along the wetted perimeter and exerting a braking effect on the flow. This factor is often regarded as the only one when choosing a roughness coefficient, but in reality it is one of many important factors.

2) Vegetation — can be considered as a kind of roughness surfaces [7], which at the same time reduces the channel capacity and inhibits movement.

Based on experimental and field studies [8], a formula for determining the roughness of an overgrown canal without weed was proposed:

$$n = n_0 \sqrt{1 + \frac{R^{\frac{4}{3}}}{2gn_0^2} \left[C_D \bar{d} \left(\bar{N} + \sigma_n \sqrt{2ln \frac{\overline{V_0}}{P_N}} \right) \right]},$$

where

n_0 – roughness coefficient without vegetation;
R – hydraulic flow radius;
d – average plant diameter;
C_D – vegetation drag coefficient;
σ_n – mean square deviation of plant density;
N – average plant density per 1 m^2;
P_N–vegetationGaussdistribution.

1) Channel heterogeneity — consists in the heterogeneity of the shape of the wetted perimeter and changes in the shape of the channel cross-section (size and shape) along its length. Depending on the heterogeneity of the channel, the roughness coefficient may increase to 0.0005 or more.

2) Channel alignment — a smooth curvature with a large radius of curvature gives a low value of *n*, while a sharp curvature increases the value of *n*. Based on the tests in trays, it was established [9] that the value of *n* is increased by 0.0001 for every 200 curvature per 100 meters of the channel length.

3) Siltation and erosion — silting can turn a very homogeneous channel into a relatively homogeneous and reduce *n*, while erosion, on the contrary, can increase *n*.

4) Hindrance — the existence of ridge agglomerate, bridge abutments and other hindrances causes an increase of *n*.

5) Dimensions and shape of the channel — the hydraulic radius increase may increase or decrease n, depending on the channel's condition.

6) Water level and flow rate — in most channels, an increase in water level and flow rate cause a decrease of *n*.

For channels without canals lining n coefficients vary in a relatively small range of 0.02–0.04.

The table of coefficients n is famous in the USA — compiled [10], it is a generalization of data available in the United States of various authors. For channels in the soil, a very large range of n is given, in particular, 0.016–0.14. This is explained by the fact that n values are given for channels that are not supported under normal operating conditions, for example, when both grass and bushes are not cleared.

A major contribution to the establishment of norms of resistance to movement made [11]. He proposed a table of coefficients n, which became the most popular in the USSR; later the he published new tables of roughness coefficients for rivers, calling them a classification of rivers by resistance. Fundamentally correct point of view of N.N. Pavlovsky [12] about the inconsistency of the indicator y in the formula for the Chezy coefficients [13] is reflected in this tables and this is a distinctive and important feature of the tables. Values of the indicator y, varying from 0.16 to 0.50 are given in tables [11]. The international standard [14] contains a table of coefficients n, varying from 0.019 to 0.20.

The question of individual assessment of factors affecting hydraulic resistance is not removed. Forexample, [15] a dependency was proposed:

$$n = \sqrt{n_0^2 + n_1^2 + n_2^2 + n_3^2 + n_4^2}$$

where
n_0– coefficient of roughness of the channel with a uniform surface of natural material;
n_1 – coefficient taking into account the heterogeneity of the surface;
n_2 – coefficient taking into account the change in the length of the shape and size of the section;
n_3 – coefficient taking into account the influence of local resistance;
n_4 – coefficient taking into account the influence of vegetation;

The question of the turbidity effect on the magnitude of hydraulic resistance and roughness has been discussed for a relatively long time. There is a decrease in the roughness coefficient with increasing turbidity.

Special hydrobiological studies [8] were established that the presence of hard coatings (concrete linings) contributes to a more intensive development of weed (phytobenthos), due to a decrease in the self-cleaning ability of the canal. In addition, the weed growth is influenced by the following factors: water transparency, flow rate, channels cleaning, the presence of biogenic substances in water and soils.

The problem of hydraulic resistance of lined canals is closely related to the reliability of linings, their durability. Monographs [16] are devoted to the operational reliability of hydraulic structures, and, consequently, channels. The state of linings surfaces change depending on its working conditions. Cracks, which increase the roughness coefficient, may appear.

In [8], there is a significant variability of the roughness coefficient of the lined canals, which, in his opinion, is associated with disorder of joints, destruction, siltation under the influence of vegetation and weed lead to a significant loss of energy. These changes in hydraulic efficiency really turn them into unlined canals, as mentioned above, with characteristic processes of siltation and overgrowing.

To establish the minimum flow rate that does not cause the siltation of concrete canals, the formula [8] was proposed by the author:

$$V_{\max} = V_{no\_silt}\left[1 + 0,141\sqrt{2ln\left(\frac{0,707\overline{\lambda_0}T}{-lnP}\right)}\right]$$

where
$V_{\text{no\_silt}}$– water speed with does not causes the siltation;
T – channel operation time;
P – siltation probability of the channel.
λ– coefficient changes in the calm and turbulent equidistant flows are different and change with increasing Reynolds number Re in different ways. For quiet flows, with increasing Re, the values of λ decrease, while for turbulent flows, on the contrary, they increase.

Thus, the analysis of many dependencies to determine the values of n and λ showed that, despite large number of methods, there are areas in which the changes in roughness are still studied not so deeply. A large number of dependencies for determining n non-uniform around the perimeter are given by authors, but in practice the values of n may vary along the length of the channels.

3. Simulation of hydrological regimes of the river in MIKE11 software

The stretch of the Eshkakon River is considered in the paper [17]. The canal of the river is located 123 km along the right bank of the Podkumokriver. The length of the river is 344 km. At this site in the area of the dam, it is necessary to increase the efficiency of surface water treatment sent for water supply to the Kislovodsk city.

In case of unsteady water movement along a river channel or along a channel, the flow characteristics in any section change over time, and at each timepoint they are not the same along the length of the section. Unsteady movement is observed in the regulation of water flow rate in hydraulic project, and in unregulated rivers — during the passage of a flood or flood wave. Unsteady motion is a general case of motion in rivers and channels, and steady-state (uniform and non-uniform) is its particular case. The task of calculating the unsteady motion is to determine two characteristics describing the state of a one-dimensional flow (flow and depth) as a function of length and time. Other characteristics, such as flow rates, can be determined from the flow and depth values.

In this paper, the model will describe the situation of the flow augmentation of variable flow rate, fed from above. In the lower site of the river stretch is described the confluence of the channel into a large reservoir (reservoir), where the fluctuations of the water level during the estimated time (1.04.2017–25.04.2017) are minor.

In the Mike 11 software [18], the Eshkakon River stretch (figure 1), cross sections (figure 2) were simulated based on: the outlines of the boundaries of the river bed, the topography of the river bed and floodplains, hydrometric data for boundary conditions, and data on water consumption (figure 3) as initial data.

The hydraulic radius was calculated by the formula

$$R = \frac{F}{P} = 24.18\,m$$

where
F – square of effective cross-section, $F = 1584.28$ m^2;
P – wetted perimeter, $P = 65.52$ m.

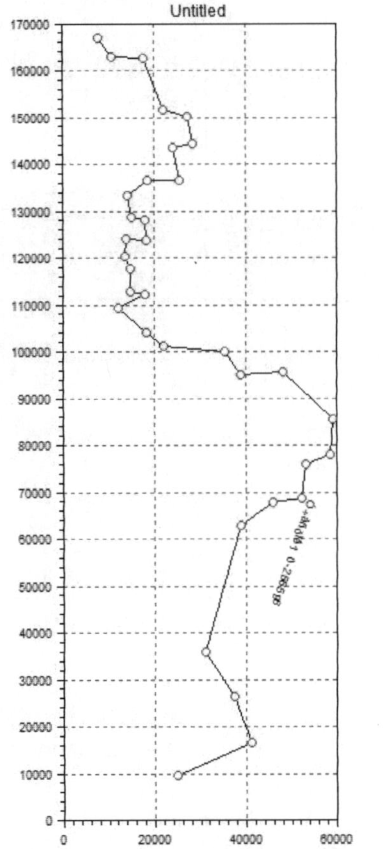

Figure 1. River stretch.

IOP Conf. Series: Materials Science and Engineering **492** (2019) 012018 doi:10.1088/1757-899X/492/1/012018

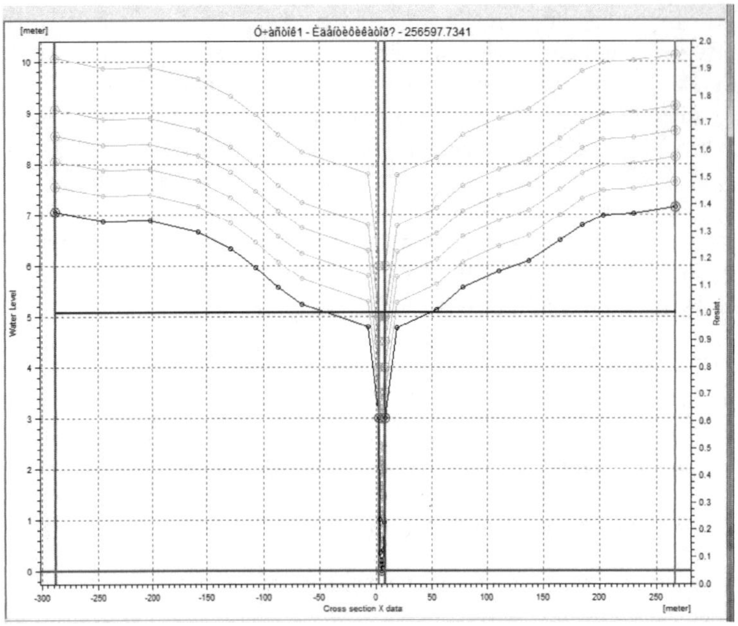

Figure 2. Profile of the canal.

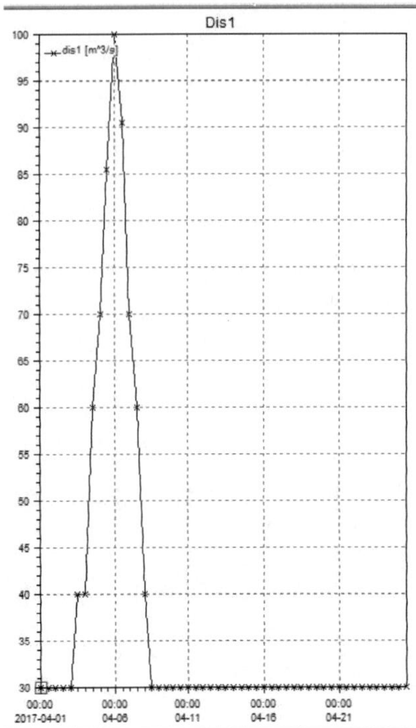

	Time	1:dis1 [m^3/s]
0	01.04.2017 00:00:00	30
1	01.04.2017 12:00:00	30
2	02.04.2017 00:00:00	30
3	02.04.2017 12:00:00	30
4	03.04.2017 00:00:00	30
5	03.04.2017 12:00:00	40
6	04.04.2017 00:00:00	40
7	04.04.2017 12:00:00	60
8	05.04.2017 00:00:00	70
9	05.04.2017 12:00:00	85.5
10	06.04.2017 00:00:00	100
11	06.04.2017 12:00:00	90.5
12	07.04.2017 00:00:00	70
13	07.04.2017 12:00:00	60
14	08.04.2017 00:00:00	40
15	08.04.2017 12:00:00	30
16	09.04.2017 00:00:00	30
17	09.04.2017 12:00:00	30
18	10.04.2017 00:00:00	30
19	10.04.2017 12:00:00	30
20	11.04.2017 00:00:00	30
21	11.04.2017 12:00:00	30
22	12.04.2017 00:00:00	30
23	12.04.2017 12:00:00	30
24	13.04.2017 00:00:00	30
25	13.04.2017 12:00:00	30
26	14.04.2017 00:00:00	30
27	14.04.2017 12:00:00	30
28	15.04.2017 00:00:00	30
29	15.04.2017 12:00:00	30
30	16.04.2017 00:00:00	30
31	16.04.2017 12:00:00	30
32	17.04.2017 00:00:00	30

Figure 3. River water flow change.

One of the main parameters is the coefficient of roughness, which is laid in the model in MIKE11. This coefficient can be calculated in different ways, for example by the Manning's Formula [19]

$$C = \frac{1}{n} R^{\frac{1}{6}} = \frac{1}{0.021} 24.18^{\frac{1}{6}} = 81.90\,;$$

IOP Conf. Series: Materials Science and Engineering **492** (2019) 012018 doi:10.1088/1757-899X/492/1/012018

Pavlovsky formula [12]

$$C = \frac{1}{n} R^{y} = \frac{1}{0,021} 24.18^{0.08} = 61.44$$

$$y = 2.5\sqrt{n} - 0.13 - 0.75\sqrt{R}\left(\sqrt{n} - 0.10\right) = 0.08;$$

Agroskinformula[20]

$$C = \frac{1}{n} + 17.72 \lg R = \frac{1}{0.021} + 17{,}72 \lg 24.18 = 72.13$$

4. Results and discussion

Based on the roughness coefficient calculations, the river stretch was modeled in the MIKE11 software. Manningsformula was taken as basic one. The change in water level in the canal profile with the calculated roughness coefficient at the extreme sections of the selected river stretch is schematically shown on figure 4. It is obvious that along the course of the river the difference in water level can reach 1.5 meters: at the initial river stretch the water level was 7.8 m, and at the final river stretch — 6.4 m.

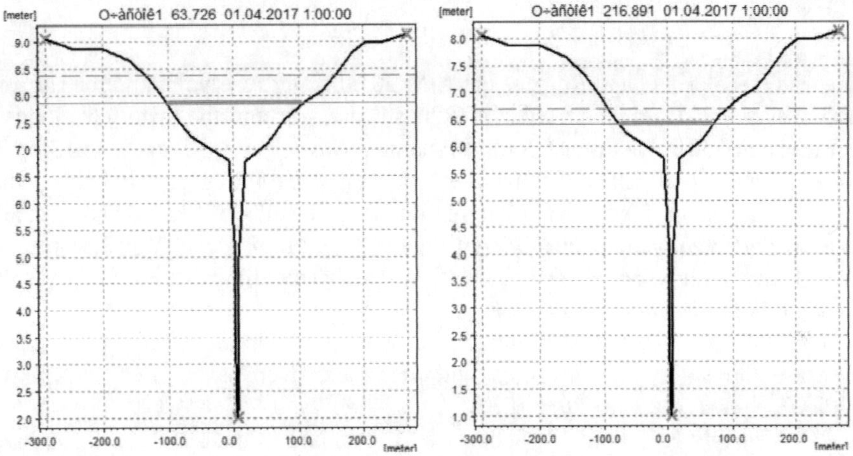

Figure 4. Water flow in the canal profile with a calculated roughness coefficient.

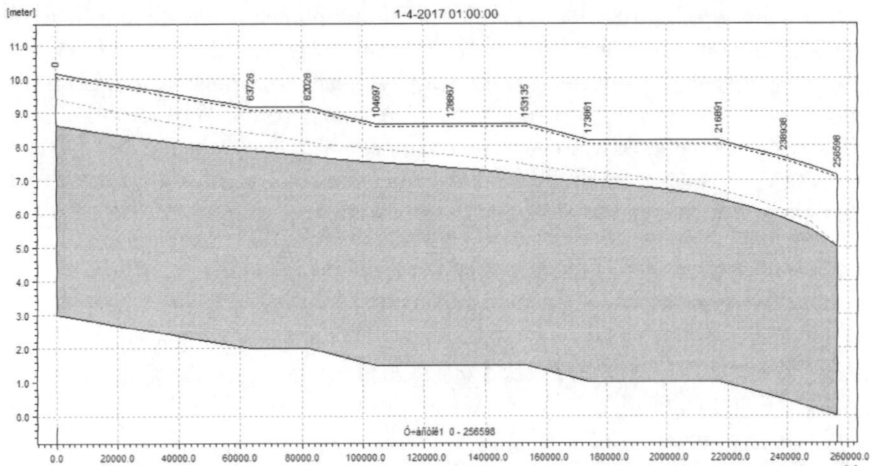

Figure 5. Water flow with a calculated roughness coefficient.

IOP Conf. Series: Materials Science and Engineering **492** (2019) 012018 doi:10.1088/1757-899X/492/1/012018

The change in the water level in a river with distance along its course in the canal is more clearly shown in figure 5. A sharp decrease in the current water level closer to the canal is the results of a significant increase in the roughness coefficient in this river stretch.

5. Conclusions

Thus, in this paper, the Eshkakon River stretch was simulated, the boundary and initial conditions that characterize the distribution of the flood wave during flood in April were set. The results obtained in the process of simulation are correlated with the hydrograph, in which the level of water in different sites was determined during field observations, and the initial discharge of water was recorded. Based on the foregoing, it can be concluded that the created model of the river stretch is correct.

It is shown that the water level in the river is directly proportional to the roughness coefficient. The highest water level in the lower reaches of the river will be achieved in the case of calculating the roughness coefficient using the Manning formula.

References

[1] Romanescu G, Stoleriu C and Romanescu A-M 2011 Water reservoirs and the risk of accidental flood occurrence. Case study: Stanca–Costesti reservoir and the historical floods of the Prut river in the period July–August 2008, Romania *Hydrological processes* **25.13** pp 2056–70

[2] Alaghmand Sina, et al. 2012 Comparison between capabilities of HEC-RAS and MIKE11 hydraulic models in river flood risk modelling (a case study of Sungai Kayu Ara River basin, Malaysia) *International Journal of Hydrology Science and Technology* **2.3** pp 270–91

[3] Ivanova O and Ivanov M 2016 1-D Mathematical Modelling of Flood Wave Propagation *Chemical Engineering***53** pp 139–44

[4] Revetria R, et al. 2017An hybrid simulator for managing hydraulic structures operational modes to ensure the safety of territories with complex river basin from flooding *Simulation Conference (WSC), Winter 2017* IEEE

[5] Ivanova O, Neusipin K, Ivanov M, Schenone M, Damiani L and Revetria R 2017 Optimization Model of a Tandem Water Reservoir System Management In: *Proceedings of the World Congress on Engineering and Computer Science***2**

[6] Olsen N RB and Stokseth S 1995 Three-dimensional numerical modelling of water flow in a river with large bed roughness *Journal of Hydraulic Research***33.4** pp 571–581

[7] Demidov AB, et al. 2017 Modelling Kara Sea phytoplankton primary production: Development and skill assessment of regional algorithms *JournalofSeaResearch***125** pp 1–17

[8] Kosichenko Yu M Hydraulic resistance and roughness of concrete channel bed *Water Resources***2**

[9] MosharovaI V, Mosharov SAandIlinskiy VV 2017 Distribution of bacterioplankton with active metabolism in waters of the St. Anna Trough, Kara Sea, in autumn 2011 *Oceanology***57.1** pp 114–121

[10] Te Chow, Ven 1959 *Open-channel hydraulics* **1** (New York: McGraw-Hill)

[11] Sgribnyj M F 1960 Average river flow velocity formula and classifications of rivers by resistance to motion *Studies and complex utilization of water resources* M

[12] Pavlovskii N N 1937 *Hydraulics handbook* (ONTI)

[13] Limerinos J T and California Dept. of Water Resources 1970 *Determination of the Manning coefficient from measured bed roughness in natural channels*

[14] Code SP. 32.13330.2012 2012 *Sewerage. Pipelines and wastewater treatment plants*(Moscow)

IOP Conf. Series: Materials Science and Engineering **492** (2019) 012018 doi:10.1088/1757-899X/492/1/012018

[15] Zhelezniakov G V 1981 *Throughput capacity of rivers and channels* (Leningrad: Gidrometeoizdat)
[16] MirtskhulavaTs E 1981 *On reliability of large channels* (Moscow: Kolos)
[17] AbazaR*1959*TheAbaziniansCaucasian Review**8**p 34-40.
[18] Havnø K, Madsen M N and Dørge J 1995 MIKE11–a generalized river modelling package. *Computer models of watershed hydrology* pp 733–82
[19] Jarrett R D 1984 Hydraulics of high-gradient streams *JournalofHydraulicEngineering***110.11** pp 1519–39
[20] Agroskin I I 1949 Formula for calculating Chézy coefficient for quadratic drag law *Hydrotechnical Construction***2**

IOP Conf. Series: Materials Science and Engineering **492** (2019) 012019 doi:10.1088/1757-899X/492/1/012019

Estimation of the environmental damage of floods in Russia at the end of the 20th century

O Ivanova[1,3], M Ivanov[1] and R Revetria[2]

[1] Bauman Moscow State Technical University, 5 Second Baumanskaya Street, Moscow, 105005, Russian Federation
[2]University of Genoa, Genoa, Italy

[3]E-mail: olgaivanova@bmstu.ru

Abstract. This paper deals with outlining the main causes of floods. We analyze statistical data on the number of largest floods and the number of people in Russia affected by this phenomenon. We also classify floods by their causes and reveal the defining relationship between the number of floods and the number of spring high waters caused by melting snow and ice. This paper states the positive and negative aspects of using reservoirs as tools of changing nature for human benefit. We justify the need of combining engineering and non-engineering means of flood protection and evaluate the environmental damage of floods.

1. Introduction

Right now the issues of ensuring safe operation of hydrotechnical installations are becoming increasingly important.

According to [1], the number of emergencies on hydrotechnical installations on river basins in Russia and worldwide has significantly increased over the last 100 years. These emergencies caused large economic and environmental damage and also human casualties (figure. 1).

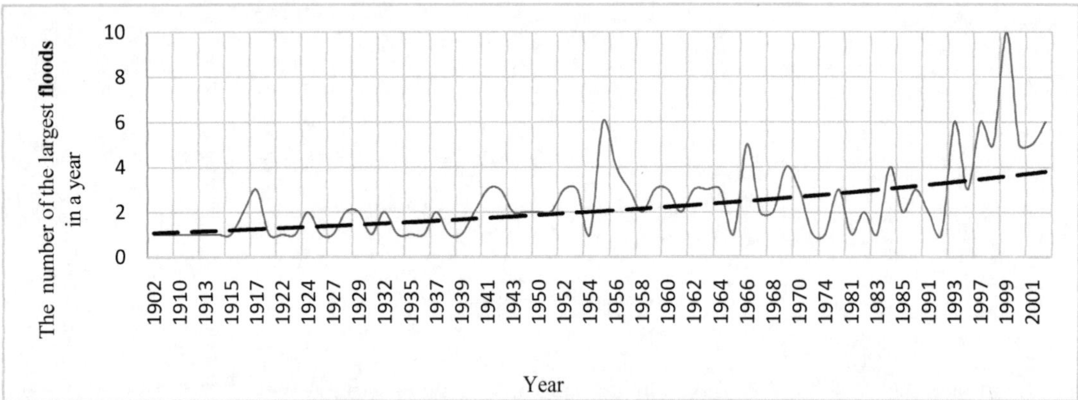

Figure 1. The tendency of the growing number of largest floods per year in Russia from 1902 to 2001.

Analysis of these emergencies enables us to formulate the main reasons causing them [2, 3]:

- Poor technical condition of hydrotechnical facilities
- Obsolescence of the systems for monitoring condition of hydrotechnical facilities
- Human factor
- Change of meteorological conditions.

The first two reasons can be eliminated to taking certain organizational measures requiring different capital and operational expenses. However, special attention should be paid to the human factor and the overall precipitation volume increase caused by the global change of meteorological conditions.

According to [4], some of the reservoirs of the Techa cascade of water bodies have shown stable water level growth. The level has grown by one meter between 1989 and 2001. The analysis of water balance components indicates that the main reason of reservoir water level growth is the change of meteorological conditions in the region in the last 10-20 years.

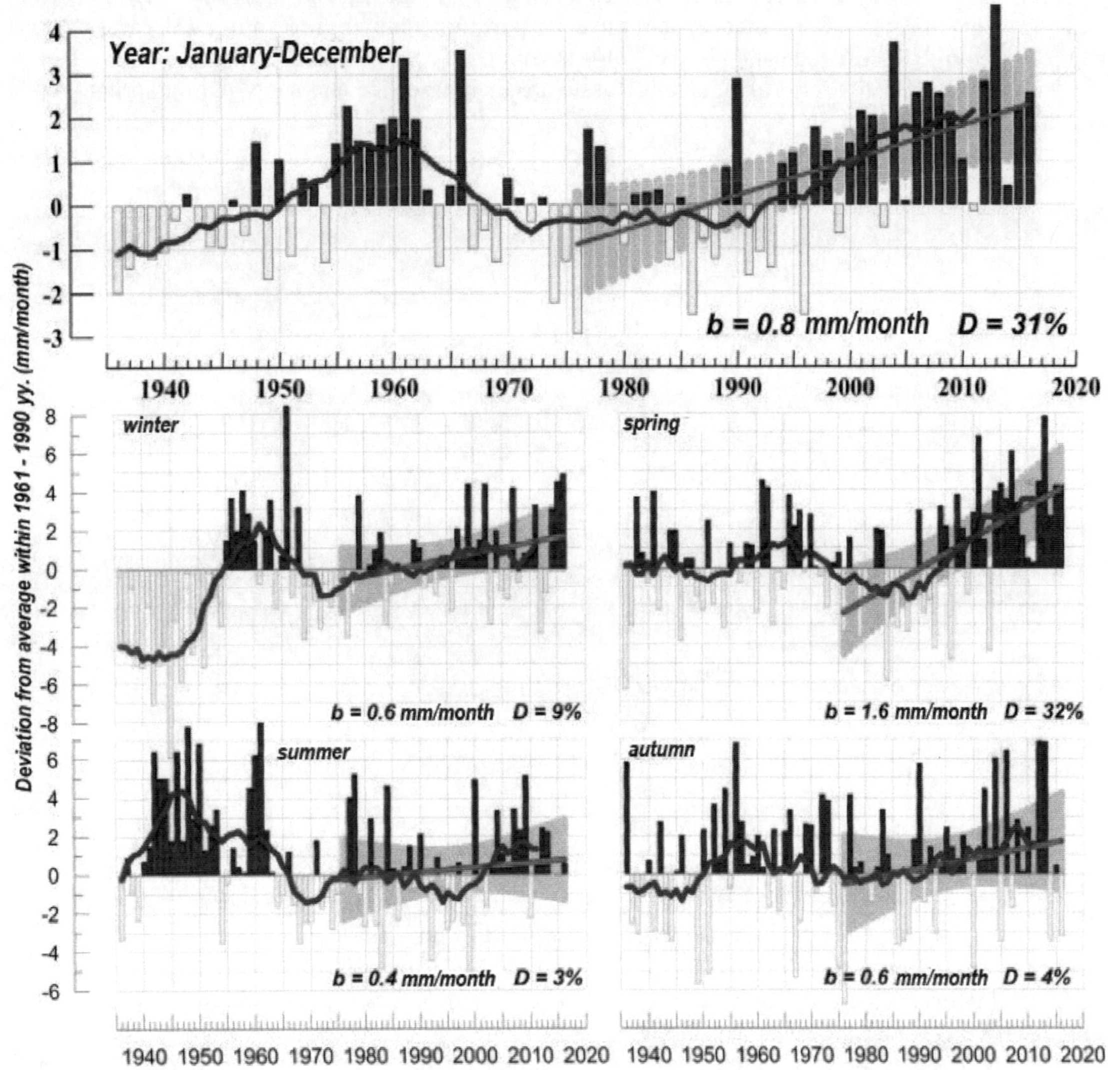

Figure 2. Mean annual and seasonal precipitation anomalies (mm/month), averaged for the territory of Russia, 1936-2016: b – trend slope (mm/month/10 years); D – trend contribution to the total variance (%).

The frequency and (or) intensity of the extreme weather phenomena may change due to global warming [5]. Figure 2 presents the time histories of the average annual and seasonal precipitation anomalies in Russia and worldwide (11 year moving average is displayed for all time histories and the linear regression trends with the 95% confidence interval are shown for the years 1976-2016).

Anomalies have been calculated as deviations from the mean value for the years 1961-1990.

The smoothed curve is obtained by an 11-year moving average. The linear trend is evaluated for the period 1976-2016; b – trend slope (mm/month/10 years), D – trend contribution to total variance (%). The chart analysis shows that the mean spring precipitation in the south of Russia in the year 2015 was 119% of the normal value, especially in the central and southern districts of the European part of Russia (the precipitation was 143% of the normal value in the Central federal district).

The human factor is playing an increasingly significant part in the natural water regime and composition change. These changes are often caused by wanting to get instant benefits. Here the human factor includes building reservoirs, channels, reclamation works, water diversion systems and so on. These interferences in the natural water exchange cycle can lead to significant increase of the wave propagation time. For instance, the wave propagation time from Rybinsk to Yaroslavl has increased tenfold because regulation of the Volga stream flow[6].

The main causes (figure 3) of floods in Russia are melting of ice and snow during spring – 51% and rainfall – 42% [7].

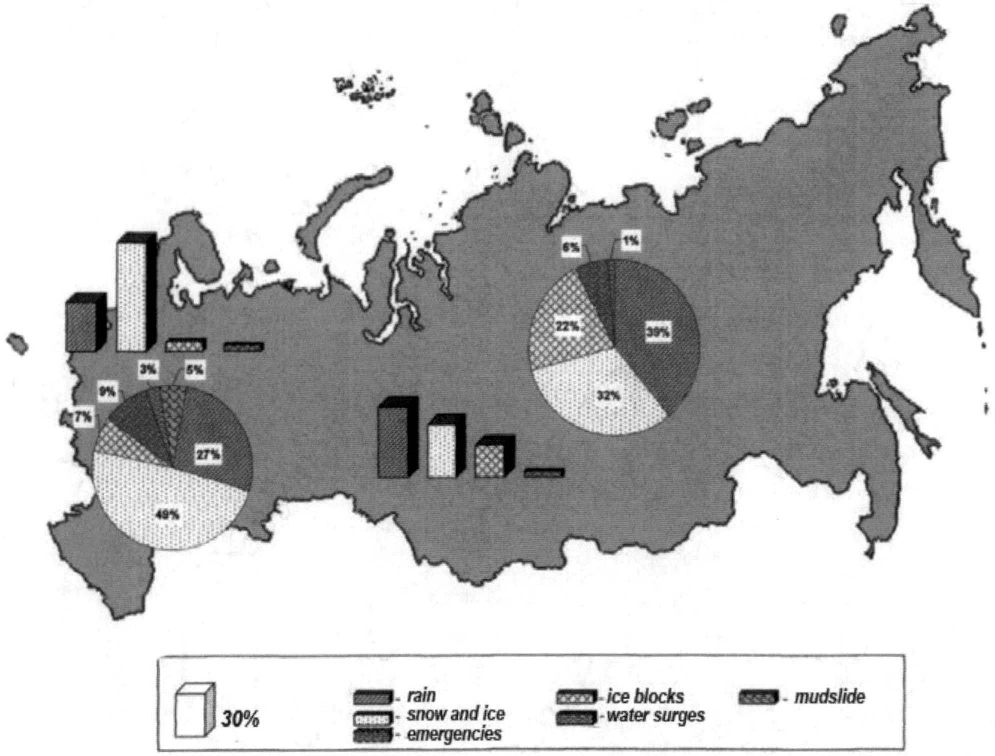

Figure 3. Causes of floods in Russia.

Several types of floods occur: high waters, floods, ice-jam and clogged drains cause floods; at the Far East even tsunamis take place sometimes.

The predicted climate warming and the inevitable growth of river valley development will undoubtedly lead more frequent and more destructive floods. Therefore developing effective measures of flood protection and prevention is an urgent issue that has to be addressed by the national governments and local authorities. One should not forget that emergency prevention can reduce the flood recovery expenses by a factor of 50-70.

A reservoir is defined as a mean of transforming the nature in the interests of humans and their economic activities. Positive changes for the environment and the economic activity from using reservoirs include:

- Reduction or full elimination of such negative natural phenomena as floods, water shortages;
- Concentration and accumulation of water resources;
- Redistribution of streamflow between years and seasons of different dryness, between days and hours, creation of water areas, improvement of conditions for water transport;
- Hydrological regime transformation for the purpose of regular land irrigation, improved usage of bottomlands below dams and bottomland protection from overly frequent and lengthy floods;
- Involving the otherwise unproductive land in economic use by accumulating water resources on that and creating a water environment that can be productive in some cases (fish farming and fishing)
- Improvement of water supply for the population, especially in dry years and periods;
- Improving environmental conditions of the adjacent territories: water amenities, softening of climate.

The powerful upward trend of flood damage in all countries, including Russia, is primarily caused by economic development intensification of flood-prone territories. The experience of many years of all countries shows that engineering measures do not provide total safety from floods, as they are not capable of preventing the main causes of flood damage growth. It is necessary to develop and approve a unified flood protection concept including both the engineering and non-engineering measures. The latter should be aimed at stimulating the landowners to organize their activity in the flood-prone areas in such a way that the damage would be at a minimum in case of a flood.

2. Evaluation of environmental damage caused by river flooding

The river Eshkanon is the right affluent of the river Podkumok. It originates in mountains, it is a typical mountain river with rapid flow and steep slopes. It is classified by the water regime as a North Caucasus type river. It is characterized by intensive summer high waters and a stable winter low water. Frequent rainfalls and downpours significantly increase the number of floods. In this case the number of floods per year can exceed tens of times. Mountain rivers carry a very large amount of sediments. The river is carrying not just small soil particles but even stones and sometimes large stone blocks. The destructive force of the mountain flow is very large. During a flood it can not only deform the stream bed but even form the new one by moving from one side of the valley to the other. The slopes and speeds are reduced downstream and the river deposits the larger sediments and carries downstream only the smaller fractions [8, 9].

Mudflows are present on mountain rivers. The mudflows are occasional mud rock flows having large destructive force. The process of mudflow forming has to do with lengthy accumulation of the rock weathering products in the upstream mountain river water. The products are carried by the river affluents to their mouths and deposited there in the form of "evacuation cones".

During floods caused by strong downpours the rapid river flow carries these cones along with bank devastation products. Temporary jams can be formed at narrow points or at river turns. When they are broken a wave full of sediments, mud, stones, rocks, washed away trees and other debris is formed. All this mass goes crashing downstream by destroying all the obstacles in its way.

The Eshkanon river water regime is determined by the long spring-summer flood and heavy rains. The main flow takes place in the spring-summer season. It is formed by solid and liquid precipitations and also by melting high-mountain snows in the upper part of the river basin.

The water quality study was conducted along the whole reservoir length and also across it. To that end water samples were collected and analyzed at the water sections shown in figure 4.

The low water period turbidity was 10 mg/L, while during the flood it was reaching the maximum value of mg/L (figure 5). Therefore during low turbidity period lasting from 7 to 9 months the water is purified only by direct filtering in the water supply system.

IOP Conf. Series: Materials Science and Engineering **492** (2019) 012019 doi:10.1088/1757-899X/492/1/012019

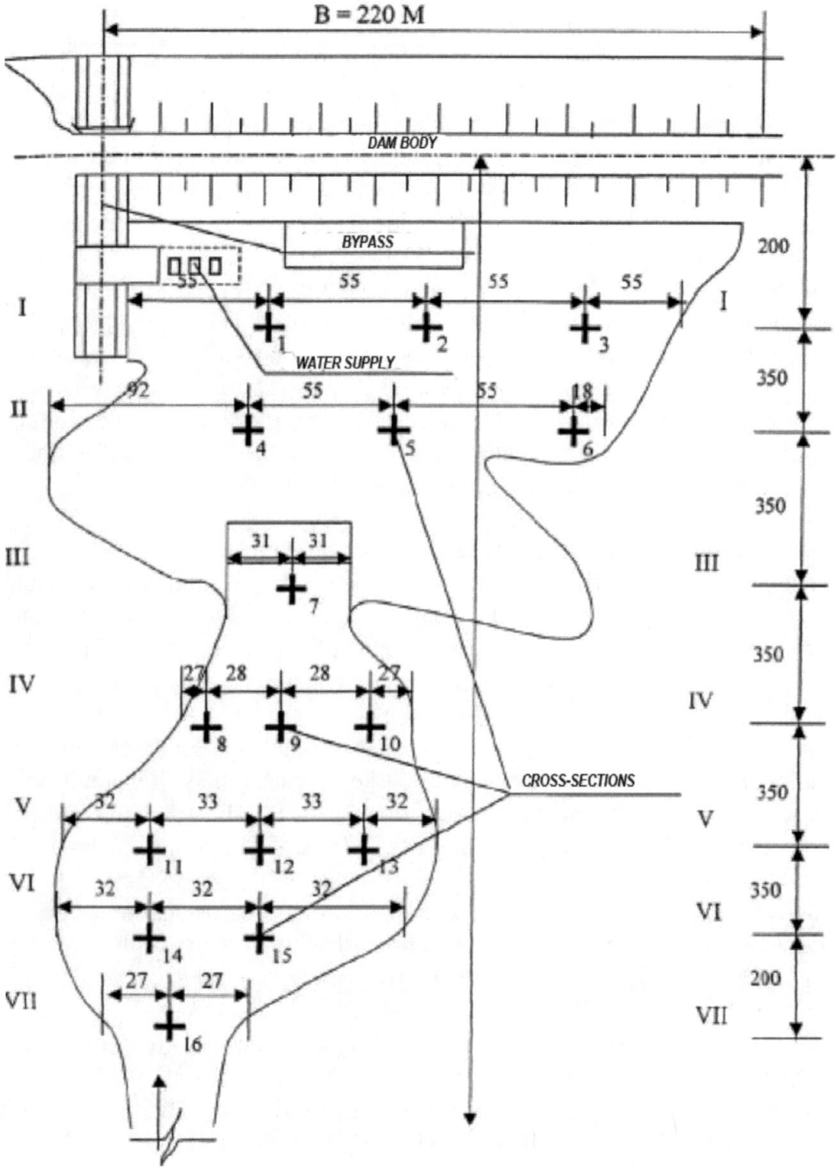

Figures 4. Cross-sections for collecting and analyzing water samples:I–VII – reservoir cross-sections; 1–16 – numbers of samples in cross-sections.

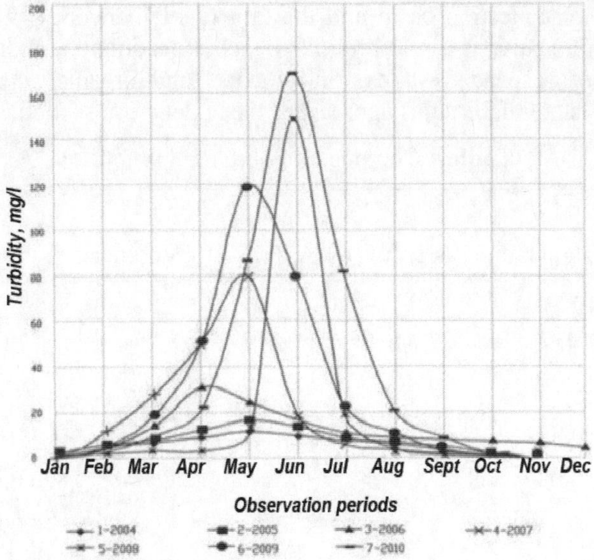

Figure 5. Reservoir water turbidity for the years of observation.

3. Qualitative flood impact analysis

Waterlogging is the rise of groundwater level and the increase of soil moisture on built-up areas and on areas being built up. Waterlogging is caused by water balance change (water inflow is greater than outflow) due to the influence of the aggregate man-made factors at certain environmental conditions. It complicates construction and exploitation of a built-up territory and leads to environmental deterioration [10, 11, 12].

According to data provided by the State Committee of the Russian Federation on environmental protection right now waterlogging is the most widespread hazardous process. Its consequences can be dangerous or catastrophic.

The waterlogging process is mostly determined by the environmental conditions of the territories being built up: lithological composition, topography, soil filtration properties, groundwater infiltration recharge character.

Territory terrain determines the flow condition of rain and meltwater, therefore it strongly influences the groundwater infiltration recharge levels. Having no system of surface flow regulation, weak terrain roughness, presence of depressions, different kinds of cavities and lows create favorable conditions for groundwater accumulation [13].

Most of the studies point out that the main cause of waterlogging is losses of water due to poor operation and dilapidation of the water networks (gravity and pressure pipelines, sewage system, storm runoffs systems) reaching 15-38%.

The latest studies have shown that the surface waters contain a large mass of toxic organic and non-organic chemicals (table 1), as well as pathogenic microorganisms having a detrimental effect on the water body flora and fauna.

The problem of polluted surface flow monitoring on urbanized territories is being studied by many researchers [14, 15].

Quantitative and qualitative composition of rain and thaw water can only be predicted with a certain degree of probability. It is governed by a whole range of factors: geographic location, precipitation intensity and duration at different seasons, character of flow basin, sanitation condition, the predominant wind direction, proximity to industrial zones, presence of highways and transport loads [16, 17]. The concentration of adulterants in rainfall runoff is largely determined on precipitation intensity, dry weather period duration, and the previous rain [18]. This can be explained by the fact that the rainfall runoff is polluted due atmospheric air substance sorption, dilution and washing away

IOP Conf. Series: Materials Science and Engineering **492** (2019) 012019 doi:10.1088/1757-899X/492/1/012019

contaminants of the water collection basin and the rain runoff network [19]. The concentration of adulterants changes significantly; as a rule it rapidly reaches maximum and diminished by the end of a rain. The content of suspended solids, oil products and organics expressed in BOD units significantly changes as the rainfall runoff approaches the outlet cross-section.

Table 1. Composition of rainfall runoff.

Indicator Value. mg/L	Volgogram	Moscow	Astrakhan	Voronezh	Penza	Maximum concentration limit
pH	7.41-7.8	7-8	8.6-9.4	7.8-8.0	7.0-8.0	6.5-8.5
Lead	–	–	–	–	–	0.03
Copper	0.022	–	–	–	0.001-0.016	1.0
Zinc	0.039	–	–	–	0.02-0.12	1.0
Nickel	0.009				0.02-0.05	0.1
Suspended solids	113-5049	65-245	215-281	39-90	50-1544	20.0
Solidresidue	445.5-5718	1088-1935		200-266		1000
COD	75-239.1	15.3-21.6	57.6-195.0	60.2-120		30
BOD_5	39.2-71.9				5.2-316	6
Oil products	0.75-3.0	12-17.5	95-197	45-52	0.125-47.5	0.3
Sulfates	126.8-215.8		195-212	48-55	63.4-792.0	500
Chlorides	98.1-149.4	51-190	323-451	91-111	36.9-86.7	350
Phosphates					0.06-5.44	3.5
Ammonia nitrogen	0.45-1.85				3.8-11.2	2
Nitrates	0.85-3				4.9-25.5	45
Nitrites	0.014-0.173				0-0.78	3.3

The experiments established that the concentration of suspended solids in the rainfall runoff close to rainwater sumps can be very volatile ranging from tens of milliliters per liter to 20000 milliliters per liter. The oil products content ranges from 0.25-2 mg/L to 80-100 mg/L [20]. The analysis shows that the suspended solids content is mainly influenced by rain intensity and the duration of the preceding dry period. The level of rainfall runoff pollution by oil products is greatly influenced by traffic intensity.

4. Main results and conclusions

It is necessary to combine non-engineering and engineering flood protection methods (flow regulation using reservoirs, building wall dams of near-river territories, straightening and deepening the river beds in order to accelerate flood water flow, building channels for channeling water waters in the natural terrain cavities, terrain level elevation). The former methods include developing economic and legal rules that take into account the specifics of using flood-prone territories. These norms include: restriction or full prohibition of the economic activities that can worsen the floods (deforestation, etc.), and also escalation of measures aimed at creating flow reduction conditions (autumn ploughing, switching to subsoil tillage, etc.). Besides one should choose such kind of economic activity that will suffer the least severe damage from a flood. Making and publishing maps of flood-prone territories, besides their direct application, will help significantly in rapid flood damage evaluation including estimates of ground flooding levels. These estimates do no texist form ost floods.

References

[1] NigmetovGM, FilatovYuA, Pchelkin VI and IuzbekovNS 2003 The upward trend of the disaster floods on the Russian Federation territory *Technologies of civil security* (1-2)

[2] Istomina M N, Kocharian A Gand Lebedeva I P 2005 Floods: causes, socioeconomic and environmental consequences *Water Resources***32**(4) 389–398

[3] MunnE 1979 *Environmentalimpactassessment*

[4] Sadovnikov V I, Glagolenko Yu V, Drozhko E G, Mokrov Yu Gand Stukalov P M 2002 Modern condition and ways of solving the problems of the Techa cascade of water bodies *Issues of radiation safety* (1) 3–14

[5] RoshydrometR F 2011 Report on climate specifics in the Russian Federation in 2010 *The Federal Service for Hydrometeorology and Environmental Monitoring of Russia.*

[6] Ismaiylov G Kh. O, Proshliakov I V and Perminov A V2013 Mathematical model of cross-border rivers water distribution *Environmental engineering* (1)

[7] Turner R K, Dent D and Hey R D 1983 Valuation of the environmental impact of wetland flood protection and drainage schemes *Environment and Planning A***15**(7) 871–888

[8] Kamo S L, Czamanske G K, Amelin Y, Fedorenko V A, Davis D W and Trofimov V R 2003 Rapid eruption of Siberian flood-volcanic rocks and evidence for coincidence with the Permian–Triassic boundary and mass extinction at 251 Ma *Earth and Planetary Science Letters***214**(1-2) 75–91

[9] Lehner B, Döll P, Alcamo J, Henrichs T and Kaspar F 2006 Estimating the impact of global change on flood and drought risks in Europe: a continental, integrated analysis *Climatic Change***75**(3) 273–299

[10] Burdakov A V, Ukharov A O, Myalkin M P and Terekhov V I 2018 Forecasting of Influenza-like Illness Incidence in Amur Region with Neural Networks *International Conference on Neuroinformatics* Springer pp 307–314

[11] Schepaschenko D, Moltchanova E, Shvidenko A, Blyshchyk V, Dmitriev E, Martynenko O, ... and Kraxner F 2018 Improved Estimates of Biomass Expansion Factors for Russian Forests. *Forests***9**(6) 312

[12] Lesiv M, Schepaschenko D, Moltchanova E, Bun R, Dürauer M, Prishchepov A V, ... and Kussul N 2018 Spatial distribution of arable and abandoned land across former Soviet Union countries *Scientific data***5**

[13] Ksenofontov B, Kozodayev A, Taranov R, Vinogradov Mand Senik E 2018 Floatation Sewage Treatment of Galvanic Productions *Ecology and Industry of Russia***22**(11) 10–13

[14] Ksenofontov B S, Antonova E S, Ivanov M V, Kozodaev A S and Taranov R A 2015 The influence of oil contaminated soil on the quality of surface waste water *Water Practice and Technology***10**(4) 814–822

[15] Gray R, Collison D and Bebbington J 1998 Environmental and social accounting and reporting *Financial reporting today* 179–214

[16] Ilyin V K, Smirnov I A, Soldatov P E, Korniushenkova I N, Grinin A S, Lykov I N, and Safronova S A 2004 Microbial utilisation of natural organic wastes *Acta astronautica***54**(5) 357–361

[17] Jamali M K, Kazi T G, Arain M B, Afridi H I, Jalbani N and Memon A R 2007 Heavy metal contents of vegetables grown in soil, irrigated with mixtures of wastewater and sewage sludge in Pakistan, using ultrasonic-assisted pseudo-digestion *Journal of Agronomy and Crop Science* **193**(3) 218–228

[18] Valdes J B, Fiallo Y and Rodríguez-Iturbe I 1979 A rainfall-runoff analysis of the geomorphologic IUH *Water Resources Research* **15**(6) 1421–1434

[19] Nakken M 1999 Wavelet analysis of rainfall-runoff variability isolating climatic from anthropogenic patterns *Environmental Modelling & Software* **14**(4) 283–295

[20] Reinhold S and Korobov D S 2007 The Kislovodsk basin in the North Caucasian piedmonts-archaeology and GIS studies in a mountain cultural landscape *Preistoria Alpina* **42** 183–207

[21] Popova A Yu 2014 Strategic priorities of the Russian Federation in the area of ecology from the standpoint of preversing the nation's health *Populatin health and living environment* **2** p 251

IOP Conf. Series: Materials Science and Engineering **492** (2019) 012020 doi:10.1088/1757-899X/492/1/012020

Waste water depuration of the manufacture of mirrored surfaces for energy saving

A Kozodaev[1, 2], Yu Korpusova[1] and A Shulzhenko[1]

[1] Bauman Moscow State Technical University, 5 Second Baumanskaya Street, Moscow, 105005, Russian Federation

[2]E-mail: kozodaevas@mail.ru

Abstract. The article points out that to protect against overheating of various energy facilities including hydraulic cylinders involve the use of a large number of mirror surfaces. The article deals with the problem of insufficiently effective water treatment of the production line of mirrors of a large enterprise. The authors analysed of the composition of pollution at each stage of production and water treatment technologies, which showed that the modernization of the water purification line from silver and ammonium nitrogen is required. The authors proposed a flow chart that allows a significant improvement in water quality and provides evidence of its effectiveness. The authors believe that it is possible to indirectly reduce the negative impact on the environment of enterprises producing components for the power engineering industry.

1. Introduction

The manufacturing process of the mirror surface includes several stages, the main of which are the following: surface preparation, application of the mirror layer, and application of a protective coating.

In the first stage, the surface is subjected to washing. After this, the blank is fed to the polishing site, carried out by mechanical brushes using a suspension of cerium dioxide. Then the process of sensitization is carried out by applying a solution of tin chloride. After sensitization, the glass is rinsed and it enters the stage of surface activation. The activation process is carried out by spraying a solution of palladium chloride on the glass. Rinse follows again. The prepared glass is sent to the silvering chamber, where the silver solution and the reducing agent are applied. This is followed by surface preparation for painting, rinsing and drying. The dried mirror is covered with a layer of protective paint and sent to high-temperature drying.

The main contribution to environmental pollution by mirror production is made by heavy metals. The main methods of wastewater treatment from heavy metals and their salts: reagent, ion-exchange, sodium cation-exchanger, sorption, etc. [1-10]. Silvering lines are the most of things to environmental pollution [11].

Sewage from the silvering line is contaminated with silver and nitrogen compounds. In many enterprises, the main aspect of the purification process is the removal of dissolved silver from wastewater, while the removal of nitrogen compounds is not provided.

2. Methods

Cleaning technology is as follows. Sewage waters contaminated with silver and nitrogen compounds enter the receiving tank (pos. 1 in figure 1), where they are treated with hydrochloric acid to a pH value of 2.0 ... 2.2. As a result, the reaction proceeds to form silver chloride, which is insoluble in water and precipitates. To intensify the reaction, the receiving tank is equipped with a mechanical stirrer. Then the water enters the precipitation chamber (see figure 1, pos. 2).

After settling, the flow flows into the chamber (pos. 3), where the precipitation of silver chloride continues. Then the wastewater enters the settling tank with a sloping bottom (pos. 4). After settling, the water from the sump with a pump is fed to the fine filters (pos. 5 and 6). Filtration fineness is 1 micron. As filters, cartridge filters with a non-woven filter element are used. Then the flow of water enters the chamber (pos. 7), where the final neutralization is performed with hydrochloric acid (pos. 8) and sodium hydroxide (pos. 9). In addition, a flow from the silanization line is supplied to the neutralization tank. The wastewater from the silver recovery line, mixed with the flow from the silanization line, is neutralized in the chamber (pos. 7) and discharged into the domestic sewage plant. Then they go to the biological treatment unit of the plant-wide sewage treatment plants or city sewer.

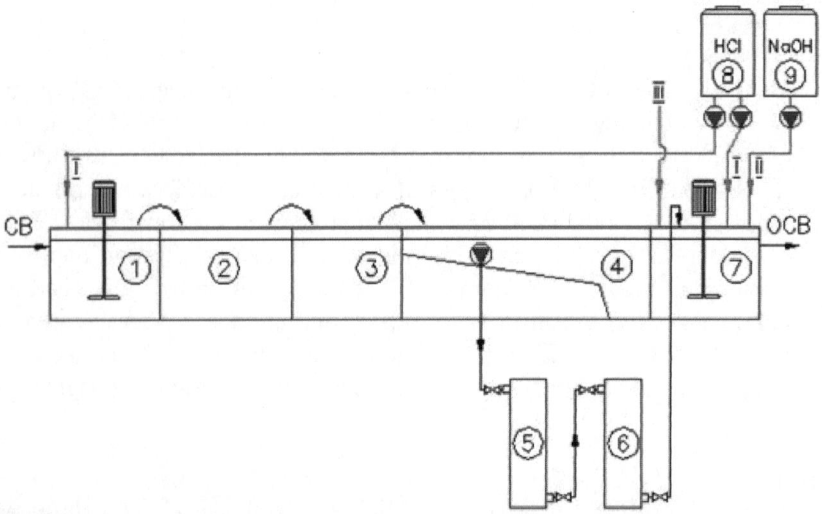

Figure 1. The scheme of wastewater treatment from the silver line:1 – tank for receiving and treating wastewater with hydrochloric acid; 2 – settling zone; 3 – the capacity of the reagent treatment of wastewater with hydrochloric acid; 4 – sump; 5, 6 – cartridge filters; 7 – neutralization capacity; 8 – hydrochloric acid supply unit; 9 – sodium hydroxide supply unit. I – supply HCl; II – NaOH feed.

During the purification process, sewage sludge is formed, the main part of which is silver chloride. Periodically, it is manually removed from the structures. In addition, in the process of purification formed waste bag filters contaminated with fine silver chloride particles. Waste generated on the silver recovery line is surrendered to refineries. Wastewater analysis results are showed in table 1.

Table 1. Qualitative composition of wastewater before and after treatment at existing facilities.

No	Name	Enter	Exi end	Notice
1	Ammoniumion	97…261	58…201	
2	Silver	0.2…3.38	0.03…0.49	

As can be seen from the results presented in the table, the purified water has a mediocre quality, especially if it is to be discharged to local biological treatment facilities. In this case, silver ions, as

IOP Conf. Series: Materials Science and Engineering **492** (2019) 012020 doi:10.1088/1757-899X/492/1/012020

well as nitrogen compounds in such high concentrations will have a toxic effect on activated sludge aerotanks.

In this regard, it is necessary to modernize the presented technological scheme.

3. Analysis result

Analysis of the literature data showed that the use of reagent methods based on pH adjustment and subsequent coagulation can significantly improve the cleaning efficiency. Based on the analysis [12–15], an improved purification scheme was developed, based on the existing technology, enhanced by the post-treatment stage and the use of a flotation tank with a new aeration system [16]. The design of a flotation tank should be carried out taking into account the initial concentration of pollutants [17, 19]. An improved scheme is showed in figure 2.

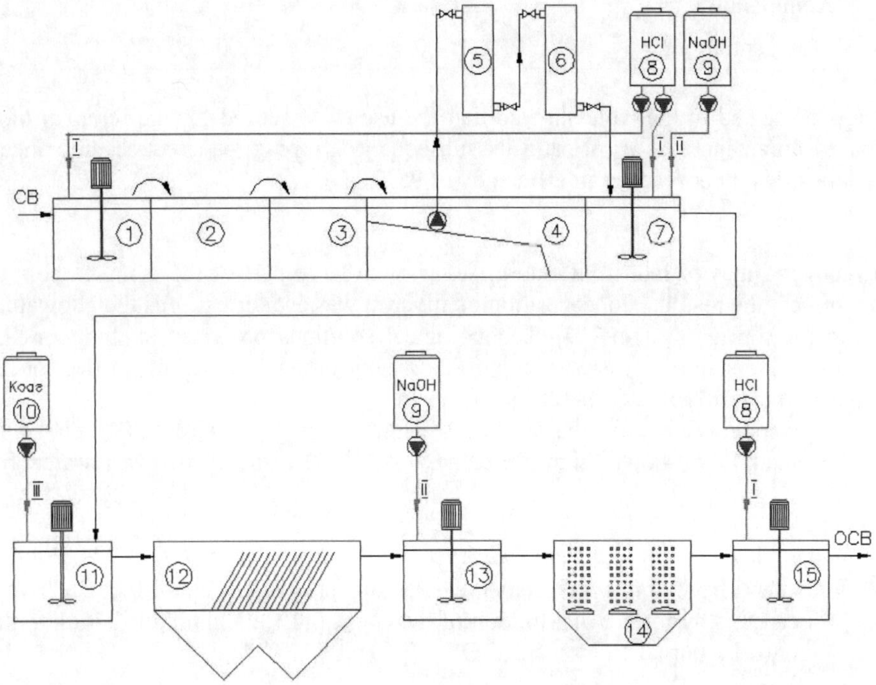

Figure 2. Improved wastewater treatment scheme: 1 – tank for receiving and treating wastewater with hydrochloric acid; 2 – settling zone; 3 – capacity of the reagent treatment of wastewater with hydrochloric acid; 4 – sump; 5, 6 – cartridge filters; 7 – pH adjustment chamber; 8 – hydrochloric acid supply unit; 9 – sodium hydroxide supply unit; 11 – camera for coagulation of water; 12 – flotation sump; 13 – pH increase chamber; 14 – aeration chamber; 15 – neutralization capacity. I – supply HCl; II – supply of NaOH; III – coagulant feed.

Wastewater contaminated with silver and nitrogen compounds enters the receiving tank (figure 2, pos. 1), where they are treated with hydrochloric acid to form and precipitate silver chloride, which is insoluble in water and precipitates. Then the water enters the precipitation chamber (pos. 2). After settling, the flow flows into the chamber (pos. 3), and then into the settling basin with a sloping bottom (pos. 4), where the precipitation of silver chloride continues. After settling, the water from the sump with a pump is fed to the fine filters (pos. 5 and 6). Filtration fineness is 1 micron. As filters, cartridge filters with a non-woven filter element are used. Then the flow of water enters the pH adjustment chamber (pos. 7), where the pH is raised. Then the purified water enters the coagulation chamber (pos. 11) and is treated with an aluminum-containing coagulant. There is a process of formation of flakes on which dissolved contaminants are sorbed. The treated water is directed by

IOP Conf. Series: Materials Science and Engineering **492** (2019) 012020 doi:10.1088/1757-899X/492/1/012020

gravity into a flotation tank (pos. 12) for clarification. The clarified water enters the pH increase chamber (pos. 13), where the pH rises to 11 units. At this value, ammonium goes into molecular nitrogen. Then the water passes into the aeration chamber (pos. 14), where it is blown off to the atmosphere, and the purified water is neutralized in the tank (figure 2, item 15).The proposed technological scheme was tested on the field drains of a glass factory. The results are showed in table 2.

Table 2. Qualitative composition of waste water after removal efficiency.

No	Name	Existed process scheme	Suggested scheme	Efficiency final purification, %
1	Nitrogen ammonium	175	32.6	81
2	Ammonium ion	226	42	81
3	Argentum	0.13	0.0065	95

The results presented in the table showed that the use of a method for the removal of ammonium nitrogen, based on an increase in pH and its subsequent stripping, can reduce the concentration by 80%. Silver after-treatment reaches an efficiency of 95%.

4. Conclusion

The silver recovery lines of industrial enterprises allow a large part of the metal to be returned back into production, but the residual concentration of silver in wastewater exceeds the allowable values for discharge into the sewage system [19]. Technological solutions proposed in this work have shown high efficiency at the stage of wastewater purification, ensuring a reduction in the concentration of silver ions below the allowable values [20] for fisheries.

The currently known method for the removal of ammonium nitrogen, based on pH adjustment and subsequent stripping, has shown high efficiency for the treatment of wastewater from mirror production

References

[1] Henze M, Harremoes P, la Cour Jansen J and Arvin E 2001 Wastewater treatment: biological and chemical processes *Springer Science & Business Media*

[2] Liu H, Ramnarayanan R and Logan B E 2004 Production of electricity during wastewater treatment using a single chamber microbial fuel cell *Environmental science & technology***38**(7) 2281–2285

[3] Tchobanoglous G, Burton F L and Stensel H D 1991 Wastewater engineering *Management* 7 1–4

[4] Patterson J W 1985 *Industrial wastewater treatment technology*

[5] Gogate P R and Pandit A B 2004 A review of imperative technologies for wastewater treatment I: oxidation technologies at ambient conditions *Advances in Environmental Research***8**(3-4) 501–551

[6] Barakat M A 2011 New trends in removing heavy metals from industrial wastewater *Arabian journal of chemistry***4**(4) 361–377.

[7] Ramalho R S 1983 *Introduction to wastewater treatment processes*

[8] Dakiky M, Khamis M, Manassra A and Mer'Eb M 2002 Selective adsorption of chromium (VI) in industrial wastewater using low-cost abundantly available adsorbents *Advances in environmental research***6**(4) 533–540

[9] Hammer M J 1986 *Water and wastewater technology*
[10] Eaton A D, Clesceri L S, Greenberg A E and Franson M A H 2005 *Standard methods for the examination of water and wastewater* American public health association **1015** 49–51
[11] Fabrega J, Luoma S N, Tyler C R, Galloway T S and Lead J R 2011 Silver nanoparticles: behaviour and effects in the aquatic environment *Environment international***37**(2) 517–531
[12] Ksenofontov B S, Kozodayev A S, Taranov R A, Vinogradov M S, Senik E V 2018 Floatation sewage treatment of galvanic productions *Ecology and Industry of Russia***22**(11) 10–13
[13] Ksenofontov B S, Butorova I A, Kozodaev A S, Afonin, A V, Taranov R A 2017 Problems of toxicity of ash and slag waste*Ecology and Industry of Russia***21**(2) 4–9
[14] Ksenofontov B S, Antonova E S, Ivanov M V, Kozodaev A S, Taranov R A 2015 The influence of oil contaminated soil on the quality of surface waste water *Water Practice and Technology***10**(4) 814–822
[15] Ksenofontov B S, Kozodaev A S, Taranov A, Balina A and Vinogradov M S 2013 Technology of flow using reagent flotation *Ekologiyaproizvodstva* 60–63.
[16] Sazonov D, Ksenofontov B, Ushinov V and Antonova E 2014 Wastewater Treatment By Flotation With Modern System Of Aeration *14th SGEM GeoConference on Ecology, Economics, Education And Legislation, 2* (SGEM2014 Conference Proceedings, ISBN 978-619-7105-17-9/ISSN 1314-2704, June 19-25, 2014) **1** 815–820
[17] Ksenofontov B S and Ivanov M V 2013 A novel multistage kinetic modeling of flotation for wastewater treatment *Water Science and Technology***68**(4) 807–812
[18] Revetria R, Damiani L, Ivanov M and Ivanova O 2017 December An hybrid simulator for managing hydraulic structures operational modes to ensure the safety of territories with complex river basin from flooding *Simulation Conference (WSC) 2017 Winter* 2717–28 IEEE
[19] Gvozdez V D and Ksenofontov B S 1986 Waste water treatment in an electroflotation apparatus with a fluidized media *KhimiyaiTekhnologiyaVody***8**(4) 70–72
[20] Csuros M and Csuros C 2016 *Environmental sampling and analysis for metals*. CRC Press.

IOP Conf. Series: Materials Science and Engineering **492** (2019) 012021 doi:10.1088/1757-899X/492/1/012021

Research of reagent-free waste water treatment methods of the chrome plating line for mineralizing the rod of hydraulic cylinders

A Kozodaev[1,2], A Andrushko[1] and A Firsova[1]

[1] Bauman Moscow State Technical University, 5 Second Baumanskaya Street, Moscow, 105005, Russian Federation

[2]E-mail: kozodaevas@mail.ru

Abstract: The article is devoted to the analysis of the subject literature on the existing methods of non-reagent treatment of electroplating wastewater produced on the lines of deposition of chromium and nickel coatings, electrochemical methods of cleaning. The authors showed and determined the advantages and disadvantages of each of the considered methods. The authors determined the most effective wastewater treatment method based on the review.

One of the most important parts of the hydraulic cylinder is. A rod is a component that works at elevated loads, even in aggressive chemical environments. To increase the life of the rod, it is necessary to chrome. In the case of marine use, nickel-chrome plated rods are used. Chrome-plated or nickel-chrome rods are made from high-grade calibrated round steel by machining the surface of the bar and applying electroplating, followed by polishing. As a result of the use of water in the electroplating process, a huge amount of wastewater is formed, containing various harmful substances that must be removed before being discharged into the sewer system or water bodies. Industry accounts for 25% of the total discharge of polluted wastewater [1]. To solve the problems of wastewater treatment, various methods (mechanical, chemical, and physicochemical) are used with the use of modern water treatment equipment with high cleaning efficiency [23-33]. One of these areas are wastewater treatment methods using electrochemical processes.

Electrochemical methods are among the most effective methods of wastewater treatment from various types of pollution. The development of electrochemical methods of wastewater treatment, improvement and creation of new technical solutions will solve the problem of preventing high anthropogenic pressure on water bodies, create conditions for the use of substandard water in water circulation, and also ensure the environmental safety of water users.

Electrochemical methods make it possible to extract valuable products from wastewater with a relatively simple treatment scheme without the use of chemical reagents. Wastewater treatment by electrochemical methods can be carried out periodically or continuously.

Of the electrochemical methods of wastewater treatment, the electrocoagulation method has become the most widespread [2]. Electrochemical coagulation is the process of purifying water from colloidal and fine impurities using coagulants as metal hydroxides (aluminum, iron), carried out by passing a directed electric current through water.

The process of electrocoagulation occurs in the apparatus is called electrolyzers. The electrolyzer (figure 1) is a vessel (or several vessels) filled with electrolyte, with electrodes placed in it — the cathode and the anode [3]. In the electric field, which is created by electrodes, there is an orderly movement of ions in conductive liquids. Metal anodes under the action of direct current ionized and passes into the purified water. Hydroxides of aluminum or iron formed in water coagulate dispersed substances.

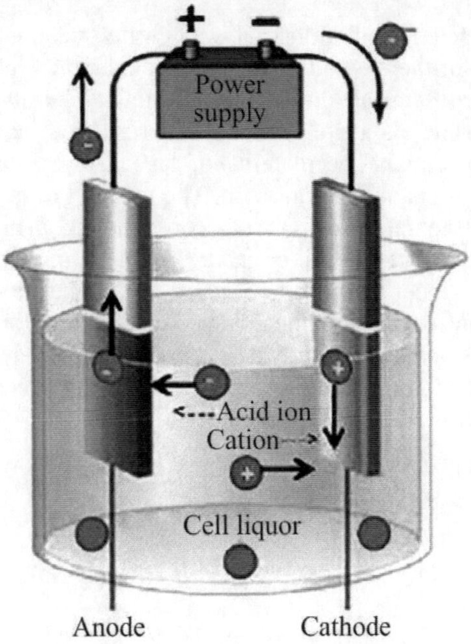

Figure1. Electrolyticprocess.

It is also possible to use insoluble electrodes. When used, coagulation can occur as a result of electrophoretic processes and the discharge of charged particles on the electrodes, the formation in solution of substances that destroy the solvation shells on the surface of particles of pollution (chlorine, oxygen).

With an increase in the concentration of suspended substances more than 100 mg/l, the efficiency of electrocoagulation decreases. With a decrease in the distance between the electrodes, the energy consumption for anodic dissolution of the metal decreases. Electrocoagulation is recommended to be carried out in a neutral or slightly alkaline medium at a current density of not more than 10 A/m^2, the distance between the electrodes is not more than 20 mm and the speed of movement is not less than 0.5 m/s.

The main advantages of electrocoagulation are: no additional water pollution, no need to use chemical reagents, low sensitivity to abrupt changes in the conditions of the cleaning process, has a high bactericidal effect, waste has good structural and mechanical properties. The main disadvantages of this method are: high energy and metal intensity of the process, relatively low productivity.

In the process of electrochemical flotation, the main role is played by bubbles of hydrogen and oxygen, which are formed at the cathode and anode, respectively, during the electrolysis of water. Gas bubbles rise in water, collide with suspended solids, stick to them, and carry them to the surface. When using soluble electrodes, the coagulation process also occurs, this leads to an intensification of the electro-flotation process due to the aluminum or iron salts formed during this process.The size of hydrogen bubbles during electrocoagulation is much smaller than with other flotation methods. Their diameter varies from 20 to 100 microns. Small bubbles of hydrogen have a greater solubility than large ones, which contributes to their release on the surface of dirt particles, intensifying the flotation process.

Electrochemical flotation is often used to treat wastewater from refineries, food processing plants, as well as in the separation and compaction of activated sludge after aerotanks in biological sewage treatment plants. [4] The optimum value of current density is 200…260 A/m^2. The main advantage of the electroflotation method is the high degree of extraction of the treated wastewater from insoluble impurities. The disadvantages of electro-flotation are the insufficiently high productivity of electro-flotation units, the release of H2 bubbles, the cost of electrodes and maintenance, and the formation of sludge by volume.

One of the widely used methods of industrial wastewater treatment is the method of contact coagulation. The development of the method of contact coagulation is electrochemical filtration. The essence of the method of electrochemical filtration is as follows. The filter (figure 2) is loaded with at least three layers of granular materials. Layer 3 and 5 materials must be electrically conductive, have different values of standard potential. The material of layer 3 must be electronegative, capable of forming insoluble hydroxide, for example, aluminum. The material of layer 5 must be electropositive. Layers 3 and 5 are spatially separated by layer 4, consisting of a non-conducting granular filter material. When water is passed through the filter charge, an electromotive force arises between layers 3 and 5 and a galvanic current. Under the action of current, electronegative material 3 is dissolved, the ions of which form a coagulant on the loading grains 4. In layers 3 and 5, oxidative and reducing reactions occur, respectively, which significantly affect the effect of water purification in the presence of organic pollutants. In layer 5, activated carbon is used, which performs, in addition to those indicated, the function of a sorbent [5].

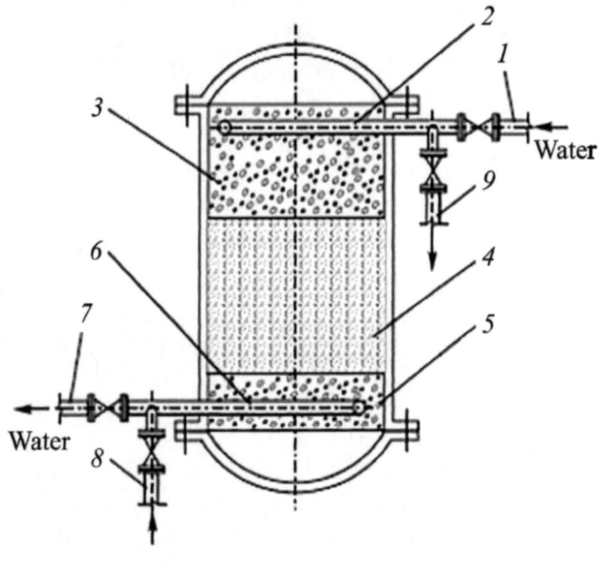

Figure 2. Electrochemical filter circuit: 1 – feed water supply; 2 – distribution system; 3 – granulated aluminum; 4 – filtering granular material; 5 – activated carbon; 6 – modular system; 7 – drainage of purified water; 8 – supply of wash water; 9 – flushing water outlet.

An important advantage of electrochemical filters is the lack of energy consumption for coagulation and redox processes, the simplicity of instrumentation. The only serious negative - the filter does not purify water from heavy metals and radioactive elements. To this it should be added that all beneficial mineral substances are preserved in water.

An important step in the development of electrocoagulation is galvanocoagulation [6]. The working area of the apparatus is loaded with a mixture of aluminum particles 20...70% and copper 80...30%. Upon contact of dissimilar particles, a short-circuited galvanic cell is formed, in which dissolution of the more electronegative metal occurs, i.e. aluminum. This forms a colloidal hydroxide, which is the main water treatment tool. Minor distances between particles contribute to an increase in current density, which, along with the developed surface of the particles, contributes to the intensification of metal dissolution.

The choice of galvanic coupler material is determined by the specific task of cleaning solutions. Usually, coke-iron is used to remove toxic metals, aluminum is used as an anode to remove anions, aluminum is also used to remove sulfates in the form of aluminum oxysulfates, and copper is used to remove precious metals. The galvanic couples with activated carbon work most effectively, which can be explained by a higher intensity of oxidation processes and, accordingly, higher output values for iron or aluminum. Such galvanic couples made it possible to significantly expand the range of application of galvanocoagulation in the practice of wastewater treatment from heavy metals, as well as to remove hardness salts and sulfates from the treated fluids, extract valuable components from ash-slag dumps [7, 8].

Galvanocoagulation method is advisable to use in the treatment of wastewater from heavy metal ions, non-ferrous and noble metals, inorganic anions (sulfates, chlorides, etc.), flotation agents, oil products and various organic impurities [9, 11-17, 22]. Also, the process of water softening occurs due to the formation of complex salts with the participation of liquid salts [10]. The advantages of this method are: lack of reagents, lack of rectifying equipment, reduction of energy costs, metal granules are used instead of sheets, sediments occupy a small volume, high reliability, high unit productivity, has the property of automatic regulation, does not require pH correction [26, 26].

The main disadvantages of the galvanocoagulation method are: scaling, difficulty in servicing due to clogging, non-recyclable wastes (iron and chromium hydroxides) are formed. As a result of the review, it was found that a galvanocoagulation method of purifying wastewater from heavy metal ions, suspended solids and other pollutants has become widely used. It is worth noting that galvanocoagulation is included in the UNESCO Register as the recommended newest method of wastewater treatment.

This method allows reducing the cost of the cleaning process due to the absence of the need to use chemical reagents, reducing the cost of energy consumption, lower cost of loading.Also galvano-coagulation is a universal method of wastewater treatment, as it allows you to purify water from the main common types of pollution.

The method of galvanocoagulation in many technological characteristics is preferable to electrocoagulation. At the same time, the method cannot be rationally applied to purify large volumes of wastewater due to the significant size of the equipment and premises for its placement.It is also necessary to create innovative solutions to intensify the process of galvanocoagulation wastewater treatment, as it is becoming more and more difficult for modern galvanic-coagulation plants to clean up drains up to the established regulatory requirements.

In this paper, experimental studies of wastewater treatment using galvanocoagulation treatment were carried out. The task of the experimental studies was to determine the effectiveness of water purification from nickel under various operating conditions.

In the course of experimental studies, the most effective types of loading for the galvanocoagulation filter, the modes of mixing, the height of loading, the time of filtration were determined.

The results of the study showed that the effect of mixing on the galvanopair prevents passivation of the surface of its elements. The optimal load ratio for the most efficient cleaning of these heavy metal ions is, with constant load mixing, 3: 1 (75% activated carbon, 25% granulated aluminum), without load mixing, the optimum load ratio is also 3 : 1. This load ratio was used further. Also it was determined the flow rate, which was changed in the range from 0.7 to 5.0 m/h. The initial concentration of nickel is 0.538 mg/l. The results are shown in table 1.

The results in table 1 showed that with increasing filtration rate, the efficiency of purification from nickel ions decreases. Optimal results were achieved at a filtration rate of 0.5…1.5 m/h. Under these conditions, it was possible to lower the concentration of heavy metals in water with an efficiency of over 80%.

IOP Conf. Series: Materials Science and Engineering **492** (2019) 012021 doi:10.1088/1757-899X/492/1/012021

Table 1.Results of water purification from nickel ions with constant stirring filter loading and different filtration rates.

Stock concentration NiC_0, mg/l	Residual concentration Ni, mg/l	Effect purification ofNi, %	Time purification, min	Filtration rate *v*, m/h
0.538	0.500	7.063	5	5.00
	0.398	26.022	10	2.40
	0.365	32.156	15	1.60
	0.212	60.594	20	1.20
	0.111	79.368	25	0.96
	0.087	83.829	30	0.68

The loading height is important; in this connection, the filtration capacity of the galvanocoagulation unit was investigated at different loading thicknesses (galvanic couples) in the galvanocoagulation chamber. Filtration was carried out at speeds of 0.7…5 m/h, the filter column was loaded with electroplating (aluminum and AG-3 coal in a ratio of 3:1, respectively), mass: 10, 20, 30 and 40 g. With these mass values, the height of the filter loading amounted to 0.07 m, 0.14 m, 0.21 and 0.28 m, respectively. The research results are showed in table 2.

Table 2.Water purification from heavy metal (nickel) at different heights of the filtering load.

Stock concentration NiC_0, mg/l	Residual concentration Ni, mg/l, to loading height h, m				Filtration rate *v*, m/h
	0.07	0.14	0.21	0.28	
0.538	0.500	0.481	0.460	0.437	5.00
	0.398	0.358	0.333	0.301	2.40
	0.365	0.345	0.325	0.300	1.60
	0.212	0.200	0.286	0.259	1.20
	0.111	0.092	0.077	0.057	0.96
	0.087	0.077	0.057	0.057	0.68

The results presented in table 2 show that increasing the height of the layer of filter material in the galvanocoagulation chamber enhances the coagulation of pollutants in the water.Summarizing, it is to note the following:

1. The effect of mixing on the galvanopair prevents passivation of the loading surface. The effect of mixing also increases the rate of dissolution of the oxide film, while the resulting flakes of formed aluminum hydroxide begin to move more actively during the filtration process, while encountering a large amount of pollutants, thereby accelerating the process of their coagulation.
2. The efficiency of cleaning from heavy metals with increasing filtration process increases.
3. With increasing filtration rate, wastewater treatment efficiency decreases. Optimum filtration rate for high cleaning efficiency is 0.5…1.5 m/h.
4. It was established experimentally that with an increase in the height of the layer of filter material, the cleaning efficiency increases.

References

[1] *Order of the Government of the Russian Federation of August 27, 2009 No. 1235-r* (revised April 17, 2012) "On approval of the Water Strategy of the Russian Federation for the period up to 2020".

[2] Stakhov Ye A 1983 *Purification of oily wastewater from the storage and transportation of petroleum products* Leningrad NedraPubl 263 p

[3] Miklashevsky N V and Korolkova S V 2000 *Clean water. Cleaning systems and household filters* Saint Petersburg BHV–Saint Petersburg Arlit Publishing Group 240 p

[4] Ksenofontov B S, Kozodayev A S, Taranov P A, Balina A and Vinogradov M S 2013 Wastewater treatment technology using reagent flotation *Ecology of production*(4) 60–63

[5] Nazarov V D, Gurvich L M and Rusakovich A A2003 *Water supply in oil production* Ufa Virtual LLCPubl 508 p

[6] Przhegorlinsky V I and Ivanishvili A.I. 1982 *Method of wastewater treatment* Patent 952756 USSR, VRBC02F 1/46 declared 09/02/1980; publ.23.08.1982

[7] Solozhenkin P M 2007Theoretical foundations and practical aspects of galvanochemical wastewater treatment. Part 1. Theoretical foundations of galvanochemical wastewater treatment *Water and Ecology: Problems and Solutions*(2) 3–17

[8] Ksenofontov B S, Kozodayev A S, Taranov R A, Balina A, Vinogradov M S and Petrova E V 2013 Development of the technology for extracting valuable components from ash-slag dumps of heat and power generation facilities *Ecology of production*(8) pp 9–12

[9] Aksenov V I, Ladygichev M G, Nichkova I I, Nikulin V A, Klein S E and Aksenov E V 2005 *Water Management of Industrial Enterprises: Reference book* In 2 books Book 1 Moscow TeplotechnikPubl 640 p

[10] Klyavlin M S, Karamov I V and Talipov R A 2002 Determination of the effective combination of galvanic couples in the practice of water softening *Problems of the building complex of Russia* Ufa 165p

[11] Espinoza-Quiñones F R, Módenes A N, Theodoro P S, Palácio S M, Trigueros D E G, Borba C E, ... and Kroumov A D 2012 Optimization of the iron electro-coagulation process of Cr, Ni, Cu, and Zn galvanization by-products by using response surface methodology *Separation Science and Technology***47**(5) 688–699

[12] Kobya M, Erdem N and Demirbas E 2015 Treatment of Cr, Ni and Zn from galvanic rinsing wastewater by electrocoagulation process using iron electrodes *Desalination and Water Treatment***56**(5) 1191–1201

[13] Stanisavljević M, Krstić I and Zec S 2010 Eco-technological process of glass-ceramic production from galvanic sludge and aluminium slag *Science of Sintering***42**(1) 125–130

[14] Shan M J, MIN Y G, Shen X, Zhang Y & Zhang H M 2008 Degeneration treatment of wastewater with micro-electrolysis *Journal of Harbin Institute of Technology* **8** 033

[15] Sokolova L P, Skornyakov V V, Beklemisheva G S, Ivanova M I, Kipriyanov A A and Veksler G B 2003 Process of waste water galvanocoagulation treating intensification *Chemical And Petroleum Engineering* (11) 39–40

[16] Zhi-jun H U and Bao-jian L Y M L 2009 CTMP Wastewater Treatment with Compound Electrochemical-biochemical Method *China Pulp & Paper* **5** 014

[17] Ksenofontov B S, Kozodayev A S, Taranov R A, Vinogradov M S and Senik E V 2018 Floatation sewage treatment of galvanic productions *Ecology and Industry of Russia***22**(11) 10–13

[18] Ksenofontov B S, Butorova I A, Kozodaev A S, Afonin A V and Taranov R A 2017 Problems of toxicity of ash and slag waste*Ecology and Industry of Russia***21**(2) 4–9

[19] Ksenofontov B S, Antonova E S, Ivanov MV, Kozodaev A S &Taranov R A 2015 The influence of oil contaminated soil on the quality of surface waste water *Water Practice and Technology***10**(4) 814–822

[20] Ksenofontov B S and Ivanov M V 2014 Case study: Use of flotation for industrial stormwater treatment. *Water Practice and Technology***9**(3) 392–397

[21] Ksenofontov B S and Ivanov M V 2013 A novel multistage kinetic modeling of flotation for wastewater treatment *Water Science and Technology* **68**(4) 807–812
[22] Ksenofontov B S, Kozodayev A S, Taranov R A, Kapitonov S N and Morozov S D 2008 The use of jet aeration in the processes of flotation wastewater treatment *Life safety*(10) 2–5
[23] Ksenofontov B S, Kozodayev A S, Taranov R A, Vinogradov M S, Voropaeva A A and Senik E V 2016 Development of a complex technology for leaching rare earth metals from ash and slag and the problem of concentration of solutions of these metals *Modern high technologies* (3-1) 44–49
[24] Hammer M J 1986 *Water and wastewater technology.*
[25] Eaton A D, Clesceri L S, Greenberg A E and Franson M A H 2005 Standard methods for the examination of water and wastewater *American public health association* 1015 49–51
[26] Chen G 2004 Electrochemical technologies in wastewater treatment *Separation and purification Technology* **38**(1) 11–41
[27] Henze M, Harremoes P, la Cour Jansen J and Arvin E 2001Wastewater treatment: biological and chemical processes *Springer Science & Business Media*
[28] Liu H, Ramnarayanan R and Logan B E 2004 Production of electricity during wastewater treatment using a single chamber microbial fuel cell *Environmental science & technology* **38**(7) 2281–2285
[29] Tchobanoglous G, Burton F L and Stensel H D 1991 Wastewater engineering *Management* 7 1–4
[30] Patterson J W 1985 *Industrial wastewater treatment technology*
[31] Gogate P R and Pandit A B 2004 A review of imperative technologies for wastewater treatment I: oxidation technologies at ambient conditions *Advances in Environmental Research* **8**(3-4) 501–551
[32] Barakat M A 2011 New trends in removing heavy metals from industrial wastewater *Arabian journal of chemistry* **4**(4) 361–377
[33] Ramalho R S 1983 *Introduction to wastewater treatment processes*
[34] Dakiky M, Khamis M, Manassra A and Mer'Eb M 2002 Selective adsorption of chromium (VI) in industrial wastewater using low-cost abundantly available adsorbents *Advances in environmental research* **6**(4) 533–540

Improving of operating efficiency of fire brigades during the suppression of peat fires by introducing a unit for bioactivating drinking water into a water supply concept (an example of Tver region)

Yu Bogdanova[1,3], **A Rozhok**[1,2,4] **and V Tatarinov** [1,5]

[1]Bauman Moscow State Technical University, 5 Second Baumanskaya Street, Moscow, 105005, Russian Federation
[2]University of Genoa, Genoa, 16145 Italy

[3]E-mail: yul9853@yandex.ru
[4]E-mail: rozhok_anastasiya@mail.ru
[5]E-mail: tatavictor@bmstu.ru

Abstract. The article is devoted to an option to improve the concept designed for the water supply of fire departments during the suppression of peat fires (for example, Tver region) by adding to it a block of bio-energy activation of drinking water. The article deals with a review of the subject area were carried out in order to justify the relevance of the development and statistics on peat fires in the region for the last 5 years. The authors showed the methods of reagent-free water treatment and their comparative analysis. The article gives a detailed analysis of the theoretical basis for the further improvement of the water supply scheme.

1. The problem of developed peatlands

Emergency situations caused by earthquakes, floods, mudflows, hurricanes, and forest fires pose the greatest danger in the natural sphere. Peat gained notoriety due to underground fires [1], [2] [3], [4], known to mankind for thousands of years. Such fires can hardly be extinguished and pose a great danger [5].

Peat fires most often occur in places of peat extraction, usually occurring due to improper handling of fire, from lightning discharges or spontaneous combustion [6].

The main problem of the elimination of massive forest and peat fires often consists of several factors - the inaccessibility of the quenching areas [7], their remoteness from water sources, the irrationality or the inability to attract vehicles to deliver water. While the need for it to carry out fire activities can reach several thousand tons per day [8], [9].

The share of peat reserves located in the Tver region is 2 billion tons (with a moisture content of 40% [10], [11]), which is 43.5% of all the peat reserves in the Central Economic Region [12]. The territory of peat bogs occupies about 500 thousand hectares, which is 10% of the area of the region [13]. There are 3 main peat enterprises CJSC Tvertorf, OJSC Vasilyevskiymokh , CJSC Seliger-Holding.

The greatest fire hazard is developed and abandoned peat deposits [14], [15], for example, OrshinskyMokh, where, according to the head of Greenpeace Russia forest program, in summer 2014, the largest peat fire was observed in Russia, deposits of more than two meters thick were burning.

Total there are in the Tver region about 16.7 thousand hectares of peatlands are developed (81 developed and 3 flooded peat deposits). These territories were freed up after intensive industrial use of natural resources [16], as a result of which the bottom layer of peat remains insignificant in thickness (the average thickness is 0.3-0.5 m).

2. Statistics on peat fires 2014-2017 years

Statistics on peat fires in the Tver region over the past 5 years is showed in table 1.

Table 1. Statistics on peat fires 2014-2018 years.

Year	Number of fireplaces	Fire area, ha	First fireplace (date)	Last fireplace (date)
2014	12	3227,45	20 April	17 August
2015	43	0,64	16 February	19 May
2016	9	1,3	21 April	3 June
2017	2	0,002	4 May	6 May
2018	20	20,115	5 May	15 September
Total	86	3249,507	–	–

3. Projects to solve the problem of peat fires in the region

3.1 Prohibition on uncontrolled grass burning

Figure 1 clearly shows the difference in the number of thermal points found in the region. A comparison of mean annual values for 2002-2015 is showed (before the banning of grass burning), for 2016, for spring 2017. The graph is based on data from Greenpeace Russia.

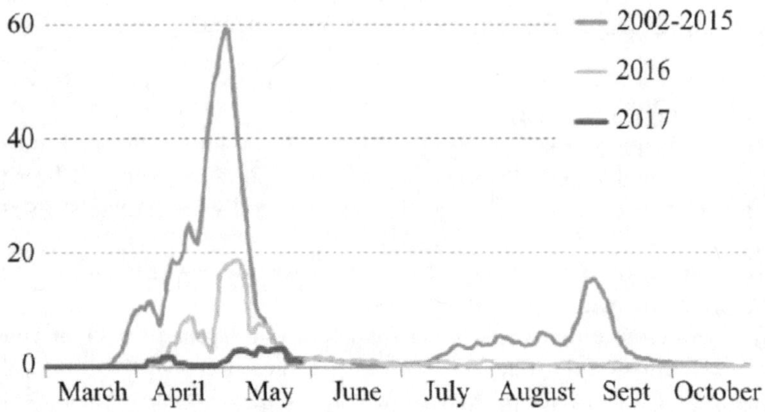

Figure 1. Comparison of the number of thermal points in the Tver region in different time periods.

It has been established that the cause for the occurrence of peat fires to a certain extent is the uncontrolled burning of dry grass. In late spring of 2017, when the first wave of fires for many regions of Russia ended, Greenpeace Russia, analyzing the results of spring, found a positive trend in the problem of peat fires in the Tver region.

In 2013, the Government of Russia was ordered by the President to impose a ban on the uncontrolled burning of grass. The ban was finally adopted in 2015.

Thus, in the region, a record decrease was achieved in the number and area of fires in the spring period and against the background of mean multiyear values and against the background of a prosperous 2016 year. As a result of successful struggle with spring burning, the total number of fires has significantly decreased.

3.2 Restoration of peat bogs in Russia
As mentioned earlier, the greatest fire danger in the forest fund of the Tver region are peat deposits, developed and previously drained. They are located on lands of various categories (forestry, agricultural, stock, industry).

To solve this problem, Tver region since 2014 has been actively involved in the Russian-German project "Restoration of peat bogs in Russia".

In 2014, Mokhovoye-II peat deposit (1.5 thousand hectares) was irrigated, in 2015, the Ozeretsko-Neplyuevskoye (900 hectares) in the Konakovsky district of Tver region and part of OrshinskyMokh deposit (1.2 thousand hectares) in Kalininsky district of Tver region.

In 2015, at the expense of the regional budget, work was carried out to water a part of the developed VasilyevskyMokh peat deposit (1,157 ha) in Kalininsky district.An analysis of the fires in 2014 showed that the largest fires were recorded at OrshinskyMokh peat deposit. The total area of this field is more than 67. thousand hectares, much of which (about 16 thousand hectares) was previously drained and developed.The watering of OrshinskiyMokh deposit was started in 2015, the area of 1,200 ha was flooded. In the first half of 2016, work was continued; the 2nd section of the field was watered on an area of 1,050 hectares.

At the time of the release of data on the project, the website of the Federal Forestry Agency (August 2016) was negotiating with the coordinators of the Russian-German project on the implementation in 2016-2017 of a pilot project for irrigation of the 3rd section of the field.

4. Development of water supply concept
One of the main problems of eliminating massive forest and peat fires is often the inefficiency or the inability to attract vehicles to deliver water. In this case, water is meant both for extinguishing the fire center itself and for food and household needs of rescue units operating in the emergency zone [17]. And, despite all the measures taken to solve this problem in the Tver region, it remains relevant due to the large number of peat deposits developed.

Thus, the relevance of developing the concept of water supply to rescue units in extinguishing peat fires was proved by the example of the Tver Region. and the circuit was designed (figure 2).

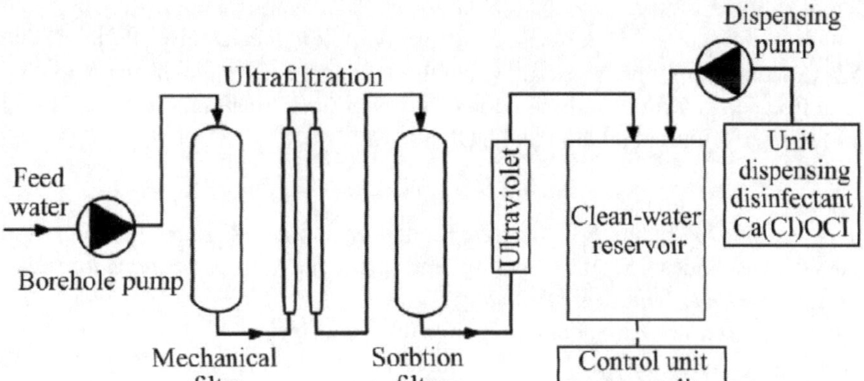

Figure 2. Diagram of the water supply of fire brigades during extinguishing peat fires.

5. Biologically active water. Subject domain overview
During the preparation of the materials and the review of the subject area, it was found out that poor-quality drinking water occupies a leading position in the pandemic of the metabolic caused morbidity

IOP Conf. Series: Materials Science and Engineering **492** (2019) 012022 doi:10.1088/1757-899X/492/1/012022

of the planet and it was decided to study the issue of non-reagent water treatment more deeply, because the mass use of water-preparation technologies water leads to the fact that water becomes pure, but biologically inert or even unsafe [18], [19].

The definition of "biologically active water" takes into account the electrophysical properties of drinking water that affect cellular metabolism: biologically active water is electrically non-equilibrium water with metastable associates of anion-radical forms of oxygen as carriers of electrons, the breakdown of which provides the flow of electrons to the cellular receptors of a living organism and maintenance in cells and organ structures of electrical nonequilibrium, necessary for the stable functioning of macroscopic quantum oscillator phase associated water control cellular processes.

This basic concept serves as a methodological basis for the biological activation of water, the activation of the vital energy of living beings and man, the formation of an electrophysically favorable environment for humans, animals and plants, and the prevention and treatment of metabolic diseases [20].

Studies show that the use of drinking water, which has reducing electron-donor properties, has a compensatory effect on the negative impact of physical environmental factors on human health. In this regard, a number of technologies for the physical activation of drinking water have been developed, such as the conditioning of drinking water with mechanochemically activated calcium micelle, membrane electrochemical activation, water activation in hypomagnetic conditions, the use of various phase modulating microadditives in drinking water. However, many of these technologies have disadvantages, of which the most important is the low integrity of the reducing properties of drinking water, due to the relatively high relaxation rates of the metastable state of water.In recent years, in the practice of water treatment, increasing attention is paid to the production of drinking water, the biological activity of which depends on the content of hydrogen isotopes. Hydrogen isotopes, whose concentration in the biosphere has increased significantly, make an important contribution to the problem of the biological activity of drinking water.

6. Conclusion

The article substantiates the relevance of improving the concept of water supply for fire departments when extinguishing peat fires by introducing into it a bioactivation unit for drinking water. In the process of writing, the main problem of the elimination of peat fires was formulated, statistics on peat fires in the Tver region for the last 5 years is given, methods of non-reagent water treatment are considered and their comparative analysis is conducted. The authors defined the basis for further improvement of the water supply scheme, which is determined by the subject of future research.

References

[1] Bonn A , Allott T , Evans M , Joosten H and Stoneman R 2016 Peatland restoration and ecosystem services: Science, policy and practice*Peatland Restoration and Ecosystem Services: Science, Policy and Practice,* 1–493

[2] Ratcliffe J L, Creevy A, Andersen R, Zarov E, Gaffney P P J, Taggart M A , Mazei Y, Tsyganov A N, Rowson J G, Lapshina E D and Payne R J 2017 Ecological and environmental transition across the forested-to-open bog ecotone in a West Siberian peatland *Science of the Total Environment,* **607-608** 816–828

[3] Yoshida H, Sazawa K, Wada N, Hata N, Marumo K, Fukushima M and Kuramit H 2018 Changes in the chemical composition of soil organic matter including water-soluble component during incubation: A case study of coniferous and broadleaf forest soils*Catena* **171** 22–28

[4] Osman K T 2013 Forest soils: *Properties and management*9783319025414 1–217
[5] Rydin H and Jeglum J K 2015 *The Biology of Peatlands* 1–400
[6] Vinogradov A V, Kuprin D S, Abduragimov I M, Kuprin G N, Serebriyakov E and Vinogradov V V 2016 Silica Foams for Fire Prevention and Firefighting*ACS Applied Materials and Interfaces***8** (1) 294–301
[7] Sister V G, Ivannikova E M, Gudkov A G, Leushin V Yu, Sidorov I A, Plyushchev V A and Soldatenko A P 2016 Detection of forest and peat-bog fire centers by means of microwave radiometer sounding *Chemical and Petroleum Engineering***52** 123–125
[8] Úbeda X, Pereira Pand Badía D 2018 Prescribed fires*Science of the Total Environment* 637–638 385–388
[9] Korotkov S A, Makuev V A, Lopatnikov M V, Nikitin V V, Sirotov A Vand Stonozhenko L V 2016 Forest-use issues in Moscow Region at the beginning of the 21st century*Bulletin of the Transilvania University of Brasov, Series II: Forestry, Wood Industry, Agricultural Food Engineering***9** (2) 17–24
[10] Gałka M, Szal M, Broder T, Loisel J and Knorr K-H Peatbog resilience to pollution and climate change over the past 2700 years in the Harz Mountains*Ecological Indicators***97** pp 183–193
[11] Lamentowicz M, Słowińska S , Słowiński M, Jassey V E J, Chojnicki B H, Reczuga M K, Zielińska M, Marcisz K, Lamentowicz Ł, Barabach J, Samso M, Kołaczek P andButtler A 2016 Combining short-term manipulative experiments with long-term palaeoecological investigations at high resolution to assess the response of sphagnum peatlands to drought, fire and warming*Mires and Peat***18** Article no 20
[12] Novenko E, Tsyganov A N, Pisarchuk N M, Volkova E M, Babeshko K V, Kozlov D N, Shilov P M, Payne R J, Mazei Y A andOlchev A V2018 Forest history, peatland development and mid- to late Holocene environmental change in the southern taiga forest of central European Russia*Quaternary Research (United States)***89** (1) 223–236
[13] Abakumov E, Maksimova E andTsibart A 2018 Assessment of postfire soils degradation dynamics: Stability and molecular composition of humic acids with use of spectroscopy methods*Land Degradation and Development***29** (7) 2092–2101
[14] Smagin A V, Sadovnikova N B, Vasenev V I andSmagina M V 2018 Biodegradation of some organic materials in soils and soil constructions: Experiments, modeling and prevention*Materials*, **11** (10) Article no 1889
[15] Bobek P, Šamonil P andJamrichová E 2018 Biotic controls on Holocene fire frequency in a temperate mountain forest *Journal of Quaternary Science***33** (8) pp 892–904
[16] Moore T R, Alfonso A and Clarkson B R2018 Plant uptake of organic nitrogen in two peatlands*Plant and Soil***433** (1-2) 391–400
[17] Bagrov V V, Kamrukov A S, Kostritsa V N andKrylov V I 2017 Autonomous ecologically safe installation for sea water desalination*Eurasian Journal of Analytical Chemistry***12** (7) 1045–1057
[18] Chaudhary N, Miller P A and Smith B2017 Modelling past, present and future peatland carbon accumulation across the pan-Arctic region*Biogeosciences***14** (18) 4023–4044
[19] Ivanov M V andKsenofontov B S 2017 Intensification of chemical agents mixing by vibroacoustical agitation *Ecology and Industry of Russia* 21 (9) 4–9
[20] Dean J F, Middelburg J J, Röckmann T, Aerts R, Blauw L G, Egger M, Jetten M S M, Meisel O H, Rasigraf O, Slomp C P and Dolman A J 2018 Methane Feedbacks to the Global Climate System in a Warmer World*Reviews of Geophysics* **56** (1) 207–250

IOP Conf. Series: Materials Science and Engineering **492** (2019) 012023 doi:10.1088/1757-899X/492/1/012023

Regulation of bubbles ascending speed o in water

B Ksenofontov[1], M Ivanov[1,4], S Chauhan[2] and I Schaikhiev[3]

[1] Bauman Moscow State Technical University, 5 Second Baumanskaya Street, Moscow, 105005, Russian Federation
[2]Traffic Planning and Environment Division, Central Road Research Institute (CSIR), New Delhi, India
[3]Kazan National Research Technological University, Kazan, Russian Federation

[4]E-mail: mivanov@bmstu.ru

Abstract. The paper presents the result of the study of various methods of controlling the rate of ascent of air bubbles in water, such as the addition of surfactants, as well as the imposition of vibration. The ascent rate was estimated by the photometric method. It is shown that with increasing bubble sizes, an increase in the ascent rate is first observed, and then after reaching a maximum, the speed may begin to fall. This is due to the increase in resistance to the movement of the bubble due to the distortion of its shape. The additional imposition of vibroacoustic effects significantly improves the hydraulic properties of air bubbles as they ascend. It is shown that there is a vibration-acoustic effect and a mechanical effect of the piston on air bubbles and flotation complexes. The authors show that at the same time, a key role in the efficiency of the process of intensification of ascent is rendered by the time of acoustic impact.

1. Introduction

The movement of gas bubbles in a liquid is used in many modern technological processes [1, 2, 3]. In particular, the ascent of air bubbles plays an important role in flotation water purification [4, 5]. One of the main parameters describing the motion of a bubble in a liquid is the ascent rate. This follows from the multistage model of flotation [6, 7]. According to it, the flotation process, taking into account the coalescence of flotation complexes, proceeds as shown in figure 1 and is described by the following system of equations:

$$\begin{cases} \frac{d}{dt}C_A = -K_1C_A + K_2C_B - K_6C_A + K_5C_C \\ \frac{d}{dt}C_B = K_1C_A - K_2C_B - K_3C_B + K_4C_C - K_7C_B + K_8C_D \\ \frac{d}{dt}C_C = K_3C_B - K_4C_C + K_6C_A - K_5C_C + K_9C_D - K_{10}C_C \\ \frac{d}{dt}C_D = K_7C_B - K_8C_D - K_9C_D + K_{10}C_C \end{cases}, \quad (1)$$

IOP Conf. Series: Materials Science and Engineering **492** (2019) 012023 doi:10.1088/1757-899X/492/1/012023

withinitialconditionsC_A (t) + C_B (t) + C_C (t) + $C_D(t)$ = C_0; C_A (0) = C_0; C_B (0) = 0; C_C (0) = 0; $C_D(0)$ =0, where *A* — theinitialstateoftheparticles; *B* — thestateofstickingandfixingoftheparticlesonthebubbles; *D*isaconditioncharacterizingtheenlarged (co-coordinated) flotationcomplex; *C* — stateofparticlesinthefoamlayer; C_A, C_B, C_D, C_C — concentration of particles in the states *A*, *B*, *D* and *C*, respectively; K1 is the constant of formation of the flotation complex; K2 is the decay constant of the flotation complex; K3 — the transition constant of the flotation complex in the foam layer; K4 is a constant characterizing the loss of the flotation complex from the foam layer; K5 is a constant characterizing the precipitation of particles of the solid phase from the foam layer; K6 is a constant characterizing the probability of the transition of particles of a solid phase from a liquid to a foam; K7 is the transition constant of flotation complexes to the enlarged flotation complex, due to their coalescence, and K8 is a constant characterizing the disintegration of the enlarged flotation complex into smaller ones; K9 — the transition constant of the enlarged flotation complex in the foam layer, and K10 — a constant characterizing the loss of the enlarged flotation complex from the foam layer.

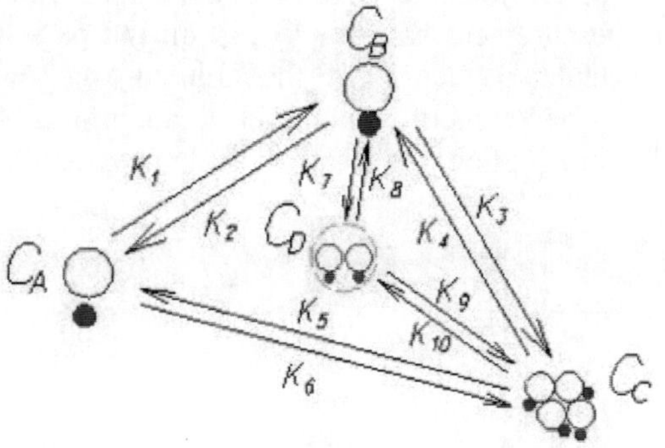

Figure 1

One of the most difficult tasks in this case is to determine the kinetic constants of the flotation process and for each of them there are different ratios for different cases [8]. The constant K3 in the general case can be determined on the basis of the following relationship:

$$K_3 = \frac{\upsilon_{BC}}{h},$$

where

υ_{BC} – float complex ascent rate [9];

h – the distance from the aeration zone to the foam layer (the depth of the flow chamber).

Thus, ultimately, it is the ascent rate that determines such significant characteristics of the flotation apparatus as the overall dimensions and the total time of the water purification process [10].

At the present time, quite often in the practice of water purification, various reagents are used [11, 12, 13, 14], in particular, surface acting agents (SAA), which may affect the rate of ascent of bubbles. For example, ion flotation is based on the process of raising a flotation complex formed by an SAA molecule, an air bubble, and an ion of a pollutant [15].

Literature devoted to the consideration of this issue quite a bit [16]. In addition, published data are controversial. In this regard, an urgent task is the experimental determination of the rate of ascent of air bubbles in water containing SAA.

However, it should be noted that this method has several significant drawbacks, the main ones being the additional pollution of SAA water, as well as the difficulty in selecting the type of surfactant and dosing the required amount of it. Therefore, it is proposed to investigate the method of vibroacoustic effects [17, 18, 19] to achieve the desired result.

2. Methods and materials

2.1. SAA research impact

To determine the nature of the effect of SAA on the rate of ascent of air bubbles of different sizes, it is necessary to solve the following problems:

1. Getting a solution of water with a given concentration of SAA.
2. Measuring bubble size.
3. Measure the speed of ascent of the bubble.

The listed tasks determined the choice of means and methods of measurement. An electronic balance and a set of volumetric flasks of various sizes were used to prepare the SAA solution of the desired concentration. The size of the bubble was determined using photo and video. Measurement of the ascent rate was carried out indirectly through the time the bubble passes a specified distance. The time was recorded using an electronic stopwatch.

The experimental setup is a transparent cylindrical column 1, mounted on a stand 2 (figure 2). At the bottom of the column installed aerators 3, connected to the compressor 4 through the valve system. Inside the column there is a measurement scale of 5, and on the outer part there are levels of the beginning and end of the calculated section of the column (the distance between the marks H = 600 mm). On a tripod 6, a video camera 7 connected to a computer 8 is installed in front of the column. To improve the quality of shooting, the installation also has an additional light source 9.

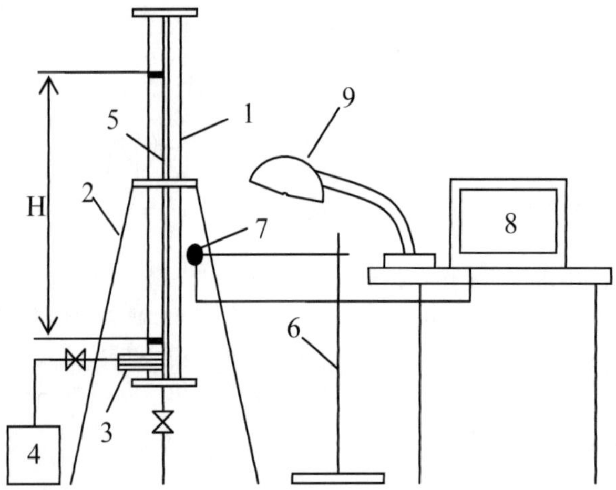

Figure 2. Scheme of the experimental setup:1 – cylindrical flask; 2 – stand; 3 – aerator system; 4 – compressor; 5 – measurement scale;6 – tripod; 7 – video camera; 8 – computer; 9 – lamp.

In order to analyze the effect of SAA on the rate of ascent of bubbles, it seems advisable to compare the values of velocity in distilled water, tap water and, finally, in water containing SAA in various concentrations. As SAA in this work, relatively frequently used substance OP-10 according to GOST 8433–81 is used. The tests were carried out for the following concentrations: 1, 5, 10, 15, 20 mg/l. In this case, the preparation of the SAA solution of the desired concentration should be carried out immediately before the start of the experiment in order to avoid undesirable consequences of its aging (for example, micelle formation).

The experiment was carried out as follows. The column was filled with the prepared liquid. A compressor was turned on, and with the help of valves, a mode is set in which bubbles of approximately the same size were formed. Using different aerators and adjusting the gas flow, we managed to get bubbles of several sizes. The ascent time measurements were taken separately for each

IOP Conf. Series: Materials Science and Engineering **492** (2019) 012023 doi:10.1088/1757-899X/492/1/012023

bubble size obtained. Moreover, to improve the accuracy of measurement was repeated ten times. With the help of a stopwatch, the time of movement of the bubble selected for observation from the lower to the upper mark was detected. A chamber connected to a computer was used to determine the size of the bubble. It allows not only to monitor the movement of bubbles in real time, but also to take pictures of them against the background of the measurement scale inside the column. A visual comparison of a bubble in a photo with a unit of measurement gives the desired size. Regulation of distance, angle of view and light makes it possible to achieve better quality photos, and therefore increase the accuracy.

After carrying out the required number of experiments, the obtained data were processed. According to the known distance and time, the speed of ascent of the bubble was calculated, and according to the photo - the size and shape. Examples of photographs obtained during the experiments are shown in figure 3.

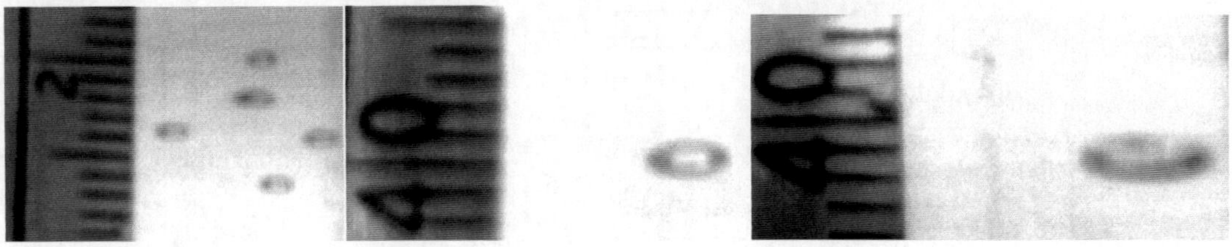

(a) (b) (c)

Figure 3. Photos of pop-up bubbles of various sizes: a) 1.5 mm; b) 3 mm; c) 4 mm.

2.2. Investigation of the effects of vibroacoustic effects

To determine the effect of vibroacoustic effects on the rate of ascent of air bubbles of various sizes, the following tasks were solved:

1. The choice of the optimal design of the laboratory setup.
2. Measuring bubble size.

The installation consists of a column 3 with a diameter of 110 mm, made of fibreglass. An aerator 5 is fixed in the column. The vibrator 1 by means of thrust drives the piston 4, which in turn is immersed to some predetermined depth in the column. The piston is a plate of various shapes, made of fiberglass. The vibrator is powered by a power amplifier, the signal to which is fed through a generator. The signal is sinusoidal.

The level of filling the column with water was equal to the number multiple of the diameter of the column and ranged from 3 to 5 diameters of the column.

In total, the installation was assembled in four different ways: when assembling according to scheme A, the aerator is located at the very bottom of the column; the piston is either directly above the aerator, or at some distance from it. The principal difference of scheme B is that the aerator is located on top of the piston. When the installation is switched on according to the scheme, the aerator is directly subjected to vibration, which is connected to the vibrator with the help of thrust. The piston in this scheme is missing. Scheme D is fundamentally different from the previous three; in this case, the column assembly with the aerator is attached directly to the vibrating table. In this case, the entire column with a liquid and a working aerator in it is subjected to vibroacoustic effects. When conducting experiments, aerators of various different types and designs were used: ceramic aerators of three sizes, two self-made aerators assembled from rubber hoses with pores of different sizes. The diameter of the bubbles ranged from 0.5 to 10 mm, depending on the type of aerator. Air was supplied to the aerators by means of two types of compressors differing in power. Both compressors have the ability to adjust the amount of air supplied. Since at this stage of this work there was no task to quantify the results obtained, for each type of aerator three levels of air flow were established:

IOP Conf. Series: Materials Science and Engineering **492** (2019) 012023 doi:10.1088/1757-899X/492/1/012023

• Maximum air flow rate — the flow rate that could be maximally obtained for this type of aerator.

• Average air flow is the flow at which the bubbles are as evenly distributed as possible over the cross-sectional area of the column.

• Minimum air flow rate - flow rate at which bubbles rise in one jet with a minimum distribution over the sectional area.

The level of consumption was limited not only by the capacity of the compressor and the pore diameter of the aerator, but also to the fact that there was some uniformity in the size of the bubbles for each level of air flow.

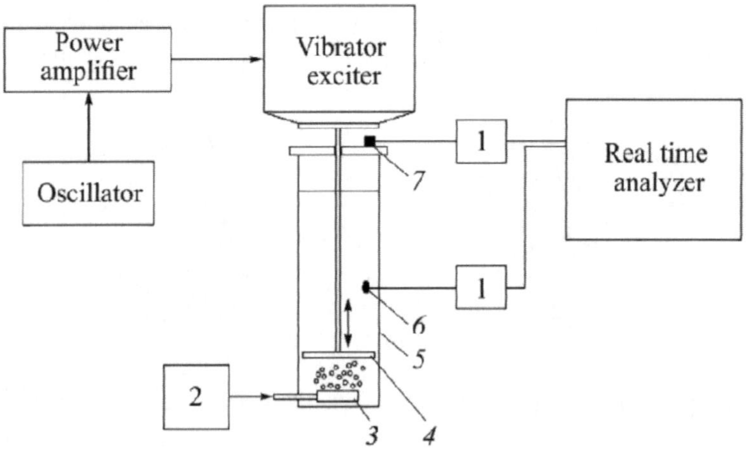

Figure 4. General installation scheme:1 – pre-amplifiers; 2 – compressor; 3 – aerator; 4 – piston; 5 – column; 6 – hydrophone; 7 – accelerometer.

The general setup diagram is presented in figure 4. The results of the experiments were recorded. As in the case of surface-active substances, photometry was used. But in addition, with the help of an accelerometer mounted on a column and a hydrophone immersed in the water layer, its own amplitude-frequency spectrum was measured. For this, a 2-channel narrow-band real-time analyzer from Bruel&Kjaer was used. One channel, with an accelerometer 7, mounted on the body of the column, was the driver, and the second (with hydrophone 6) was measuring.

3. Results and discussion

Thus, at first, the ascent rate of bubbles of various sizes in distilled water, tap water, and water containing different concentrations of SAA were obtained. The results of measurements and calculations are showed in table 1.

Analyzing the results of the experiments, it can be noted that with an increase in the bubble size, an increase in the ascent rate is first observed, and then after reaching the maximum, the speed may begin to fall. This is due to the increase in resistance to movement of the bubble due to the distortion of its shape. With an increase in the concentration of SAA in water, the rate of ascent of bubbles of the same size decreases. This is due to the fact that SAA molecules are adsorbed on the surface of the bubbles, thereby increasing resistance.

It is worth noting that when exposed to vibration with frequencies from 10 to 10,000 Hz, the wavelength changes from 150 m to 15 cm, respectively, and the direction of sound propagation is two-sided (both downward from the piston and upward). Thus, from this point of view, the change in the relative position of the piston and the aerator, as well as the distance between them, should not have any significant effect. However, during the experiments, various effects were discovered. This is due to the fact that the air bubble, leaving the aerator, falls on a vibrating piston, which has on them in addition to the acoustic mechanical effect. Thus, there is a vibration-acoustic effect on air bubbles and flotation complexes and the mechanical action of the piston. Changing the distance between the piston and the aerator increases the time it takes for a bubble or flotation complex to travel from the aerator to the piston. And the duration of the acoustic impact may be decisive to achieve the desired result.

Table 1. Experimental values of the ascending speed of bubbles

No[a]	Ascending speed, m/sec, bubble diameter, mm									
	0.5	0.9	1	1.5	2	2.5	3	4	5	6
1	–	0.099	0.117	–	–	–	0.221	–	0.205	–
2	–	–	0.106	–	0.233	0.227	–	–	–	0.236
3	0.105	–	–	0.174	–	–	0.193	–	–	–
4	–	–	–	0.155	0.168	–	–	0.203	–	–
5	0.062	–	–	0.160	–	–	–	–	–	0.209
6	–	–	0.130	–	0.193	–	–	–	–	0.205
7	–	–	0.085	–	0.173	–	–	0.185	–	–

[a]The number sin the first column correspond to the experiment sin the following fluids: 1 – distilled water; 2 – tap water; 3 – water with a concentration of SAAC = 1 mg/l;4 – the same with C = 5 mg/l; 5 – the same with C = 10 mg/l; 6 – the same with C = 15 mg/l; 7 – the same with C = 20 mg/l.

References

[1] Mishima K and Hibiki T 1996 Some characteristics of air-water two-phase flow in small diameter vertical tubes *International journal of multiphase flow***22**(4) 703–712

[2] Bendiksen K H 1984 An experimental investigation of the motion of long bubbles in inclined tubes *International journal of multiphase flow***10**(4) 467–483

[3] Vasiliev N V, Zeigarnik Y A and Khodakov K A 2017 Evolution of steam-water flow structure under subcooled water boiling at smooth and structured heating surfaces *Journal of Physics: Conference Series***891**(1) 012008 IOP Publishing

[4] Zabel T F 1992 Flotation in water treatment *Innovations in Flotation Technology* 431–454 Springer, Dordrecht

[5] Fakhru'l-Razi A, Pendashteh A, Abdullah L C, Biak D R. A, Madaeni S S and Abidin Z Z 2009 Review of technologies for oil and gas produced water treatment *Journal of hazardous materials***170**(2-3) 530–551

[6] Ksenofontov B S and Ivanov M V 2013 A novel multistage kinetic modeling of flotation for wastewater treatment *Water Science and Technology***68**(4) 807–812

[7] Sazonov D, Ksenofontov B, Ushinov V and Antonova E 2014 Wastewater Treatment by Flotation with Modern System of Aeration 14th SGEM GeoConference on Ecology, Economics, Education And Legislation 2 June 19-25, 2014*SGEM2014 Conference Proceedings***1** 815–820 ISBN 978-619-7105-17-9/ISSN 1314-2704

[8] Polat M and Chander S 2000 First-order flotation kinetics models and methods for estimation of the true distribution of flotation rate constants *International Journal of Mineral Processing***58**(1-4) 145–166

[9] White E T and Beardmore R H 1962 The velocity of rise of single cylindrical air bubbles through liquids contained in vertical tubes *Chemical Engineering Science***17**(5) 351–361

[10] Talaia M A 2007 Terminal velocity of a bubble rise in a liquid column *World Academy of Science, Engineering and Technology* **28** 264–268
[11] Bel Fdhila R and Duineveld P C 1996 The effect of surfactant on the rise of a spherical bubble at high Reynolds and Peclet numbers *Physics of Fluids* **8**(2) 310–321
[12] Watcharasing S, Kongkowit W and Chavadej S 2009 Motor oil removal from water by continuous froth flotation using extended surfactant: effects of air bubble parameters and surfactant concentration *Separation and Purification Technology* **70**(2) 179–189
[13] Alves S S, Orvalho S P and Vasconcelos J M T 2005 Effect of bubble contamination on rise velocity and mass transfer *Chemical Engineering Science* **60**(1) 1–9
[14] Rodrigue D, De Kee D and Fong C C M 1996 An experimental study of the effect of surfactants on the free rise velocity of gas bubbles *Journal of non-newtonian fluid mechanics* **66**(2-3) 213–232
[15] Korchagova V N, Kraposhin M V, Marchevsky I K and Smirnova E V 2017 Simulation of droplet impact onto a deep pool for large Froude numbers in different open-source codes *Journal of Physics: Conference Series* **918**(1) 012037 IOP Publishing
[16] Zhang Y and Finch J A 2001 A note on single bubble motion in surfactant solutions *Journal of Fluid Mechanics* **429** 63–66
[17] Minnaert M 1933 XVI On musical air-bubbles and the sounds of running water *The London, Edinburgh, and Dublin Philosophical Magazine and Journal of Science* **16**(104) 235–248
[18] Ellenberger J and Krishna R 2003 Shaken, not stirred, bubble column reactors: Enhancement of mass transfer by vibration excitement *Chemical Engineering Science* **58**(3-6) 705–710
[19] Ivanov M V and Ksenofontov B S 2014 Intensification of flotation treatment by exposure to vibration *Water Science and Technology* **69**(7) 1434–1439

IOP Conf. Series: Materials Science and Engineering **492** (2019) 012024 doi:10.1088/1757-899X/492/1/012024

Analysis of hydraulic modes of operation of the divergent plates block of the flotation-sedimentation tank

B Ksenofontov[1,2], E Senik[1] and M Ivanov[1]

[1] Bauman Moscow State Technical University, 5 Second Baumanskaya Street, Moscow, 105005, Russian Federation

[2] E-mail:ksenofontov@bmstu.ru

Abstract. This article shows the results of hydraulic characteristics analysis of operating modes of divergent plates of the thin-layer clarification block of the flotation-sedimentation tank for wastewater treatment. A method for calculating the velocity gradient and the geometric parameters of the block of divergent plates is proposed and tested by a computer modeling in this article, and besides the distance between the plates varies from 12.5 mm to 141 mm. The particles are assumed to be spherical in shape; its modeling is carried out by using the Lagrangian multiphase model. By the results of modeling it has been discovered, that with the set parameters of the plates block, the time for particles to reach the highest point of the sedimentation trajectory is 140 seconds. Moreover, the plate length, required for particles sedimentation, turns out to be 590 mm. Whereas the sedimentation time and the plate length, calculated by the suggested method, are almost the same with the results acquired by modeling. In connection with that the suggested formulas appliance becomes possible for calculation of the sedimentation time and the plate length.

1. Introduction

The most complete wastewater treatment from both hydrophobic and hydrophilic particles is achieved by the flotation-sedimentation process. The most important condition of highly efficient treatment, and for the other such methods too, is optimization and matching of the main technological parameters [1-19], and in this particular case the velocities of the treated water flow at on the flotation and sedimentation stages [1]. For that purpose we have developed the thin-layer clarification blocks in the form of a set of divergent plates (or blocks of divergent plates) [2]. Apart from that in dependence of its application the divergent plates block may be applied for other purposes:

- hydrodynamic modes matching in the flotation and sedimentation tanks;
- contaminating particles sedimentation on the lower plate of an element of the divergent plates block (by "an element of the divergent plates block " it is meant the two plates next to each other) [3];
- safety provision for the flakes of contaminating particles.

At the same time, destruction of the flakes of contaminating particles is characterized by the velocity gradient [3], an important parameter for the particles sedimentation is the plate length. And for hydrodynamic modes matching in the flotation and sedimentation tanks the geometrical parameters of the divergent plates block are essential (the angle between the plates). Moreover for the device operating volume calculation the residence time of treated water at the divergent plates block is needed [5].

2. Modeling of the hydraulic modes of a divergent plates block

A method to calculate the velocity gradient and the geometric parameters of the block of divergent plates is suggested and tested by computer modeling [6].

The divergent plates block scheme is shown in the figure 1.

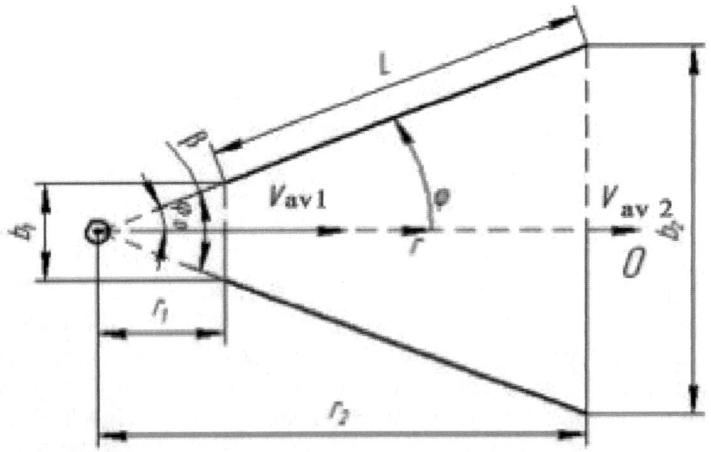

Figure 1. The divergent plates block scheme: β – angle between the plates; b_1 иb_2 – distance between the plates at the input and output from the plates block accordingly; v_{av1} and v_{av2} – r_1 and r_2 distances from the diffusor center accordingly; L – block plate length.

In accordance with [4, 5, 6] in thin-layer clarification blocks, which are composed of parallel plates, it is recommended to take the operation depth of the sedimentation zone h_{ti} in the range from 25 to 200 mm, so consequently distance between the plates at the input of the divergent plates block $b_1 = h_{ti} \cdot \cos\alpha$, where α – is a tilt angle to the horizon.

In regard of the water flow velocity at the input and output from the diffusor the following is applicable: at the input to the diffusor [10] the average velocity has to match the velocity in the sedimentation tank (form 5 to 10 mm/sec), and at the output - in the flotation tank (no more than 5 nn/sec).

Symmetrical and all-over divergent flow has to be developed in the plates block element for the safety of contaminating particles flakes [10]. Wherein the developing flow type strongly depends on the angle between the plates. Maximum angles between the plates of the divergent plates block, at which symmetrical and all-over divergent flow are developed, for various velocities of the water flow in the sedimentation tank and distances between the plates are shown in table 1.

In such a manner, if the angles of divergence are smaller than those mentioned in the table 1, only symmetrical and all-over divergent flow will occur in the divergent plates block.

Thus, if the block of divergent plates functions only for the hydrodynamic modes matching, then the distance between the plates of the thin-layer clarification block from the flotation tank side b_2, in accordance with the sequence from the continuity equation [11, 12, 13, 14], will be:

$$\frac{v_{av1}}{v_{av2}} = \frac{r_2}{r_1} = \frac{b_2}{b_1}, consequently\, b_2 = \frac{v_{av1} \cdot b_1}{v_{av2}} \tag{1}$$

Then the angle between plates β has to be chosen, and at the set parameters it should not be greater than the values in the Table 1. After that the plate length L_{match} is calculated, which is required for the hydrodynamic modes matching, by the formula:

$$L_{match} = \frac{b_2 - b_1}{2\sin\frac{\beta}{2}} \tag{2}$$

Table 1. Maximum angles between the plates of the block of divergent plates at various distances between the plates and the velocities of water flow at the input of the divergent plates block.

Velocities of the water flow in the sedimentation tank, mm/sec	Distance between the plates (from the sedimentation tank side), mm								
	140	120	100	80	60	40	20	15	12,5
10	2º	2º	3º	4º	5º	8º	16º	22º	27º
9	2º	3º	3º	4º	6º	9º	19º	25º	3ºº
8	3º	3º	4º	5º	7º	10º	21º	28º	30º
7	3º	4º	4º	6º	8º	12º	24º	30º	30º
6	4º	4º	5º	7º	9º	14º	28º	30º	30º
5	4º	5º	6º	8º	11º	17º	30º	30º	30º

To calculate the operating volume of the whole device, it is necessary to know the residence time of water at the block element [10] as well:

$$t_{match} = \frac{r^2 - r_1^2}{2r_1 \cdot v_{avl}} = \frac{L(b_1 + b_2)}{2b_1 \cdot v_{avl}} \tag{3}$$

If the plates block, apart from the hydrodynamic modes matching at the sedimentation and flotation tanks, has to provide the sedimentation of the particles with the specific hydraulic thickness, then the plate length matters, besides the sedimentation time is important too:

$$\begin{cases} t_{sed} = \dfrac{r_0}{2v_{f0}}\left[\left(1 - \dfrac{v_{f0}}{u}\left[\ln\dfrac{\left(tg\left(\dfrac{\pi}{4}+\dfrac{\alpha}{2}\right)\right)}{tg\left(\dfrac{\pi}{4}+\dfrac{\alpha+\beta}{2}\right)}\right]\right)^2 - 1\right]; \\ r_{oc} = \sqrt{2r_0 v_{f0} t_{oc} + r_0^2} \end{cases} \tag{4}$$

Herewith the plate length L_{sed}, required for sedimentation is:

$$L_{sed} = r_{sed} - r_0 \tag{5}$$

Calculation of the main parameters of the divergent plates block is made with the input parameters: u = 0,7 mm/sec, β= 5º, α= 60º, b_1 = 20 mm, v_{f0} = 10 mm/sec , by the formulas (4) and (5). As a result the following data has been acquired: r_0= 229 mm; t_{oc}= 146 sec; L_{oc}= 620 mm.

3. The results and consideration

To verify the acquired data computer modeling of the particles motion in the element of the divergent plates block has been performed [11]. The particles were assumed to be spherical in shape, modeling has been carried by using of the Lagrangian multiphase model [17, 18].

By the results of modeling it has been discovered, that with the set parameters of the plates block, the time for particles to reach the highest point of the sedimentation trajectory is 140 seconds. Moreover, the plate length, required for particles sedimentation, turned out to be 590 mm.

IOP Conf. Series: Materials Science and Engineering **492** (2019) 012024 doi:10.1088/1757-899X/492/1/012024

Wherein the sedimentation time and the plate lenght, calculated by the suggested method are almost the same with the results acquired by modeling. In connection with that using of the suggested formulas becomes possible for calculating of the sedimentation time and the plate length.

If the plates block serves for both the hydrodynamic modes matching and for water clarification, then the greatest calculated value is used. I.e. if $L_{match}>L_{sed}$, then $L = L_{match}$, $\tau = \tau_{match}$; if $L_{match}<L_{sed}$, then $L = L_{sed}$, $\tau = \tau_{sed}$.

In the present case, $L_{match}<L_{sed}$, then $L= L_{sed}$= 620 mm, $\tau = \tau_{sed}$= 146 sec.

Overall, the suggested method allows to calculate the main parameters of the divergent plates block, which in turn makes it possible to apply this method for calculating of the floatation-sedimentation tanks and define the optimum conditions to increase efficiency of the purification by 20-30%.

References

[1] Isaev S A, Leont'ev A I, Mil'man O O, Sudakov A G, Usachov A E and Gul'tsova M E 2018 Intensification of Heat Exchange in Laminar Vortex Air Flow in a Narrow Channel with a Row of Inclined Oval Trenches *Journal of Engineering Physics and Thermophysics* **91**(4) pp 963–974

[2] Ksenofontov B S and Petrova E V 2013 *Flotation-sedimentation tank*: patent 132434 Russian Federation Reg 20 09 2013

[3] Blott S J and Pye K 2001 GRADISTAT: a grain size distribution and statistics package for the analysis of unconsolidated sediments. *Earth surface processes and Landforms* **26**(11) pp 1237–1248.

[4] Ivanov M V and Ksenofontov B S 2014 Intensification of flotation treatment by exposure to vibration *Water Science and Technology* **69**(7) pp 1434–39

[5] Clara M, Kreuzinger N, Strenn B, Gans O and Kroiss H 2005 The solids retention time — a suitable design parameter to evaluate the capacity of wastewater treatment plants to remove micropollutants *Water research* **39**(1) pp 97–106

[6] Womersley J R 1955 Method for the calculation of velocity, rate of flow and viscous drag in arteries when the pressure gradient is known *The Journal of physiology* **127**(3) pp 553–563

[7] SNiP 2.04.03-85 1986 *Canalization. External Supply Lines and Constructions*

[8] A reference book to SNiP 2.04.03-85 *Construction Design for waste water treatment* 1990 Moscow Stroyizdat Publ

[9] Arefyev K Y, Prokhorov A N and Saveliev A S 2018 Study of the breakup of liquid droplets in the vortex wake behind pylon at high airspeeds *Thermophysics and Aeromechanics* **25**(1) pp 55–66

[10] Kim M I, Kim O S, Lee D H and Kim S D 2007 Numerical and experimental investigations of gas–liquid dispersion in an ejector *Chemical Engineering Science* **62**(24) pp 7133–39

[11] Moghaddam S S, Moghaddam M A and Arami M 2010 Coagulation/flocculation process for dye removal using sludge from water treatment plant: optimization through response surface methodology *Journal of hazardous materials* **175**(1-3) pp 651–657

[12] Landau L D and Lifshits E M 1986 *Theoretical physics: Manual* In 10 vol Vol VI*Hydrodynamics* 3rd edition, revised Moscow Nauka Publ Chief editorial Physics-mathematics literature736 p

[13] Slezkin N A 1955 *Dynamics of viscous incompessable liquid* Moscow State Publishing Company of technical-theoretical literature 521 p

[14] Targ S M 1955 *The main concerns of the Laminar Flows theory* Moscow State Publishing Company of technical-theoretical literatur 420 p

[15] McDonald M G and Harbaugh A W 1988 *A modular three-dimensional finite-difference ground-water flow model* Vol 6 p. A1 Reston, VA: US Geological Survey

[16] Ksenofontov B S and Ivanov M V 2013 A novel multistage kinetic modeling of flotation for wastewater treatment *Water Science and Technology* **68**(4) pp 807–812

[17] Kulik V V, Parkin A N and Navasardyan E S 2016 Numerical Modeling Procedure for Micromachined Cryogenic Cooler Elements Using ANSYS Fluent Software and Viscous Flow in a Small-Diameter Channel with Heat Transfer as an Example *Chemical and Petroleum Engineering* **52**(7-8) pp 531–538

[18] Marocco L and Inzoli F 2009 Multiphase Euler–Lagrange CFD simulation applied to wet flue gas desulphurisation technology *International Journal of Multiphase Flow* **35**(2) pp 185–194

[19] Yeganeh A, Gotoh H and Sakai T 2000 Applicability of Euler-Lagrange coupling multiphase-flow model to bed-load transport under high bottom shear *Journal of Hydraulic Research* **38**(5) pp 389–398

IOP Conf. Series: Materials Science and Engineering **492** (2019) 012025 doi:10.1088/1757-899X/492/1/012025

Hydrodynamic filters in hydraulic fluid cleaning system of hydraulic drive

V Devisilov[1] and E Sharai[1,2]

[1]Bauman Moscow State Technical University, 5 Second Baumanskaya Street, Moscow, 105005, Russian Federation

[2]E-mail: e9.sharay@yandex.ru

Abstract. The paper deals with prospects for the use of hydrodynamic filters in the hydraulic fluids' system for the preparation and regeneration based on a review of the current state in the field of hydrodynamic filtering. The authors show the classification of known design schemes for hydrodynamic filters according to two characteristics: presence or absence of sludge liquid and methods for creating an additional force field. The authors note the advantages and disadvantages of each type's hydrodynamic filter. In this paper, the authors establish the dependence of the particles' separation efficiency caused by the rotation of the filter septum on the average median particle size of impurities. Upon reading the paper one realized that the prospects of using hydrodynamic filters to provide the required frequency of the hydraulic systems hydraulic fluid of hydraulic drives.

1. Introduction

One of the methods to improve the reliability of technical systems is to reduce the failure rate [1-4]. For hydraulic systems, the failure rate of equipment depends on a number of factors: the working fluid cleanliness class, operating conditions, and timely maintenance. Practice shows that increasing the reliability of the elements of the hydraulic system by 80% depends on the purity of the working fluid. The presence of mechanical impurities in the working fluids of hydraulic systems reduces the life of hydraulic actuators and can lead to sudden failures during device seizure, leakage, valve clogging, etc. Both the concentration of particles in the fluid and their size affect the wear process of hydraulic actuators. Particles of contamination enter the working fluid when pouring fluid into the tank; when worn rubbing surfaces; through hydraulic seals. Therefore, it is necessary to constantly monitor and promptly clean the working fluid of hydraulic actuators from mechanical impurities. Of the known methods of cleaning liquids from solid contaminants, only filtering can guarantee the required cleaning fineness.

Certain portions of hydraulic fluids are high viscosity mineral oils. High fluid viscosity and clogging of the filter material lead to the need to create a high-pressure drop across the filter septum and frequent filter regeneration. Therefore, it is advisable to use combined devices, such as hydrodynamic filters, providing highly efficient separation processes. Hydrodynamic filtration differs from the traditional one by the presence of an additional component of velocity v_t, tangent to the filtration surface (figures 1a, 1b). The hydrodynamic effect is the continuous removal of the accumulated layer of sediment from the surface of the filtering material [5-7]. Realization of this effect requires high tangential speed v_t flow rate relative to filtration rate v_0. Within reasonable limits of the ratio of these velocities, a hydrodynamic effect arises for large particles of pollution exceeding 300 μm. To reduce the size of particles that can be removed from the surface of the filter septum, a combination of hydrodynamic filtration with an additional force mechanism is used (figures 1c, 1d). In this case, forces, usually centrifugal and vibration have an additional effect on the particle of pollution near the filter septum and do not allow it to settle on the filter surface.

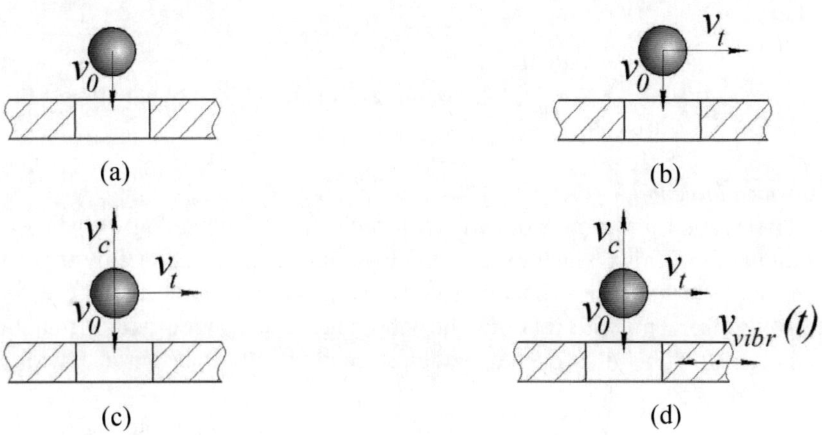

Figure 1. Filtering circuit: a) customary; b) hydrodynamic; c), d) hydrodynamic with force action.

The area of application of hydrodynamic filters is quite wide. The advantage of these devices is that with guaranteed observance of the purity of the liquid, determined by the cell size of the filter material, they have the ability to self-purification, which increases the regeneration time and service life [8].

2. Basic types of hydrodynamic filters

There are several types of hydrodynamic filters, differing in the organization of output streams, and in the method of flow swirl, which provides for the creation of an additional force field. Conditionally, they can be divided into the following groups:

2.1. Sludge liquid or without sludge liquid

The first class of hydrodynamic filters is characterized by the sludge fluid or without sludge fluid and includes two groups of filters: full-flow and part flow [8]. The full-flow filters (figure 2a) pass through the filter septum all the liquid that has arrived for purification. The advantage of this mode of operation is obvious - this is the absence of sludge fluid. However, the self-cleaning ability of such filters is lower than part-flow filters (figure 2b). They are used where full collection of impurities is necessary, with low viscosity of the medium being cleaned with low obliteration properties or with a slight difference between the density of the solid and liquid phases.

In part-flow hydrodynamic filters, a part of the filtered fluid (usually 10–15%) is by passed along the filter septum to flush accumulated sediment from its surface, and then is dumped along with the contaminants into the tank for the filtered fluid or is disposed of after proper decontamination as waste.

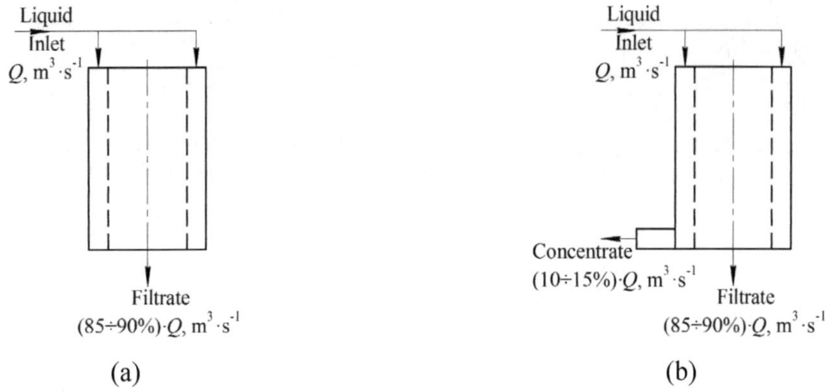

Figure 2. Hydrodynamic filter circuit: a) full-flow filter; b) part-flow filter.

2.2. Creating way a force field

The second classification group includes hydrodynamic filters containing elements of its design to create an additional swirl flow, which reduces the load on the filter material by separating part of the contamination by a centrifugal mechanism. Five types of hydrodynamic filters can be distinguished: with a rotating filter septum; with rotating and vibrating filter septum; with a rotating filter septum with a protective shell; with a filter element with a large curvature of the surface; with special twisting devices.

The greatest effect of swirling flow allows to obtain devices with a rotating filter septum, as they allow you to adjust the rotational speed depending on the properties of the medium being cleaned and thereby obtain the greatest force on the stream to be cleaned with minimum filter sizes [9] (figure 3). The combination of rotation of the filter septum with vibration along the axis of rotation increases the self-cleaning ability of the filter due to the destruction of the accumulated sediment layer and its simultaneous removal from the surface of the filter septum by centrifugal and hydrodynamic forces [10,11] (figure 3b). However, the flow in the channel formed by the casing and the filter septum is close in its structure to the Couette-Taylor flow, which, as is well known, can be unstable [12-15]. With the advent of macro-vortex structures in the working area, particle separation deteriorates due to the uneven distribution of radial and longitudinal velocity along the length of the channel; part of the filter septum can be excluded from work due to the circulation of secondary currents. In addition, the presence of vortex structures increases the hydraulic resistance in the filter. Therefore, for hydrodynamic filters with a rotating filter septum, it is necessary to choose the optimal operating conditions for which the flow in the working channel will be irrotational and stable. The proposed design with perforated protective shell, located between the housing and the filter septum. The presence of a rotating perforated shell allows the separation of not only coarse, but also fine particles of dirt, because in the annular gap between the filter septum and the perforated protective shell, the liquid rotates like a solid body with a constant angular velocity, which prevents the penetration of fine particles of dirt to the surface of the filter material [16, 17]. Thus, the rotating perforated shell allows you to increase the service life of the hydrodynamic filter by self-cleaning the filter septum and reducing the particle size separated by centrifugal forces, compared to the rotating filter septum with a similar frequency, but in the absence of a protective perforated shell. However, this design, complicated by the presence of additional moving elements, does not solve the problem of vortex formation in the working annular channel and requires the selection of operating parameters that ensure a steady flow [18-20].

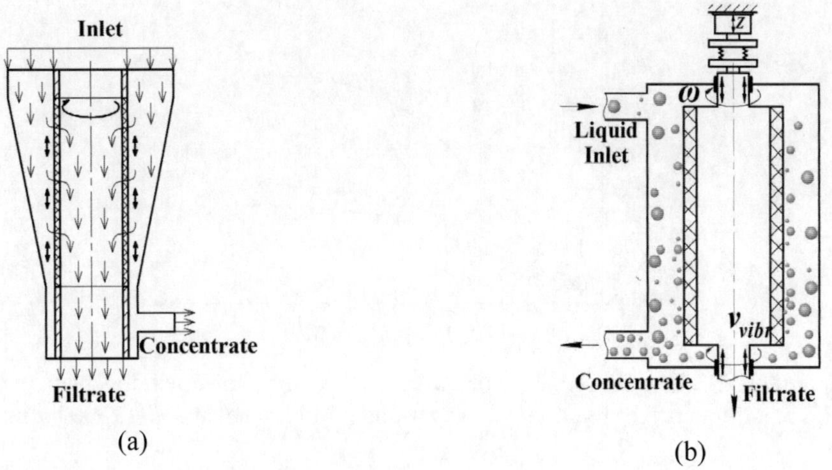

Figure 3. Hydrodynamic vibration filter circuit: a) with rotating filter septum; b) with rotating and vibrating filter system.

3. Calculation of the centrifugal separation's efficiency in a hydrodynamic filter

As a result of mathematical modeling of hydrodynamic and separation processes occurring in a hydrodynamic vibration filter using modern CAE programs, the dependence of the particle separation efficiency due to the rotation of the filter septum on the average particle size of contaminants at the rotation frequency f_{rot} = 5 Hz and particles densityρ_p = 2500 kg/m^3 for non-Newtonian pseudoplastic fluids subject to the rheological power dependence of Oswald de Ville:

$$\tau = m\left|\gamma\right|^{n-1}\gamma, \tag{1}$$

where
τ – shift voltage;
n– flow index (n< 1);
m – consistency index;
γ – sliding velocity.

Evaluation of separation efficiency was carried out according to the expression:

$$\eta_{dm} = \frac{N_{rot}}{C_p}100\% \tag{2}$$

where
C_p – particles count in the raw liquid;
N_{rot} – particles count dimension d_m, separated with centrifugal mechanism.

As a result of the calculations, it was established that at the frequency of rotation of the filter septum f_{rot} = 5 Hz centrifugal mechanism captured from 60 to 100 % medium sized particles 200…400 μm, from 20 to 60 % sized particles 150…200 μm, and from10 to 20 % sized particles 60…150 μm depending on the flow index n (figure 4). A further increase in the frequency of rotation of the septum will increase the centrifugal contribution to the overall mechanism for cleaning fluid. However, there is a limit to the growth of the cleaning efficiency of the centrifugal mechanism, due to the loss of stability of the flow in the annular gap of the filter. The purpose of further research is to determine the limits of flow stability and the choice of optimal operating parameters of the filter.

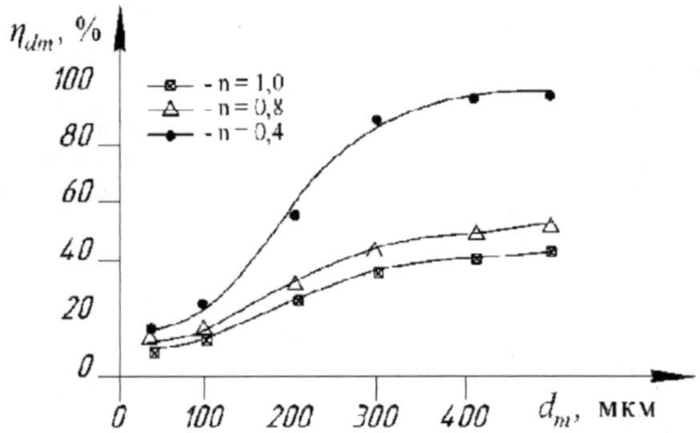

Figure 4. Separation efficiency dependence of the centrifugal mechanism on the average particle size of dirt particles at a frequency of rotation f_{rot} = 5 Hz and particles density ρ_p = 2500 kg/m^3.

4. Conclusions

To ensure uninterrupted and reliable operation of hydraulic systems, it is necessary to clean working fluids from mechanical impurities in a timely manner. The proposed mechanism for cleaning highly viscous non-Newtonian fluids by filtration with a combination of the action of force fields reduces the load on the filter material by separating some of the contaminants by centrifugal forces, and increases the service life of the filter due to its ability to self-regeneration.

Acknowledgments

This paper was made according to State Order of the Ministry of Science and Higher Education of the Russian Federation (No 10.7766.2017/8.9).

References

[1] Conrad F, Hilbrecht B and Jepsen H 2000 Design of low-pressure and high-pressure tap water hydraulic systems for various industrial applications (No. 2000-01-2614) *SAE Technical Paper*

[2] Livingstone G and Cavanaugh G 2015 The real reasons why hydraulic fluids fail *Tribology & Lubrication Technology***71**(7) p 44

[3] Badakh V, Tarasenko T, Puzik O and Krayushkina K 2013 Functional units based on cavitation effects for hydraulic systems of vehicles *Transport engineering and management* Vilnius: Lithuania pp 50–54

[4] Onishchenko D, Pankratov S, Zotov A, Osipkov A and Poshekhonov R 2017 Study of influence of hydraulic thermoelectric influence of hydraulic thermoelectric generator resistance on gasolie engine efficiency *International journal of applied engineering research***12** (5) pp 721–727

[5] Yim S and Kwon Y 1997 A unified theory on solid-liquid separation: Filtration, expression, sedimentation, filtration by centrifugal force, and cross flow filtration*Korean Journal of Chemical Engineering***14**(5) pp 354–358

[6] Kim Mand Zydney A 2006 Theoretical analysis of particle trajectories and sieving in a two-dimensional cross-flow filtration system *Journal of membrane Science***281**(1-2) pp 666–675

[7] Al-Malack M and Anderson G 1997 Use of crossflow microfiltration in wastewater treatment. *Water Research***31**(12) pp 3064–72

[8] Finkelstein Z 1986 *Primenenie I ochistka rabochikh zhidkostej dlya gornykh mashin*

[Application and cleaning fluids for mining machines] Moscow: Nedra Publ.
[9] Aleksandrov A, Devisilov V, Sharai E and Kiselyova D 2018 Effect of geometric parameters of working channel of hydrodynamic filter with protective baffle on medium flow structure *Herald of the Bauman Moscow State Technical University, Series Natural Sciences***2** (77) pp 23–38 DOI: 10.18698/1812-3368-2018-2-23-38
[10] Devisilov V and Sharai E 2018 Particle Separation in an Annular Converging Channel with an Inner Rotating Permeable Baffle *High Temperature***56** (4) pp 576–580 DOI: 10.1134/S0018151X18040053
[11] Devisilov V and Sharai E 2016 Numerical study of the flow structure in a hydrodynamic filter *Theoretical Foundations of Chemical Engineering***50** (2) pp 209–216 DOI: 10.1134/S0040579516020044
[12] Ostilla R, Stevens R, Grossmann S and Verzicco Rand Lohse D 2013 Optimal Taylor–Couette flow: direct numerical simulations *Journal of fluid mechanics***719** pp 14–46
[13] Montavon C, Grotjans H, Hamill I, Phillips H and Jones I 2000 Mathematical modelling and experimental validation of flow in a hydrocyclone In: *5th International Conference on Cyclone Technologies* Warwick, UK **31** pp 175–186
[14] Bunyawanichakul P, Kirkpatrick M, Sargison J and Walker G 2006 Numerical and experimental studies of the flow field in a cyclone dryer *ASME Journal of Fluid Engineer* no 128 (6) pp 1240–1250 DOI: 10.1115/1.2354523.47
[15] Varaksin A, Protasov M, Marinichev D and Vasil'ev N 2015 An Analysis of the Parameters of Flare Particles for Optical Diagnostics of Vortex Flows *Measurement Techniques***58** (6) pp 655–660
[16] Durand A 2003 Bioreactor designs for solid state fermentation *Biochemical Engineering Journal***13**(2-3) pp 113–125
[17] Mochalin I, Brazhenko V and Yashchuk O 2017 An experimental research of the efficiency of a fluid mechanical cleaning by a rotating filter*Journal of civil engineering and management* **23** (3) pp 43–46
[18] Yahi F, Hamnoune Y, Lecheheb S and Bouabdallah A 2015 Experimental investigation of the free surface effect on the conical Taylor-Couette flow system *Topical problems of fluid mechanics* (Prague) p 273
[19] Noui-Mehidi M 2011 Design optimization of a conical annular centrifugal contractor *FDMP* **7** (2) p 141
[20] Xiaofei X, Pu W, Lanxi X and Dapeng C 2010 Occurrence of Taylor vortices in the flow between two rotating conical cylinders *Communications in Nonlinear Science and Numerical Simulation*vol **15**(5) p 1228

IOP Conf. Series: Materials Science and Engineering **492** (2019) 012026 doi:10.1088/1757-899X/492/1/012026

The effect of design parameters of the closed type regenerative pump the energy characteristics

N Isaev

Bauman Moscow State Technical University, 5 Second Baumanskaya Street, Moscow, 105005, Russian Federation

E-mail: isaev.nikita@bmstu.ru

Abstract. The article deals with the application area and features of the regenerative pumps. The main disadvantages of the existing methods for the design of the vortex pump flow part are given. Based on the nature of the flow in the original flow part, calculated according to the classical method, a number of design parameters have been selected and a parametrized 3-D model of a closed type vortex pump has been developed on the base of it, which allows changing parameters together and separately in a wide range. Method for determining the influence of a single constructive parameter on energy characteristics using computational fluid dynamics methods is described in the article. A complex optimization of the flow section of the closed type regenerative pump using the LP-τ search method in the STAR-CCM+ software package was carried out.

1. Introduction

The application area of regenerative pumps is wide enough. They are used in the chemical industry for supplying reagents, for pumping volatile liquids at gas-dispensing stations, in public utility pumping facilities, instead of liquid-ring pumps as vacuum pumps and low-pressure compressors, on vessels for supplying industrial and drinking water, and in some other cases. For regenerative pumps, the following parameters are usually characteristic: pump flow up to 720 l/min, pump head up to 250 m, and overall efficiency does not exceed 35-38% [1]. A regenerative pump in comparison with centrifugal pump with the same dimensions and shaft rotation frequency has the head 4-9 times larger, and its design is technologically simpler and cheaper. The regenerative pump is not able to pump highly viscous liquids due to a sharp drop in pressure. It is also impossible to pump liquids with solid inclusions, since the rapid wear of the walls of the end and radial gaps leads to an increase in overflows and, accordingly, to a drop in pressure and efficiency [2].

The principle scheme of the closed type regenerative pump is shown at figure 1 The impeller 1 usually has radial or inclining blades. The impeller is in a cylindrical housing with a concentric channel starting from the suction chamber and ending in the pressure chamber. The channel is shut down by a closing dike 2, which performs the function of sealing between the suction and pressure chambers [3-6].

Nowadays there are several classic methods for regenerative pumps analysis, the main representatives of which are: O.V. Baibakov, B.I. Nahodkin, K.P. Fleuderer, V.V. Shaumyan and a number of other scientists [1-4]. All of these techniques either require a large amount of calculations, or are applicable to narrow classes of pumps, or are based on statistics from the regenerative pumps

IOP Conf. Series: Materials Science and Engineering **492** (2019) 012026 doi:10.1088/1757-899X/492/1/012026

existing at that time. In addition, now there is no comprehensive method for optimizing the flow part of regenerative pumps based on hydrodynamic modeling methods. Since the increased requirements for energy efficiency have been recently made against pumps [7–10], the question of increasing the efficiency of regenerative pumps has become particularly acute.

Therefore companies set BMSTU department "Hydromechanics, Hydromachines and Hydro-Pneumoautomatics" a task of developing a parametrized model of a closed type regenerative pump and determining which design parameters affect the energy characteristics of the regenerative pump mostly using modern hydrodynamic modeling methods.

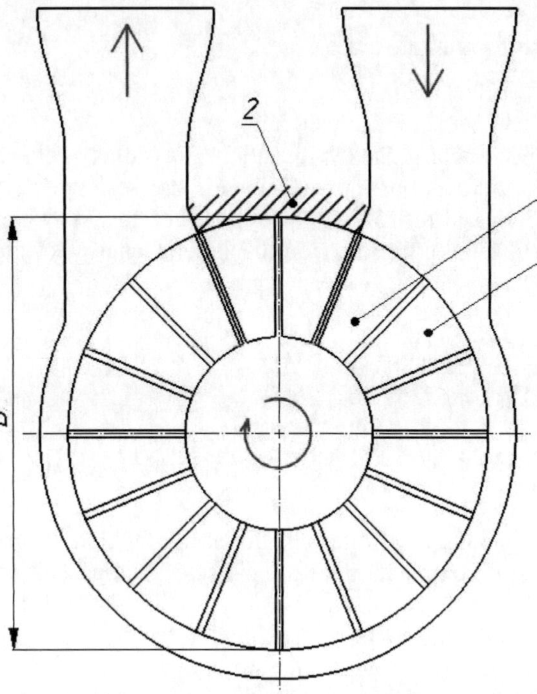

Figure 1. The principle scheme of the closed type regenerative pump.

This study had been conducted in the framework of the development of a double-stage centrifugal-regenerative pump for the following parameters:

Pressure $H = 54$ m;

Feed $Q = 32$ m^3/h;

Shaft speed n=2900 rpm.

2. Methods

The method of numerical modelling is based on solving discrete analogs of the basic hydrodynamic equations [11-14]. In the case of an incompressible fluid model (ρ = const), this is:

Mass conservation equation (continuity equation)

$$\frac{\partial \bar{u}_j}{\partial x_j} = 0$$

where

$\bar{u}_j$ – the averaged value of the fluid velocity in the projection on the j-th axis (j = 1, 2, 3);

IOP Conf. Series: Materials Science and Engineering **492** (2019) 012026 doi:10.1088/1757-899X/492/1/012026

The equation of momentum conservation (Reynolds averaging):

$$\rho\left[\frac{\partial U_i}{\partial t}+U_i\frac{\partial U_i}{\partial x_j}\right]=-\frac{\partial P}{\partial x_i}+\frac{\partial}{\partial x_i}\left[\tilde{\tau}_{ij}^{(v)}-\rho\langle u_i u_j\rangle\right],$$

where

U, P – averaged speed and pressure;

$\tilde{\tau}_{ij}^{(v)}=2\mu\tilde{S}_{ij}$ – viscous stress tensor for incompressible fluid;

$\tilde{S}_{ij}=\frac{1}{2}\left[\frac{\partial \tilde{u}_i}{\partial x_j}+\frac{\partial \tilde{u}_j}{\partial x_i}\right]$ – instant strain rate tensor;

$\rho\langle u_i u_j\rangle$ – Reynolds stresses.

The introduction of the Navier—Stokes equation, Reynolds-averaged, makes the system of equations non-closed, since additional unknown Reynolds stresses appear. A semiempirical k—ω SST model of turbulence has been used for solving the system in this task, which introduces the necessary additional equations: the transfer equations for the kinetic energy of turbulence and the relative dissipation rate of the energy:

$$\frac{\partial k}{\partial t}+U_j\frac{\partial k}{\partial x_j}=P_k-\beta k\omega+\frac{\partial}{\partial x_j}\left[(\upsilon+\sigma_k\upsilon_T)\cdot\frac{\partial k}{\partial x_j}\right]$$

$$\frac{\partial \omega}{\partial t}+U_j\frac{\partial \omega}{\partial x_j}=\alpha\cdot S^2-\beta\cdot\omega^2+\frac{\partial}{\partial x_j}\left[(\upsilon+\sigma_\omega\upsilon_T)\cdot\frac{\partial \omega}{\partial x_j}\right]+2\cdot(1-F_l)\cdot\sigma_{\omega 2}\cdot\frac{1}{\omega}\cdot\frac{\partial k}{\partial x_j}\cdot\frac{\partial \omega}{\partial x_j}$$

The flow part is divided into a set of finite cells, for each of which discrete analogues of continuous equations are composed. The combination of all discrete analogs has formed a closed system of algebraic equations.

To obtain a parametric model, as a first approximation the flow part calculated by the method of B.I. Nahodkin, was used. The modelling results are shown at figures 2, 3. Head is H = 46.2 m, hydraulic efficiency η = 39.6%.

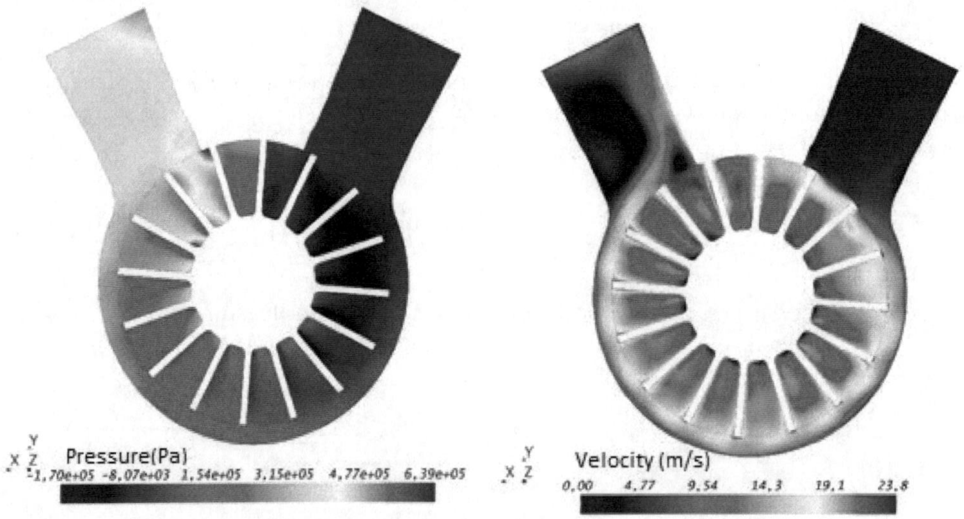

Figure 2. The fields of pressure and velocity distribution in the section, perpendicular to pump rotation axis.

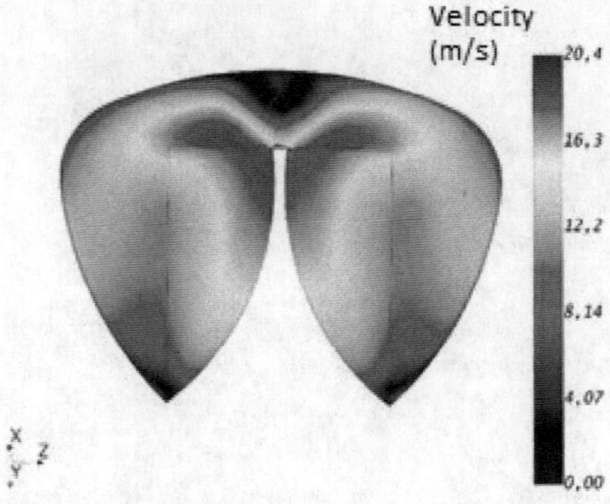

Figure 3. The field of velocity distribution in the between blades space.

After analyzing the fluid flow in the flow part of the regenerative stage, it was decided to consider the effect on the energy characteristics of the following parameters (figure 1, figure 4):
D – the outer diameter of the impeller;
b – vortex wheel width;
n – the number of vortex wheel blades (on the one side);
h – radial gap between the impeller and the casing;
Kf1 = A/B the coefficient of "roundness accuracy" of between blades space;
Kf2 = $L2/L1$ the symmetry coefficient of between blades space.

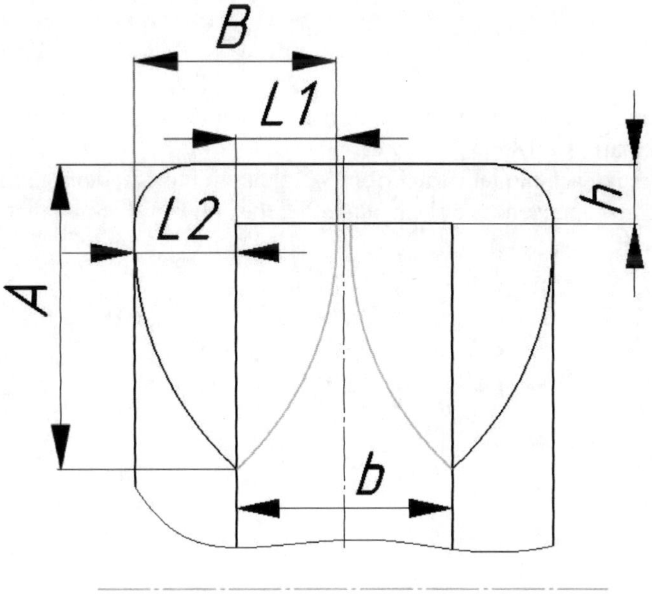

Figure 4. The parameters that have influences on the energetic characteristics of closed type regenerative pumps.

The parametrized model of the flow part of the closed type regenerative pump is shown at figure 5. It has opportunity to change the chosen parameters simultaneously or separately in a wide range of values', which makes it convenient for obtaining the optimal flow part by the LP-τ search method [15].

IOP Conf. Series: Materials Science and Engineering **492** (2019) 012026 doi:10.1088/1757-899X/492/1/012026

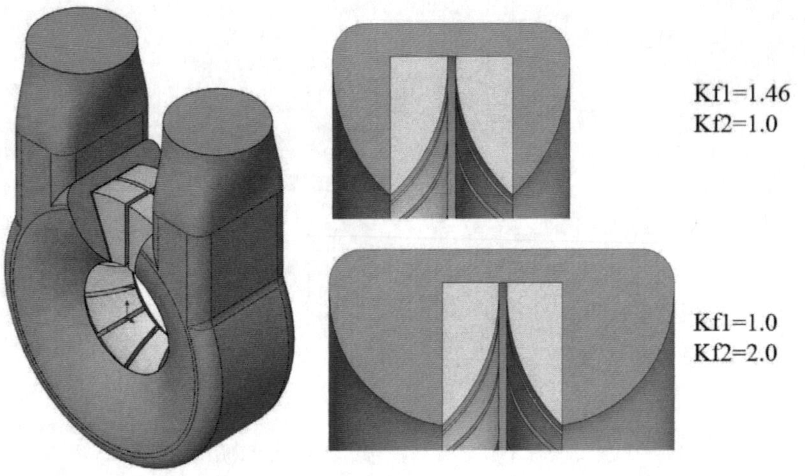

Figure 5. The parametrized model of the flow part of the closed type regenerative pump.

3. Results

In order to evaluate the effect of a single parameter on the energy characteristics of the regenerative pump with the other committed parameters, for each parameter a range of values was selected, which is divided into 15 equal intervals.

The range of parameter changing:

$D = 110\ldots130$ mm;

$b = 26\ldots40$ mm;

$n = 13\ldots24$;

$h = 5\ldots15$ mm;

Kfl = 0.8…1.2;

Kf2 = 1.0…2.0.

The boundary conditions were the total value of the outlet pressure and the fluid velocity amplitude at the outlet of the flow part [16–18].

The number of nodes in each model varied from 380 thousand to 420 thousand nodes.

Figures 6–11 show the dependences of head and efficiency on variable parameters.

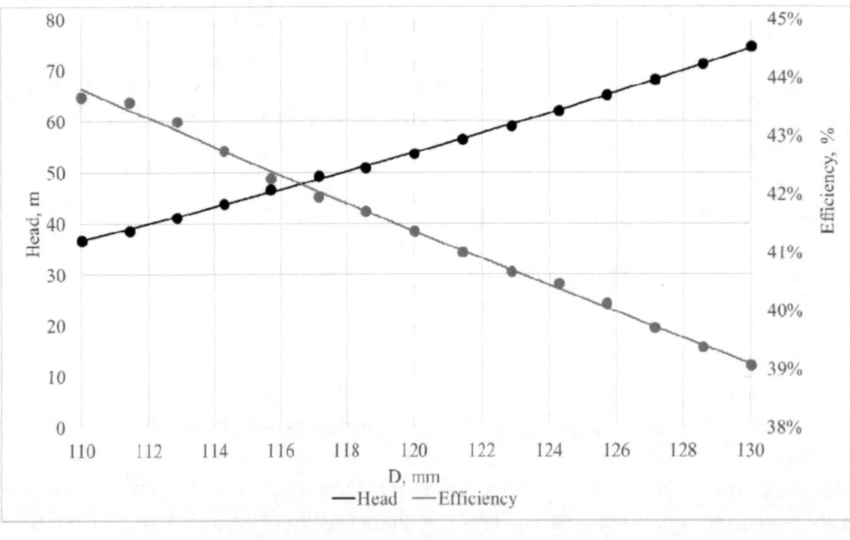

Figure 6. The dependencies of efficiency and pressure from *D*.

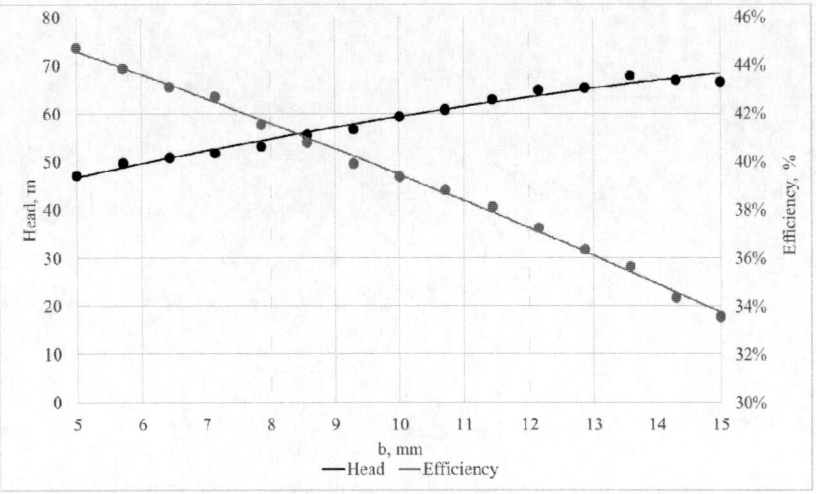

Figure 7. The dependencies of efficiency and pressure from *b*.

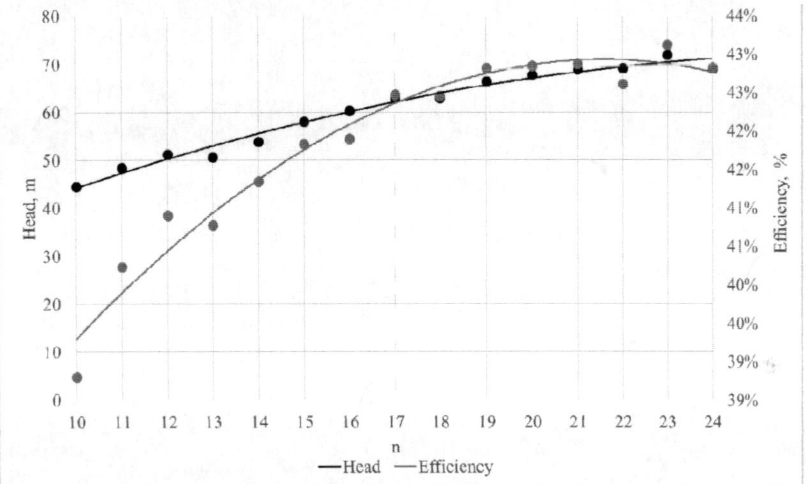

Figure 8. The dependencies of efficiency and pressure from *n*.

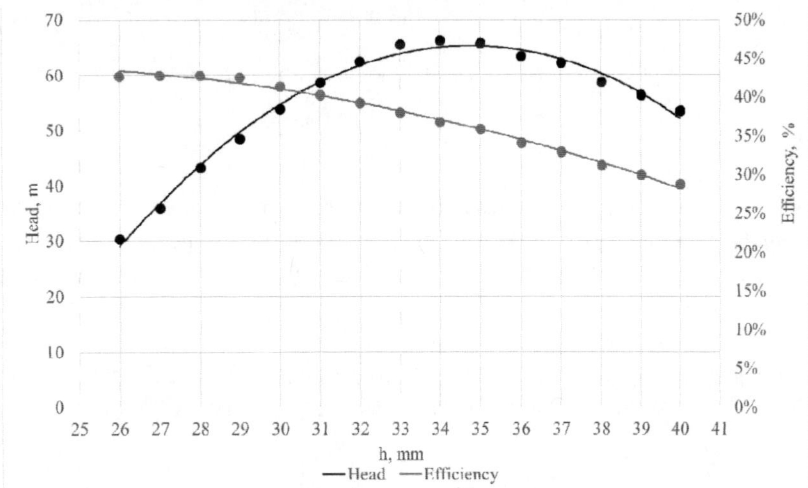

Figure 9. The dependencies of efficiency and pressure from *h*.

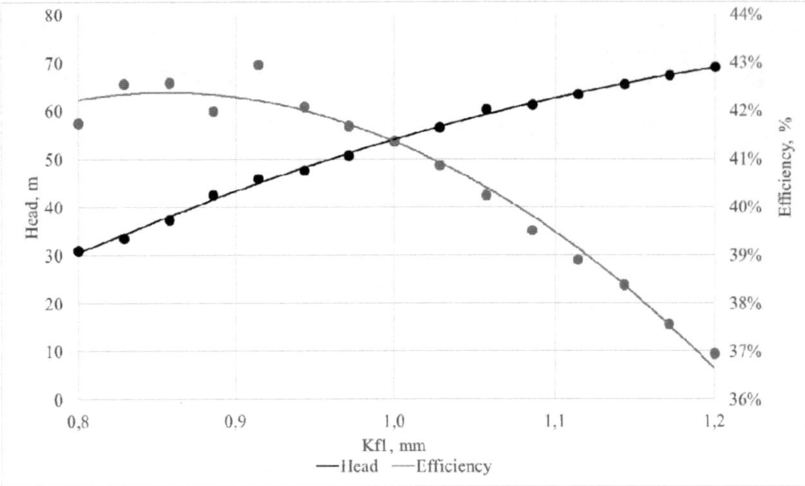

Figure 10. The dependencies of efficiency and pressure from Kf1.

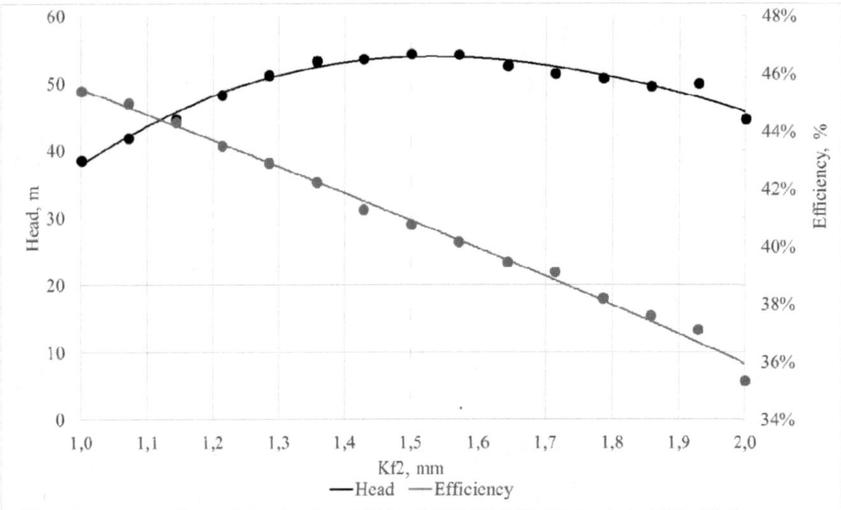

Figure 11. The dependencies of efficiency and pressure from Kf2.

On the basis of the obtained results, it was possible to narrow the range of parameters changes and carry out a comprehensive optimization of the flow section of a closed-type regenerative pump using computational fluid dynamics methods with the LP-τ sequence. The number of design models appeared to be 128 [19-20].

According to the optimization results, the best model has the following characteristics: head $H = 44$ m, hydraulic efficiency $\eta = 51\%$, which is 11.4% more than the original model has. The fields of pressure and velocity distribution in the flow part are presented in figures 12, 13.

IOP Conf. Series: Materials Science and Engineering **492** (2019) 012026 doi:10.1088/1757-899X/492/1/012026

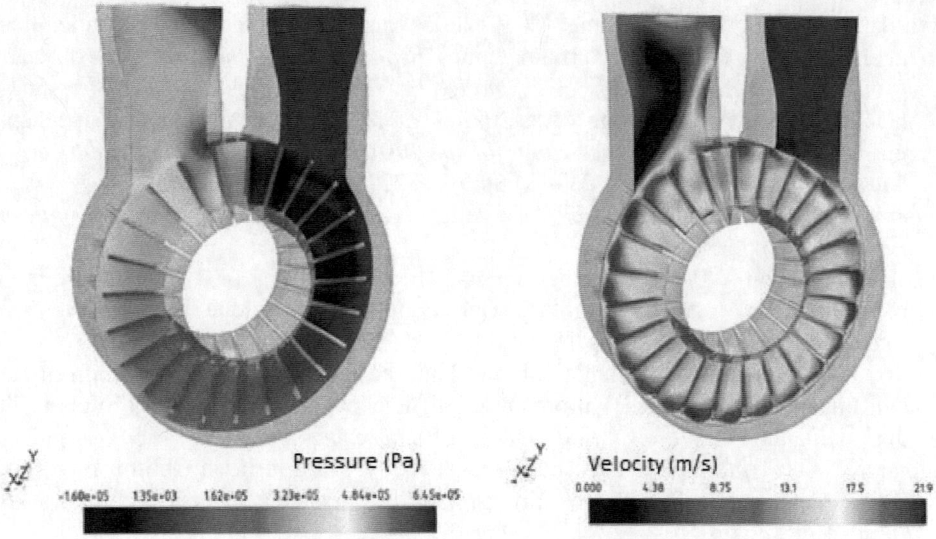

Figure 12. Fields of pressure and velocity distribution in the section perpendicular to the rotation axis of the pump.

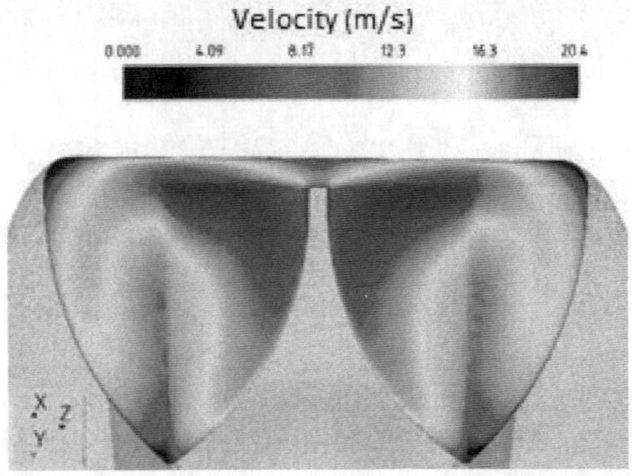

Figure 13. Velocity distribution field in between blades space.

4. Discussion

Based on the results of the work, the following conclusions may be drawn:

1. A parametrized model of the flow part of the closed type regenerative pump has been developed, which allows the design parameters changing in a wide range;

2. As it was shown by hydrodynamic modeling, the hydraulic efficiency decreases while parameters D, b, h, Kf1 are increasing. As the number of blades increases, hydraulic efficiency increases greatly. The highest value is obtained at n = 24. The coefficient of symmetry is advisable to choose from the range of Kf2 = 0.8 ... 1.

References

[1] Yoo I S, Park M R and Chung M K 2005 Improved momentum exchange theory for incompressible regenerative turbomachines *Proceedings of the Institution of Mechanical Engineers. Part A: Journal of Power and Energy*, **219** (7) Article ID A09704 pp 567–581
[2] Song J W, Engeda A and Chung M K 2003 A modified theory for the flow mechanism in a regenerative flow pump *Proceedings of the Institution of Mechanical Engineers, Part A: Journal of Power and Energy* **217** (3) pp 311–322
[3] Sixsmith H and Altmann H 1977 A regenerative compressor *Journal of Engineering for Industry* **99**(3) pp 637–647
[4] Quail F J, Scanlon T H and Baumgartner A 2012 Design study of a regenerative pump using one-dimensional and three-dimensional numerical techniques *European Journal of Mechanics* B/Fluids **31** (1) pp 181–7
[5] Guskov A M, Lomakin V O, Banin E P and Kuleshova M S 2017 Minimization of Hemolysis and Improvement of the Hydrodynamic Efficiency of a Circulatory Support Pump by Optimizing the Pump *Flowpath Biomedical Engineering***4** pp 229–233
[6] Lomakin V O, Chaburko P S and Kuleshova M S 2017 Multi-criteria Optimization of the Flow of a Centrifugal Pump on Energy and Vibroacoustic Characteristics *Procedia Engineering***176** pp 476–482
[7] Gouskov A M, Lomakin V O, Banin E P and Kuleshova M S 2016 Assessment of Hemolysis in a Ventricular Assist Axial Flow Blood Pump *Biomedical Engineering***4** pp 12–15
[8] Lomakin V O, Kuleshova M S and Bozh'eva S M 2016 Numerical Modeling of Liquid Flow in a Pump Station *Power Technology and Engineering***5** pp 324–327
[9] LomakinV O 2015 *Proceedings of 2015 International Conference on Fluid Power and Mechatronics*
[10] Lomakin V O, Kuleshova M S and Kraeva E A 2015 Fluid Flow in the Throttle Channel in the Presence of Cavitation *Procedia Engineering***106** pp 27–35
[11] Polyanin A D and Lychev S A 2016 Decomposition methods for coupled 3D equations of applied mathematics and continuum mechanics: Partial survey, classification, new results, and generalizations *Applied Mathematical Modelling***40** (4) pp 3298–3324
[12] Leontiev A I and Burtsev S A 2015 Device for separation of vortex gas-dynamic energy *Doklady Physics***60** (10) pp 476–478
[13] Grishin Y A and Bakulin V N 2015 Numerical Investigation of Flow in a Centrifugal Compressor *Journal of Engineering Physics and Thermophysics***88** (5) pp 1274–1279
[14] Kaplunov S M, Val'Es N G, Samolysov A V and Marchevskaya O A 2015 Determining the tube bundle streamlining critical parameters using the numerical experiment method *Thermal Engineering* (*English translation of Teploenergetika*) **62** (8), article A009 pp 593–597
[15] Polyanin A D and Lychev S A 2014 Various representations of the solutions of systems of equations of continuum mechanics *Doklady Physics***59** (3) pp 148–152
[16] Popov Ye P 1995 New method for composing the equations of a spline from specified points *Journal of Computer and Systems Sciences International***33** (4) pp 135–146
[17] Isaev S A, Kornev N V, Leontiev A I and Hassel E 2010 Influence of the Reynolds number and the spherical dimple depth on turbulent heat transfer and hydraulic loss in a narrow channel *International Journal of Heat and Mass Transfer***53** (1-3) pp 178–197
[18] Volkov V Yu, Belova O V, Krutikov A A and Skibin A P 2013 A coolant flow simulation in fast reactor wire-wrapped assembly *Thermal Engineering* (*English translation of Teploenergetika*) **60** (6) pp 429–433
[19] Kulik V V, Parkin A N and Navasardyan E S 2016 Numerical Modeling Procedure for Micromachined Cryogenic Cooler Elements Using ANSYS Fluent Software and Viscous Flow in a Small-Diameter Channel with Heat Transfer as an Example *Chemical and Petroleum Engineering***52** (7-8) pp 531–538

[20] Pelevin F V, Avraamov N I, Ir'yanov N Y, Orlin S A, Lozovetskii V V and Ponomarev A V 2018 Intensification of Heat Exchange in the Regenerative Cooling System of a Liquid-Propellant Rocket Engine *Journal of Engineering Physics and Thermophysics* **91** (3) pp 601–10

IOP Conf. Series: Materials Science and Engineering **492** (2019) 012027 doi:10.1088/1757-899X/492/1/012027

Determination of parameters of the water-air mixture generated by an ejection aeration system with a dispersing agent

E Antonova

Bauman Moscow State Technical University, 5 Second Baumanskaya Street, Moscow, 105005, Russian Federation

E-mail: e.s.antonova@bmstu.ru

Abstract. The ejection way of water aeration is considered in the work. The various devices used for dispersing the water-air mixture in order to obtain finely dispersed bubbles, have been analyzed, their advantages and disadvantages have been indicated. An original dispersing device, characterized by simplicity of implementation, versatility and a number of other advantages, which allows to obtain the required parameters at lower power consumption compared with analogues, has been proposed. It has been experimentally shown that the proposed dispersing device makes it possible to obtain fine bubbles with sizes less than 100 microns, and the obtained density distribution functions are polymodal. The dependences of bubble sizes and aeration intensity on the ratio of the jet diameter at the outlet of the ejector to the geometric parameters of the dispersant are obtained.

1. Introduction

There are many technological processes in which aeration systems are used, and for the most suitable behavior of such processes it is necessary to obtain and maintain certain hydrodynamic parameters. The aeration systems are used in waste-water treatment, while the effectiveness of treatment depends on the hydrodynamic regime and parameters such as the intensity of aeration and the size of the generated bubbles. Such indicators as the possibility of obtaining the required parameters, reliability, energy efficiency [1-4] should be taken into account, during the aeration system choosing.

Water aeration using ejector devices has become widespread due to the simplicity of their design, high reliability, and economical use. The ejector is a type of two-phase jet device. Aeration by ejectors is based on the use of the energy of the working fluid which moves linearly at a speed of about 15-17 m/s through a nozzle having a certain shape and size for obtaining a pressure fall creating an ejection of the gas or liquid phase [5].

Many technological processes, which use ejectors for water aerating, require a high intensity of aeration, which is achieved by the gas content increasing (the ratio of air flow to water flow) and the generation of fine bubbles of less than 500 microns. However, according to the results of experimental studies presented in [6, 7], it can be concluded that with an increase in gas content, an increase in the size of the bubbles occurs, which in some cases has negative consequences for the implemented processes. The average size of the bubbles generated by the ejectors of the simplest structures is

1000–3000 μm [8]. Massive bubbles with high ascending speeds are in close proximity to the ejector. In this regard, they are not uniformly distributed throughout the volume of the aerated liquid.

Both improved designs of the ejectors themselves and special dispersing devices are used [9-19] for improving the dispersion of the water-air mixture.

In works [10-14] the Venturi ejector is described. The dispersion of bubbles in it occurs due to the fact that a shock wave is created in the diverging part of the Venturi tube and the bubbles are broken up under the influence of pressure. The average size of the bubble is about 100-400 microns. In [10, 15], a microbubble generator with turbulence is presented, jet disintegration in which is caused by flow swirling and difference in flow velocity in different parts of the mixing chamber. The average size of the bubbles is 20–200 μm, however, fine bubbles with a size less than 100 μm can be collected only with low gas content. In the generator of microbubbles with a spherical body described in [10, 16], flow turbulization occurs, which makes it possible to obtain bubbles with sizes of 120–490 μm. The bubble generator with a porous tube allows to obtain bubbles with a size less than 500 microns due to the air suction through the porous tube. The average size of the bubbles is less than 500 microns. The disadvantage of this method is pores plugging [10, 17]. The use of a static mixer installed behind the ejector was proposed in [3]. The bubbles fragmentation to the required size (30-80 microns) occurs due to multiple reorganization of the velocity field and to changing the direction of current lines of the mixed components, thus a significant interface increase is achieved and dispersion occurs. However, the disadvantage of the method is the dependence of the dispersion quality on the water and air flow, an increase in hydraulic resistance. In papers [18, 19], it was proposed to use a disk mounted opposite the ejector outlet, while the bubbles dispersion occurs during the impact on its surface. The dispersion of bubbles in contact with a solid surface is the most technically simple solution that does not require additional energy consumption, but the presence of massive bubbles larger than 1 mm is the disadvantage.

Changing the design of the ejector and installing special dispersing devices allows to obtain the required size of the bubbles in some cases. However, a number of drawbacks should be noted: the complexity of some technologies, the insufficient degree of dispersion, and achievement of the required parameters only with certain system settings. Thus, the problem of the water-air mixture dispersing is relevant, but it requires further study of this process and the development of the most optimal solutions.

In order to ensure the required parameters using an ejection aeration system and eliminate the drawbacks of existing technologies, it is proposed to use the original dispersing device (figure 1) described in [20].

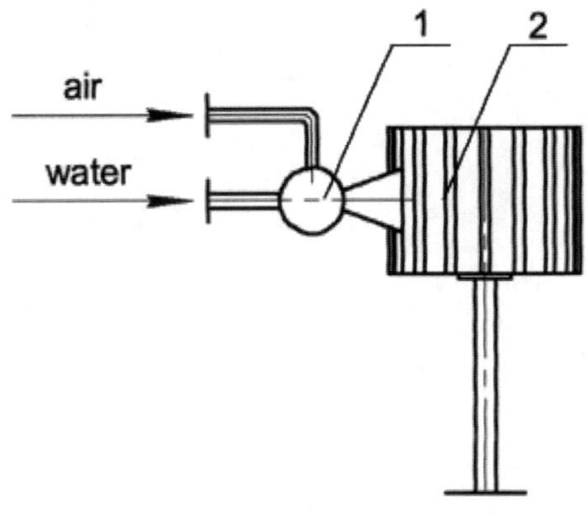

Figure 1. The dispersing device: 1 – ejector, 2 – disperser.

Bubbles are dispersed by contact of the water-air mixture jet, which leaves the ejector and passes through the guide element with the riffled disperser.

IOP Conf. Series: Materials Science and Engineering **492** (2019) 012027 doi:10.1088/1757-899X/492/1/012027

2. Materials and methods

To study the process of dispersion of bubbles, generated by an ejection aeration system with the original device for dispersing and determining the dispersion composition of the water-air mixture, a laboratory setup has been created, the scheme of which is shown at figure 2.

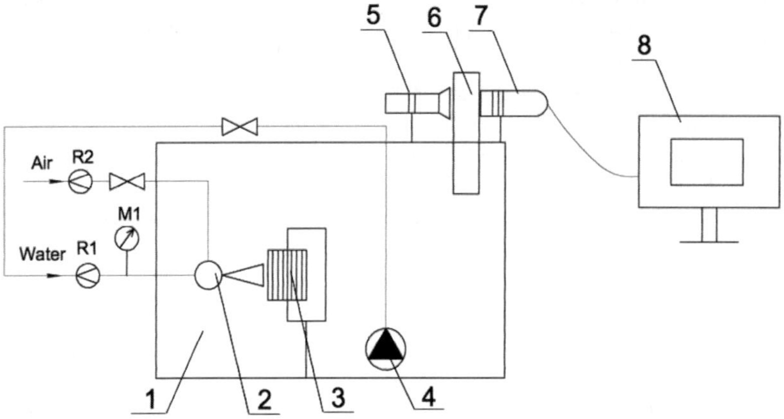

Figure 2. Laboratory setup scheme:1 – aeration chamber; 2 – ejector; 3 – disperser; 4 – pump, 5 –light source; 6 – tank; 7 – USB microscope; 8 – computer; R1 – rotameter for water flow; R2 –rotameter for air flow; M1 – manometer.

The laboratory setup consists of an aeration chamber 1, an ejector 2, a disperser 3, a drowned pump 4. To observe the bubbles and fix them on the aeration chamber 1, a small transparent tank 6 is mounted, which is partially immersed in the water. On one side of the tank, a USB microscope 7 connected to computer 8 is installed, on the opposite side opposite the microscope objective – the light source 5. Water consumption in all experiments was 360 l/h, air flow 100 l/h. The outlet jet velocity was 5.1 m/s. The diameter of the jet at the ejector outlet was 5 mm. The geometric parameters of the dispersers are presented in table 1.

Table 1. Geometric characteristics of dispersers

No	The dispersers diameterd_d(mm)	The dispersersheight h_d (mm)
1	10	20
2	15	20
3	20	20
4	30	20
5	40	10
6	40	20
7	40	40
8	75	20

When the laboratory setup was in operation, bubbles with a size of more than 500 μm floated directly above the dispersant, and fine bubbles, which are of the greatest interest, overspread through the aeration chamber. Bubbles exposure was carried out using a USB microscope in accordance with the method [21]. As a result of statistical image processing, bubble density distribution functions were obtained. The number of bubbles buoyed to the surface per unit of time was also determined. As

IOP Conf. Series: Materials Science and Engineering **492** (2019) 012027 doi:10.1088/1757-899X/492/1/012027

a result, dependences of the bubbles average size and the aeration intensity on the ratio of the disperser diameter to the jet diameter at the ejector outlet (d_d/d_j), and the height of the disperser to the jet diameter at the ejector outlet (h_d/d_j) were obtained.

Based on the obtained data, the intensity of aeration of the bubbles of each group was calculated by the formula (1):

$$q_i = f \cdot n_i \sum_{j=1}^{m} \frac{n_j \pi d_j}{6}, \quad (1)$$

where

f — the number of bubbles buoyed to the surface per second;

n_i - the fraction of bubbles of the i-th group;

n_j is the fraction of bubbles of each size;

d_j – the size of the bubbles of each group.

3.Results and discussion

As a result of statistical processing, it was found that the original density distribution functions are non-uniform (the coefficient of variation for all cases is more than 0.33). Next, homogeneous aggregates within the sample were selected, and two normal distributions were obtained. The coefficients of variation of the selected distributions were less than 0.33, therefore, the resulting samples are homogeneous. For the obtained distributions, $\chi^2<\chi^2_{кр}$ with a significance degree of 0.95, therefore, the difference from the normal distribution is random, which does not contradict the hypothesis about the normal data distribution. Thus, the distribution density function for each of the groups of bubbles will be (2):

$$f(d) = \frac{1}{\sigma_i \sqrt{2\pi}} \exp\left[\frac{-(d_i - d_{mi})^2}{2\sigma_i^2}\right], \quad (2)$$

where

d_i — the bubble size of each group, micron;

d_{mi} — average bubble size, μm; σ_i is the standard deviation of the bubble size of each group from the mean, μm.

The theoretical and experimental density distribution functions of the bubbles size for a disperser with a diameter of 40 mm (d_d/d_j = 8), height 40 mm (h_d/d_j = 8) are presented at figure 3.

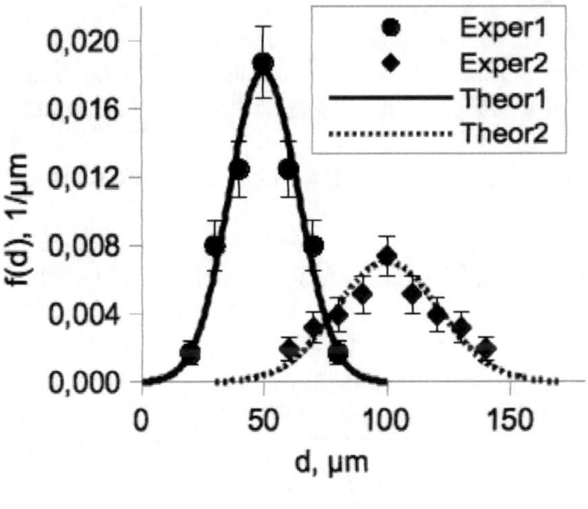

Figure 3. Theoretical and experimental functions of density distribution of bubble sizes for a dispersant with a diameter of 40 mm (d_d/d_j = 8), height 40 mm (h_d/d_j = 8).

The dispersed composition of bubbles with this disperser is characterized by two average sizes: d_1 = 50 μm, (σ = 14 μm), the proportion of bubbles 56%, d_2 = 100 μm (σ = 21 μm), the proportion of bubbles 33%. The proportion of bubbles, which have size of more than 145 microns, is about 15%. The appearance of such bubbles is random and does not have a significant impact on the process. Similarly, the distribution density functions for other dispersers have been obtained. Based on the obtained data, the dependences of the aeration intensity and the size of the bubbles of each group have been determined.

The results of determining the intensity of dispersed bubbles aeration, which are determined by the formula (1), are presented at figures 4, 5.

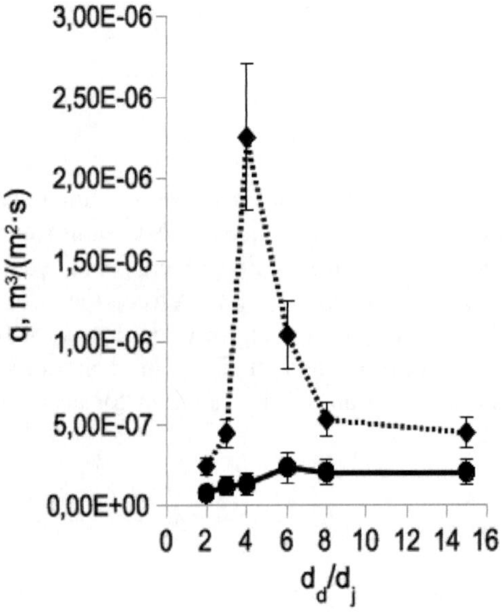

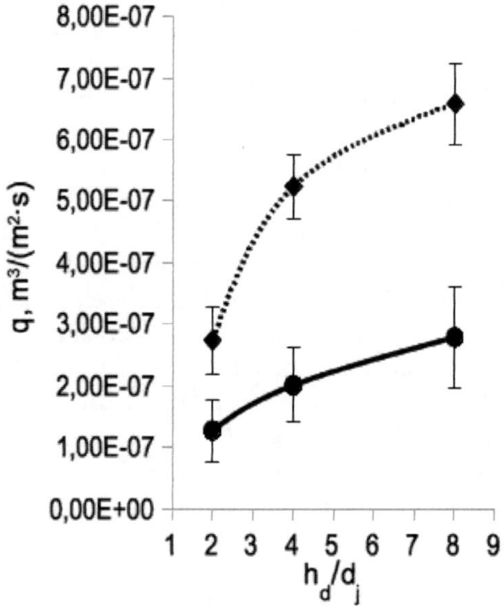

Figure 4. The dependence of the intensity of aeration on the ratio of the disperser diameter to the jet diameter: bubbles of the first group (solid line), bubbles of the second group (dashed line).

Figure 5. Dependence of the aeration intensity on the ratio of the disperser height to the jet diameter: bubbles of the first group (solid line), bubbles of the second group (dashed line).

With an increase in the $d_d/d_j$and h_d/d_j ratios, the number of dispersed bubbles increases and the intensity of aeration from them increases. The use of a disperser at $d_d/d_j \leq 3$ and $h_d/d_j \leq 2$ is impractical due to the small number of dispersed bubbles.

Thus, when designing a disperser, it is recommended to take its height $h_d \geq 4d_j$, and the diameter – $d_d \geq 6d_j$.

The dependence of the bubble size on the ratio of the disperser diameter to the jet diameter (d_d/d_j) is shown at figure 6. The dependence of the bubble size on the ratio of the disperser height to the jet diameter (h_d/d_j) is shown at figure 7.

After analyzing the obtained dependences, it can be concluded that as the ratio d_d/d_j increases to 8, the size of the bubbles decreases due to an increase in the area of contact of the jet with the disperser. The size of the bubbles of the second group varies in the range from 110 μm to 190 μm, and the size of the bubbles of the first group remains approximately constant about 55-65 μm. The size of the bubbles of the second group varies in the range from 90 μm to 125 μm, and the size of the bubbles of the first group is 50-65 μm with h_d/d_jincreasing.

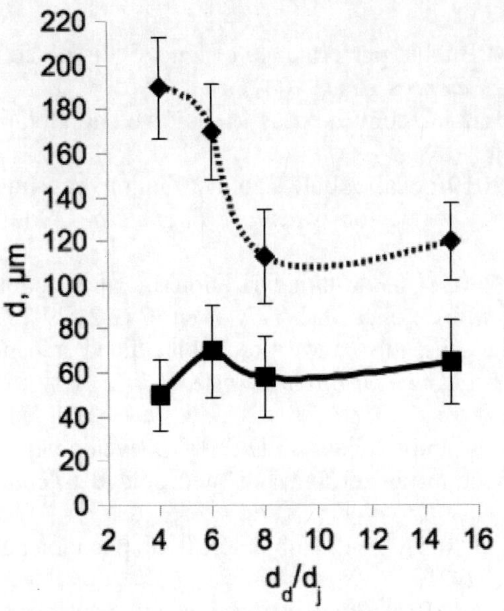

Figure 6. Dependence of the bubble size on the ratio of the disperser diameter to the jet diameter: bubbles of the first group (solid line), bubbles of the second group (dashed line).

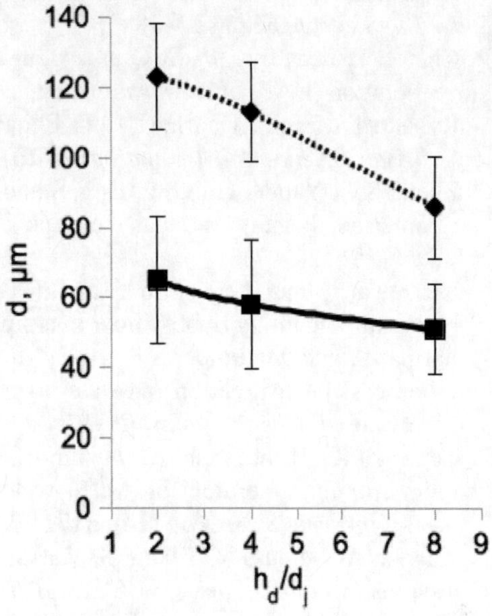

Figure 7. Dependence of bubble size on the ratio of the disperser height to the jet diameter: bubbles of the first group (solid line), bubbles of the second group (dashed line).

4. Conclusion

It was experimentally shown that the use of a disperser when using an ejection aeration system with a disperser allows to obtain fine bubbles with sizes less than 100 microns. The dispersed composition of bubbles is characterized by several aggregates with their mean sizes. The dependences of bubble sizes on the ratios of the geometric parameters of the disperser to the diameter of the ejector outlet jet allow to control the process of obtaining a water-air mixture of a given dispersion composition, taking into account the necessary distribution in the aeration chamber.

References

[1] Ksenofontov B S and Ivanov M V 2013 A novel multistage kinetic modeling of flotation for wastewater treatment *Water Science and Technology* **68 (4)** 807–812

[2] Ksenofontov B S and Ivanov M V 2014 Intensification of flotation treatment by exposure to vibration *Water Science and Technology* **69 (7)** 1434–39

[3] Ksenofontov B S and Ivanov M V 2014 Case study: Use of flotation for industrial stormwater treatment *Water Practice and Technology* **9 (3)** 392–397

[4] Ansari M, Bokhari H and Turney D E 2018 Energy efficiency and performance of bubble generating systems *Chemical Engineering and Processing: Process Intensification* **125** 44–55

[5] Tsegel'skii V G, Akimov M V, Safargaliev T D 2018 Experimental study of the effect the basic geometrical parameter and the active gas nozzle expansion ratio have on the performance

characteristics of supersonic gas ejectors fitted with a conical mixing chamber *Thermal Engineering* **65 (1)** 27–38
[6] Chen F T, Peng F X, WuX Q and Luan Z K 2004 Bubble performance of a novel dissolved air flotation (DAF) unit *Journal of Environmental Sciences* **16 (1)** 104–107
[7] Rykaart E M and J Haarhoff 1995 Behaviour of air injection nozzles in dissolved air flotation *Water Science and Technology* **31 (3)** 25–35
[8] Zheng S Q, Yao Y, Guo F F, Bi R S and Li J Y 2010 Local bubble size distribution, gas–liquid interfacial areas and gas holdups in an up-flow ejector *Chemical Engineering Science* **65 (18)** 5264–71
[9] Serizawa A, Inui T, Yahiro T and Kawara Z 2005 Pseudo-laminarization of micro-bubble containing milky bubbly flow in a pipe *Multiphase Science and Technology* **17 (1-2)** 79-101
[10] Parmar R and Majumder S K 2013 Microbubble generation and microbubble-aided transport process intensification—A state-of-the-art report *Chemical Engineering and Processing: Process Intensification* **64** 79–97
[11] Terasaka K, Hirabayashi A, Nishino T, Fujioka S and Kobayashi D 2011 Development of microbubble aerator for waste water treatment using aerobic activated sludge *Chemical Engineering Science* **66 (14)**, 3172–79
[12] Kaneko A, Nomura Y, Takagi S, Matsumoto Y and Abe Y 2012 Bubble break-up phenomena in a venturi tube. *Trans. JSME Ser. B*, **78 (786)** 207–217
[13] Gordiychuk A, Svanera M, Benini S and Poesio P. 2016. Size distribution and Sauter mean diameter of micro bubbles for a Venturi type bubble generator *Experimental Thermal and Fluid Science* **70** 51–60
[14] Yin J, Li J, Li H, Liu W and Wang D 2015 Experimental study on the bubble generation characteristics for anventuri type bubble generator. *International Journal of Heat and Mass Transfer* **91** 218–224
[15] Li P and Tsuge H 2006 Water treatment by induced air flotation using microbubbles *Journal of Chemical Engineering of Japan* **39 (8)** 896-903
[16] Sadatomi M, Kawahara A, Kano K, Ohtomo A 2005 Performance of a new micro-bubble generator with a spherical body in a flowing water tube *Experimental Thermal and Fluid Science* **29 (5)** 615–623
[17] Sadatomi M, Kawahara A, Matsuura H and Shikatani S 2012 Micro-bubble generation rate and bubble dissolution rate into water by a simple multi-fluid mixer with orifice and porous tube *Experimental Thermal and Fluid Science* **41** 23–30
[18] Kim H S, Lim J Y, Park S Y and Kim J H 2017 Effects on swirling chamber and breaker disk in pressurized-dissolution type micro-bubble generator. *KSCE Journal of Civil Engineering*, **21(4)**, 1102–06
[19] Kim H S, Lim J Y, Park S Y and Kim J H 2018 Effects of distance of breaker disk on performance of ejector type microbubble generator. *KSCE Journal of Civil Engineering*, **22(4)** 1096–1100
[20] Ksenofontov B S and Antonova E S 2016 Kinetic model for flotation treatment process using ejection system of aeration with disperser *Ecology and Industry of Russia* **20 (12)** 9–13
[21] Rodrigues R T and Rubio J 2003 New basis for measuring the size distribution of bubbles *Minerals Engineering* **16 (8)** 757–765

IOP Conf. Series: Materials Science and Engineering **492** (2019) 012028 doi:10.1088/1757-899X/492/1/012028

Ring-shaped fluorescence of mineral oil on the front edge of the choke coil

V Pilgunov[1, 2] and K Efremova[1,3]

[1] Bauman Moscow State Technical University, 5 Second Baumanskaya Street, Moscow, 105005, Russian Federation

[2] E-mail: vnp41@yandex,
[3] E-mail: efremova.k.d@gmail.com

Absctract: Hydraulic drive is a quite complicated technical system widely used in various industries: transport, agriculture, aviation, aerospace, robotics etc. During operation of the hydraulic actuators there are various modes and physical processes that occur in the working fluid to its devices. It is known that various processes like cavitation, degassing, erosion are observed in hydraulic drive devices, which are deeply studied and fully described in the literature. As follows from the analysis of the literature, the emergence of such a process as the light emission of a liquid in the form of flashes of light, the nature of which is ultrasound, is possible. The paper presents the material on the study of a rare and little-known phenomenon in the devices of hydraulic drives – the glow of the working fluid without ultrasound in it. To fill a certain gap in the study of the nature of the fluorescence of the working fluid, in particular, the fluorescence of mineral oil, the authors of this paper conducted an experimental study of cavitation processes in the transparent model of the hydraulic device – in the diaphragm choke coil.

1. Introduction

In the scientific and technical literature, two terms are widely used for fluid fluorescence, namely: "sonoluminescence" and "light emission". The first term is directly connected with the ultrasound as the reason that causes the fluorescence of the liquid; the second provides a broader range of causes for the fluorescence of the liquid.

The first report on sonoluminescence in the fluid flow was published in 1933 by Marinesco and Trillat [1], and a little later (1934) Frenzel and Schultes found a weak fluorescence in the fluid that arose under the action of acoustic oscillations [2]. In 1964, Jarman and Taylor [3, 4] registered a weak luminescence of water in the zone of closure of cavitation cavities in the Venturi pipe. In his dissertation Peterson [5] in 1966 also noted light emission in the field of bubble closure in hydrodynamic cavitation.

At the moment, there is no single explanation of the phenomenon of light emission in science. Currently popular methods for solving hydrodynamic problems [6–10] imply the numerical solution of the Navier—Stokes equations (so-called CFD-methods). However, these methods have some significant drawbacks: high costs of machine and human time, relatively low accuracy, since the results obtained by such methods are fundamentally discrete and do not provide the possibility to continuously study the physical quantities of interest to us. All this leads to the search for methods alternative to the above.

The authors of this paper have conducted a study of cavitation processes in a transparent model of the diaphragm choke coil. During the study of cavitation processes in the choke coil, a localized stable ring-shaped light emission in the blue part of the color spectrum was found at its entrance edge. The brightness of the light emission increased as the flow rate increased. Since at the time of detection of light emission in the literature there was no information about fluid fluorescence in the low pressure region, the task of conducting additional studies to determine the causes and nature of light emission in hydrodynamic cavitating flows of mineral oil in narrow slits was set.

2. Experimental study of the fluorescence process of mineral oil on the sharp inlet edge of choke coil

In hydraulic automation devices, various types of choke coil are widely used as shut–off and control elements: in diaphragm and cylindrical choke coils, the flow of the working fluid has a pronounced turbulent character, and in the compressed section, a high speed causes cavitation and the associated active release of undissolved air and steam bubbles. In the zone of high pressure bubbles are instantly closed, which causes erosion destruction of the material of the channel walls and active acoustic processes. Rapid closure of vapor-gas cavities, in accordance with the laws of thermodynamics, can cause a local temperature rise, as well as, under certain conditions, fluid fluorescence.

Research of the literature in the field of the theory of sonoluminescence [11] showed that it is possible to suppose the electrical and electrochemical nature of its appearance and existence, so the identification of the causes of light emission at the entrance edge of choke coil was the task of the study, the results of which are presented in this article.

In the 70-ies of the twentieth century A. I. Koldomasov discovered the fluorescence of distilled water in a narrow channel [12]. The nature of the fluorescence was explained by plasma discharges during water cavitation, but the cause of discharges and the source of plasma formation were unclear. Later (2007–2010) light emission was found in the study of cavitation in organic liquids flowing at high speed in narrow dielectric cracks [13-15]. Since 2004, studies of cavitation regimes of high-speed flow of highly purified deionized water in narrow profiled dielectric channels have been carried out in the Keldysh Center [16]. The detected bright fluorescence was localized in the region of the largest narrowing of the channel with a diameter of 1-2 mm at a distance of 2-3 mm from the entrance to the capillary of the Venturi pipe at a flow rate exceeding 40 MS-1. Fluorescence manifested itself in the form of sparks, galloping along the flow in the Central part of the channel.

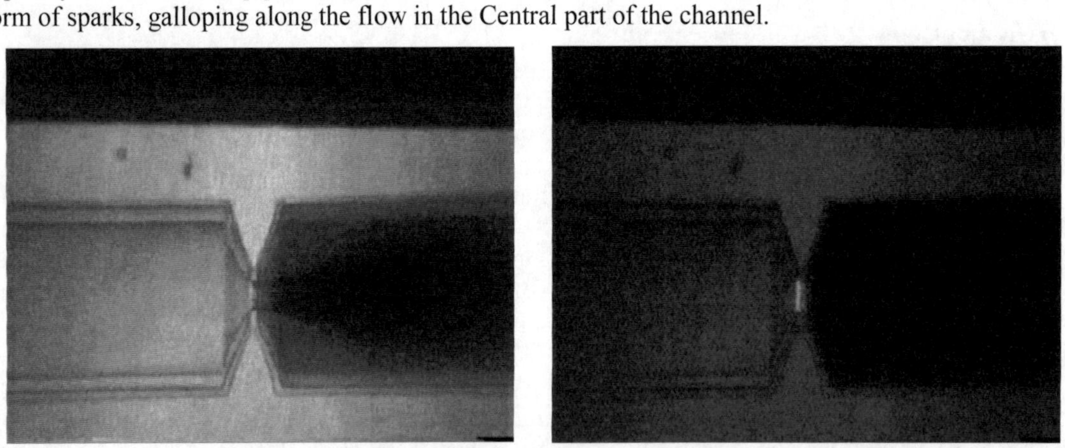

(a) (b)

Figure 1. Light emission in the cavitating flow of mineral oil in the inlet section of choke coil: a) light emission in the choke coil inlet section; b) stable light emission in the input section of choke coil.

IOP Conf. Series: Materials Science and Engineering **492** (2019) 012028 doi:10.1088/1757-899X/492/1/012028

It follows from the above that the causes of light emission in liquids have not yet been identified in many cases. In this regard, this experimental study was aimed to study the process of light emission in the cavitating flow of mineral oil at the inlet edge of the diaphragm choke coil.

During the experimental study, in the flow of mineral oil at the inlet edge of the diaphragm choke coil, made of transparent organic glass, the authors found a bright light emission in the blue part of the spectrum (figure 1).

Fragments of video filming of the stages of development of the cavitation process in the diaphragm choke coil, recorded by a digital camera with shooting speeds of 50 and 2000 FPS, are shown in the photo (figure 2):

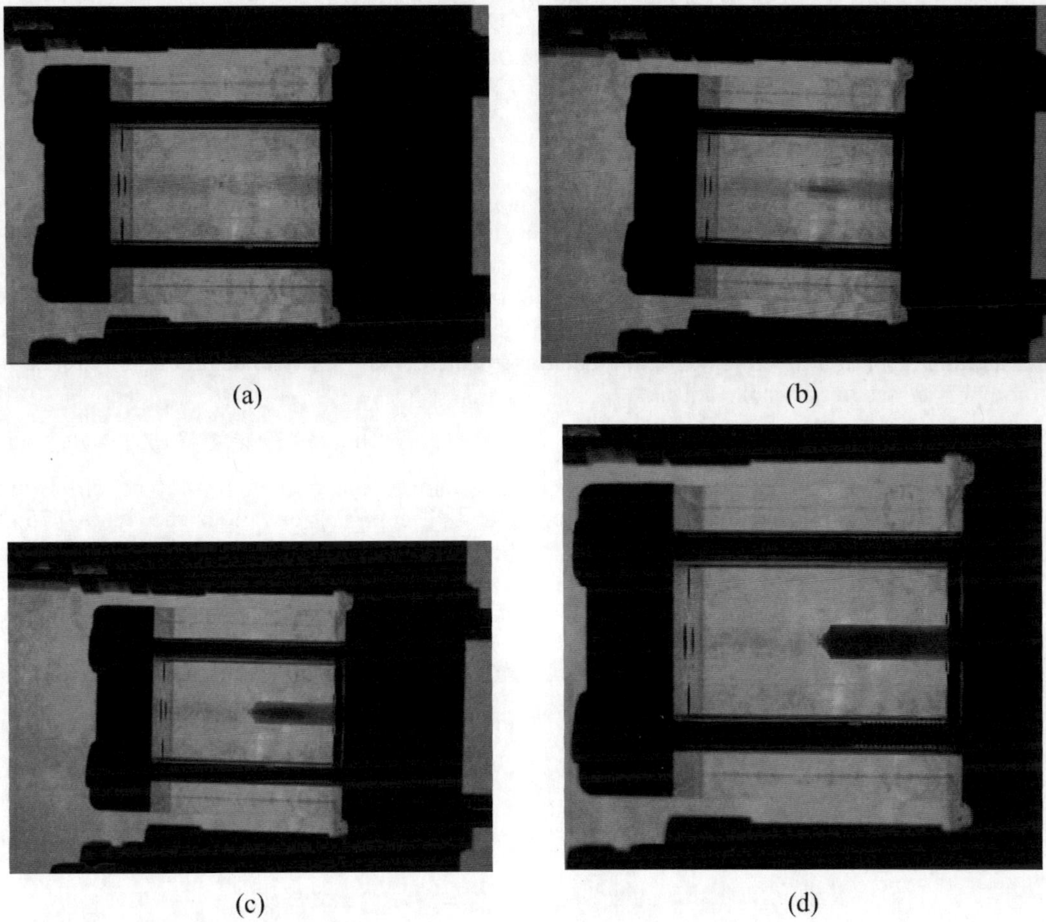

(a) (b) (c) (d)

Figure 2. The development of the process of hydrodynamic cavitation: a) origin of the cavitation ring on the inlet edge of the choke coil; b) the appearance of a cavitation torch; c) the spread of the plume in the area of the allotment; d) supercavitation mode.

In the photo (figure 2) the whole process of hydrodynamic cavitation is observed, which begins with the origin of the ring at the inlet edge of the choke coil (figure 2a), then, as the velocity of the liquid increases, the ring turns into a cavitation torch (figure 2b), expanding (figure 2c), closes with the inner surface of the tap, which results in supercavitation mode (figure 2d). It should be noted that light emission occurs at the input sharp edge of the choke coil at the time of the cavitation ring, continues until the super-cavitation mode starts and stops at the developed super-cavitation (figure 2d). At the same time, the continuity of the fluid flow is maintained on the channel axis in the high pressure zone.

IOP Conf. Series: Materials Science and Engineering **492** (2019) 012028 doi:10.1088/1757-899X/492/1/012028

The beginning and end of the light emission process are determined by the pressure levels p_1 in the inlet and p_3 in the outlet of the choke coil, determining the pressure difference of the choke coil $p_{cc} = p_1 - p_3$ and the speed in the compressed section (figure 3).

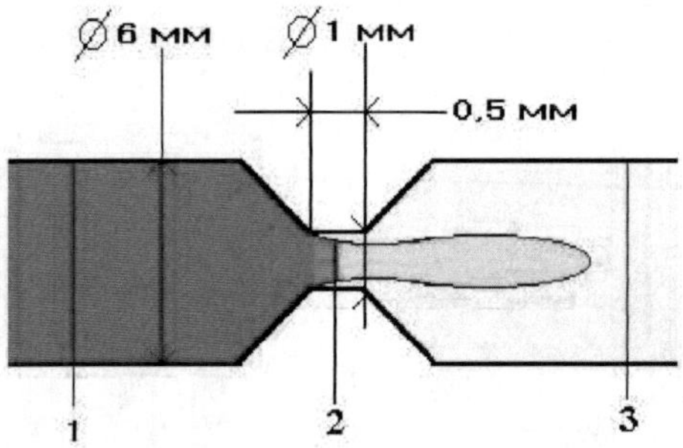

Figure 3. Flow structure and geometry of the diaphragm choke coil.

The solution of equations of Bernoulli for sections 1-2 and 1-3 in terms of continuity of fluid flow in the mode of origin of the cavitation ring and when the condition $p_{2abc} = p_{ov}$, where p_{ov} is the pressure of saturated oil vapor at a given temperature, determined the condition of the beginning of cavitation in the compressed section of choke coil:

$$p_1 - 5.43p_3 = 4.43\,(p_{at} - p_{ov}). \quad (1)$$

The coefficients in the supply of equation (1) determining the cavitation start condition are obtained for the choke coil geometry shown in figure 3, and the compression ratio of the jet ε = 0.75. It should be noted that in the mineral oil heated to 50 °C light emission was not observed.

The graph (figure 4) presents the diagram of light emission (SI) origin and termination at the change of the pressure of the backwater p_3 in the choke coil outlet and the pressure drop p_{cc} on the choke coil.

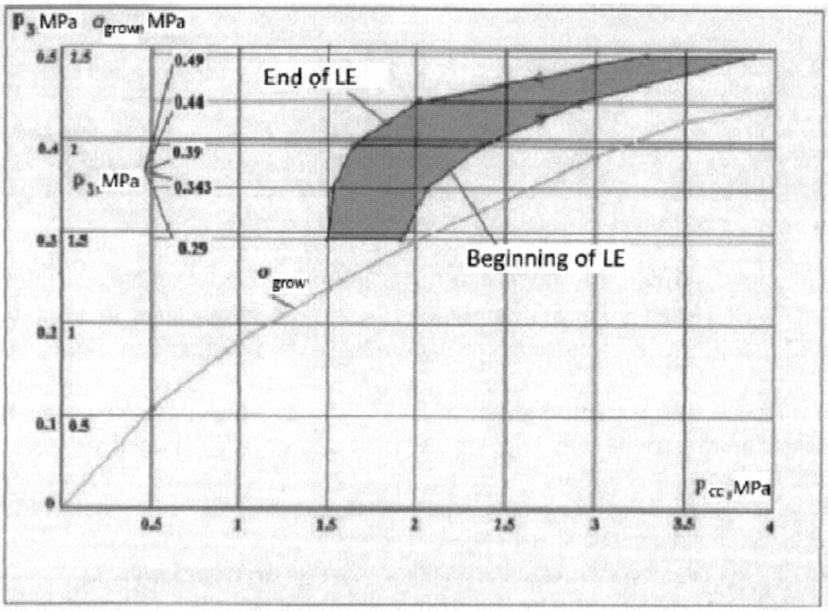

Figure 4. Influence of the backwater flow parameters on the light emission.

IOP Conf. Series: Materials Science and Engineering **492** (2019) 012028 doi:10.1088/1757-899X/492/1/012028

Horizontal sections of the diagram corresponding to the specified back pressure p_3 in the outlet of the choke coil allow to determine pressure levels p_1 in the inlet, at which the light emission begins and ends: for example, at the back pressure in the tap p_3 = 0.44 MPa, light emission occurs when the pressure difference on the choke coil p_{cc} = 2.80 MPa, that corresponds to the pressure in the inlet p_1 = 2.36 MPa. Light emission stops when the pressure difference is equal to p_{cc} = 2.00 MPa. It follows from the diagram that in the process of light emission there is a hysteresis. The digital increase in the photo of the glowing ring to the pixel level showed that in its conditional center of symmetry the color of light emission is pure white, remaining brightly blue on the periphery.

In this paper, the cavitation process in the flow of fresh mineral oil of the firm *SHELL* which has the viscosity ν_{40} = 20 sSt was initially investigated.

Cavitation in hydrodynamic flow of MGE10 mineral oil with viscosity ν_{40} = 10 sSt it was more intense, and light emission was not observed. Thus, we can assume the existence of dependence of light emission from the following factors: the properties of a basis of mineral oil, the pressure of the saturated vapor and its package structure of corrective additives.

In the course of long-term studies of the cavitating flow of mineral oil in the transparent model of diaphragm choke coil, the intensity of light emission decreased and after a certain time the light emission completely stopped. Replacement of spent mineral oil with fresh did not lead to the resumption of light emission.

In a new instance of the choke coil model of the previous geometry, light emission resumed in the same quality and at the same flow modes. This allows to make a conclusion about dependence of intensity of light emission from the sharpness of the leading edge of choke coil, is susceptible to dulling with abrasive wear of the material of the walls. A fragment of the experimental stand for the study of hydrodynamic processes in choke coil is shown in figure 5.

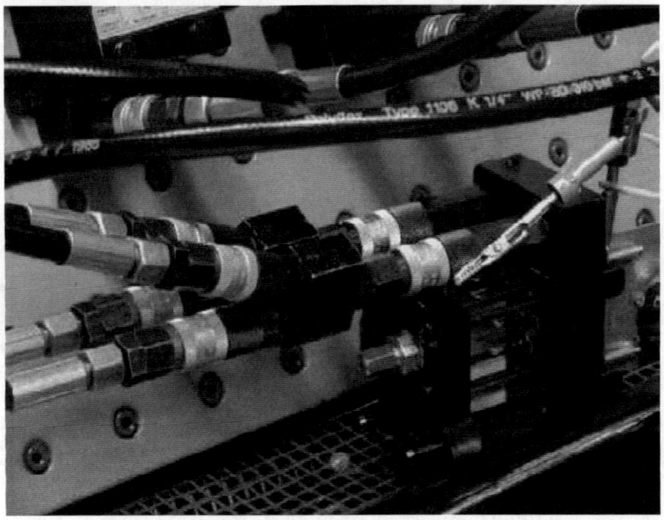

Figure 5. Experimental stand for the research of the characteristics of the choke coil.

Pressure drop in the choke coil $\Delta p = p_1 - p_2$ was estimated taking into account hydraulic losses in high-speed flexible hose connectors and in the inlet and outlet channels of choke coil. Flow parameters in diaphragm choke coil are shown in figure 6. At the highest value of the pressure drop on the choke coil Δp = 4 MPa the Reynolds number in the choke coil input section has reached the value Re = 3.5×10^3; this corresponded to the beginning of the turbulent regime of fluid motion, while the flow regime in the inlet and outlet corresponded to the laminar flow at Reynolds number Re = 500. The speed of movement of the liquid in the inlet, in the cross-section of choke coil and the compressed cross section of the flow received, respectively, the values V_1 = 1.9 m/s; V_2 = 65 m/s; V_{comp} = 100 m/s.

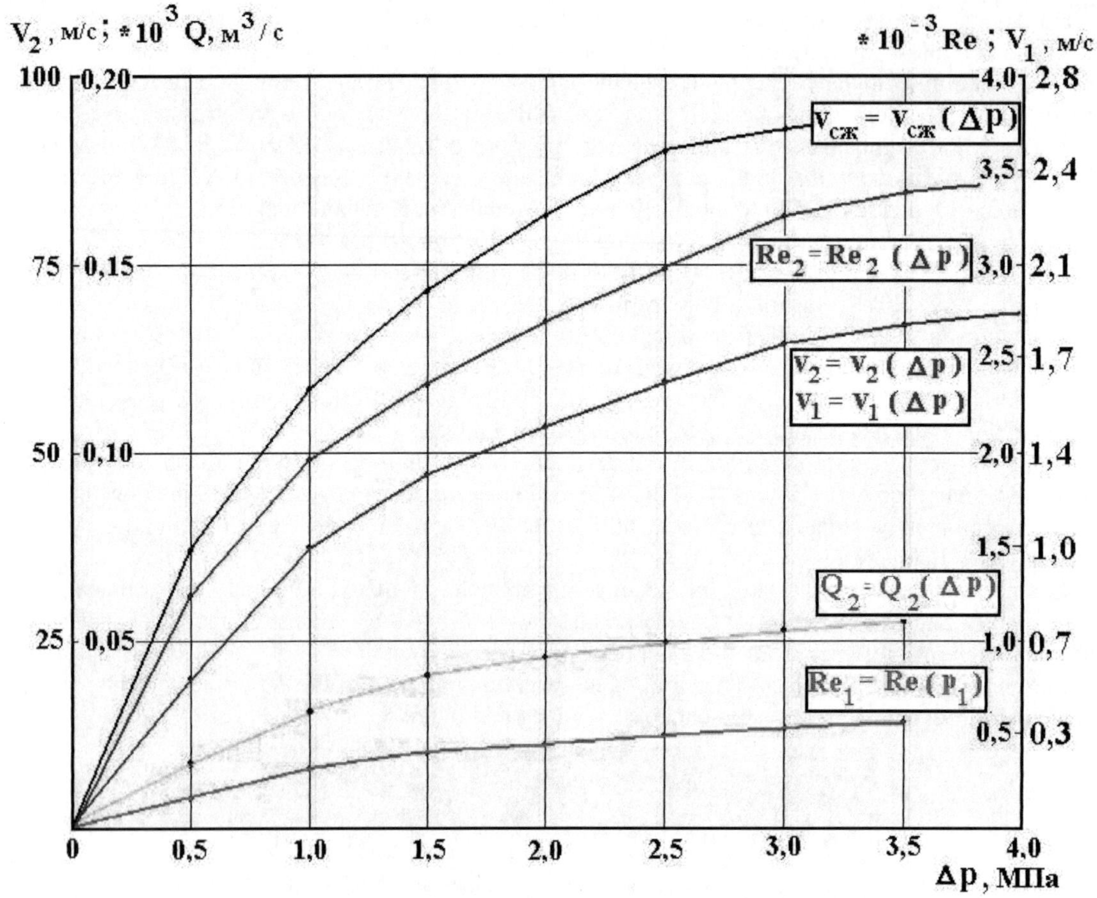

Figure 6. The parameters of the flow of mineral oil in a diaphragm choke coil.

Comparison of graphs in figure 4 and figure 6 shows that light emission occurs at fluid flow rates in a compressed section $V_{comp} \geq 75$ m/s.

3. Physical and mechanical justification of the results of the experimental study

The results of the study of the hydrodynamic flow of mineral oil (a medium with dielectric properties) in narrow channels showed that electric charges and associated complex processes may form inside the flow, causing, in the presence of additional conditions, light emission in the compressed section of the flow, when it passes through the sharp entrance edge of the diaphragm choke coil [17]-[20].

Under certain conditions, light emission may occur in the flow, the nature of which is difficult to determine, but it is most likely that it is determined by electrical rather than thermal processes. The absence of melting of the sharp entrance edge in the model of diaphragm choke coil made of organic glass is not consistent with the stated versions of “plasma discharges” [12] and “spark discharges” and “plasma breakdown” [16]. Localization of light emission at the entrance sharp edge of the diaphragm choke coil in the compressed flow section, its absence in other sections of the turbulent flow, as well as the dependence of the light emission on the degree of sharpness of the entrance edge, necessitated the analysis of the dynamics of the process of bending peripheral streams flow conditionally sharp entrance edge diaphragm choke coil (figure 7).

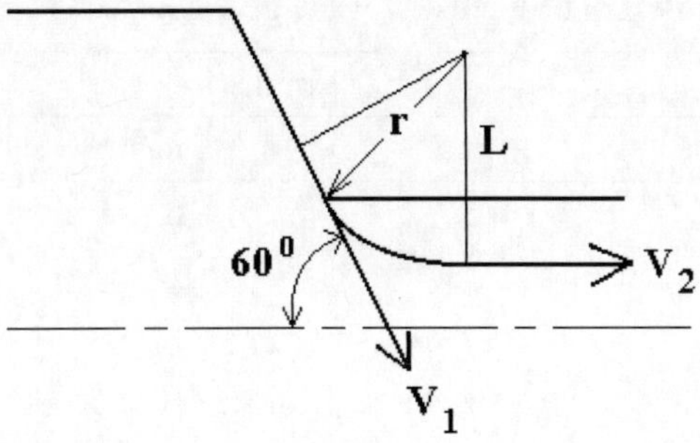

Figure 7. The scheme of diffraction of the peripheral stream sharp edge of the input choke coil.

If we assume the uniformity of motion of the liquid particle in the area of the circumference of the peripheral jet of the sharp entrance edge of the choke coil along a curvilinear trajectory with a constant radius r, such movement can be considered as uniformly accelerated, with zero local acceleration $\partial V/\partial t = 0$ and final values of convective accelerations $\partial V/\partial S \neq 0$, caused by a change in the direction of the velocity vector. In this case, the acceleration is directed to the center of the circle perpendicular to the velocity vector and is considered as a normal centripetal acceleration of the portable motion $a_{cp}= V^2/r$.

On a particle of liquid mass Δm, located on the peripheral stream and the envelope of the sharp edge, in the center of its mass acts the centrifugal force $F_{cf,}$ determined by the pressure forces on the side of adjacent particles on the outside of the trajectory (supporting force), and the tensile forces on the part of the particles on the inside of the trajectory adhering to the solid wall (attracting force). It can be assumed that in the case of non-wettability of the wall with a liquid, the attracting force will be absent.

The volume of a droplet liquid particle of spherical shape with radius R is related to its mass and density $\Delta W = \Delta m/\rho = 4\pi R^3/3$. In this case, the radius R is defined by the expression

$$R = (3\Delta W/4\pi)^{1/3} = (3\Delta m/4\pi\rho)^{1/3}. \tag{2}$$

If we assume the modulo equality of the supporting and attracting forces, the resulting surface force acting on the plane diametrical cross section of the sphere with the area of

$$\Delta S = \pi R^2 = \pi(3\Delta m/4\pi\rho)^{2/3}, \tag{3}$$

will be determined by the equation

$$\Delta m V^2/r = 2\pi p(3\Delta m/4\pi\rho)^{2/3}. \tag{4}$$

Expressing from equation (4) the value of p, which is considered to be the intensity of the surface force in "Fluid Mechanics", we obtain

$$p = \Delta m V^2/2\pi r(3\Delta m/4\pi\rho)^{2/3}. \tag{5}$$

The substitution of the average density of mineral oil ρ = 860 kg/m^3, defines an expression for the numerical evaluation of the intensity of the surface force

$$p = 37\Delta m^{0,33} V^2/r. \tag{6}$$

In the expression (6) the mass Δm is not determined, so it is advisable to estimate the intensity of the surface force reduced to the mass $\Delta m^{0,33}$:

$$p/\ \Delta m^{0,33} = 37\ V^2/r. \tag{7}$$

Table 1 presents the intensity of the reduced surface force, calculated in accordance with the equality of (7) for specific values of r and V.

IOP Conf. Series: Materials Science and Engineering **492** (2019) 012028 doi:10.1088/1757-899X/492/1/012028

Table 1. The intensity of the given surface forces

r = 0.5 mm		r = 0.1 mm		r = 0.1 mm	
V, m/s	$\Delta m^{0,33}$, m/s^2	V, m/s	$\Delta m^{0,33}$, m/s^2	V, m/s	$\Delta m^{0,33}$, m/s^2
0.5	9×10^3	0.5	18×10^3	0.5	90×10^3
1	3.7×10^4	1	7.4×10^3	1	37×10^3
2	1.48×10^5	2	2.96×10^5	2	14.8×10^5
5	9.25×10^5	5	18.5×10^5	5	92.5×10^5
10	3.7×10^6	10	7.40×10^6	10	37×10^6
50	9.25×10^7	50	18.5×10^7	50	92.5×10^7

The data in table 1 show that a mineral oil particle located on a peripheral jet, enveloping the sharp entrance edge of the choke coil, can be affected by a sufficiently large tensile force, which can cause light emission. Due to the speed of the processes, the liquid particle does not have time to turn into steam, which explains the localization of light emission in the pre-cavitation zone. According to equality (7) and table 1, the magnitude of the tensile force depends significantly on the sharpness of the inlet edge, which determines the compression ratio of the jet and the radius of curvature of the trajectory. For example, reducing the fillet radius of an edge from 1 mm to 0.05 mm increases the intensity of the reduced surface force by 20 times. This, to some extent, reveals the reason for the lack of light emission in the diaphragm choke coil, which in the process of long-term operation occurred dulling its sharp entrance edge due to abrasive wear. Quantitative assessment of the intensity of the tensile force gave the following results:

1) $R = 0.1$ mm; $\Delta W = 4.2\times10^{-12}$ m^3; $\rho = 860$ kg/m^3; $\nabla m = 0.36\times10^{-8}$ kg; $\nabla m^{0,33} = 0.0016$.
For $r = 1$ mm and $V = 50$ m/s: $\boldsymbol{p}$ **= 152 kPa**; for r = 0.5 mm and $V = 50$ m/s: $\boldsymbol{p}$ **= 304 kPa**;
2) $R = 0.05$ mm; $\Delta W = 0.52\times10^{-12}$ m^3; $\rho = 860$ kg/m^3; $\nabla m = 0.045\times10^{-8}$ kg; $\nabla m^{033} = 0.00079$.
For $r = 1$ mm and $V = 50$ m/s: $\boldsymbol{p}$ **= 75 kPa**; for $r = 0.5$ mm and $V = 50$ m/s: $\boldsymbol{p}$ **= 150 kPa.**

The above flow characteristics in the diaphragm choke coil allow us to estimate the averaged velocity of the fluid in the compressed section and the intensity of the tensile force.

D. Bernoulli equation written for the input section of choke coil channel and the compressed section of the flow (its pre-cavitation zone) in the conditions of continuum preservation $V_i \times A_i$ = inv, (where V_i is the speed averaged for the A_i flow cross section), determines the average value of the pressure in the compressed section of flow

$$p_2 = p_1 + \rho V_1^{\,2} / 2 - (1 + \zeta_{in}) \times \rho V_2^{\,2} / 2, \tag{8}$$

where
$\varsigma_{in} = 0.05$ is the coefficient of hydraulic resistance of the input sharp edge choke coil;
$\rho = 860$ kg/m^3 is the density of mineral oil.

After substitution of the value of the imperfect compression ratio of the jet $\varepsilon = 0.65\div0.70$, the equation (8) will take the form

$$p_2 = p_1 - 451\ V_2^2. \tag{9}$$

The algebraic values of pressure p_2 averaged over the flow cross-section obtained by the results of the experiment had a negative value, which indicates the presence of tensile stresses in the cross-section $p_{grow} = -p_2$.

The values of the tensile stresses averaged over the flow section calculated from the test results σgrow = σgrow (Δp), where $\Delta p = p_1 - p_3 = p_{cc}$ is the differential pressure on diaphragm choke coil; p_3 is the pressure in the outlet of the choke coil, are shown in the graphs figure 4.

From the graph figure 6 it follows that

– the average velocity of the liquid in the compressed section reached the values V_{2max} = 100 m/s,

– the average value of the tensile force intensity (tensile stress) $\sigma_{grow\ max} = 2.2$ MPa.

– tensile stress corresponding to the beginning of light emission in the input section of the choke coil lies within $\sigma_{grow} = 1.4 \ldots 2.2$ MPa and depends on the amount of pressure in the backwater choke coil.

– the light emission is terminated at lower values of tensile stress $\sigma_{grow} = 1.1 \ldots 2.0$ MPa, which indicates the presence of hysteresis effect in the process.

Taking into account that the input flow section is taken on the left edge of the cavitation ring, the pressure field along the flow section is inhomogeneous, and the tensile stress calculated from the Bernoulli equation σ_{grow} should be considered as averaging the pressure plot. The same factor can be explained by the influence of the back pressure p_3 on the intensity and stability of light emission: the termination of light emission in the regime of intense supercavitation is due to a decrease in σ_{grow} in conditions of closure of the attached and moving cavitation areas.

The time required for the origin and closure of the moving cavity in the case where the main role is played by the inertia forces is a few milliseconds, In addition, the estimates of the tensile force intensity performed on the basis of the solution of the Bernoulli equation and the analysis of the dynamics of the process of envelope of the liquid particle of the sharp entrance edge of choke coil are quite correlated.

4. Conclusions

The results of the analysis of processes in the flow of dielectric fluid and the results of the experiment on the study of the hydrodynamic flow of mineral oil in narrow channels allow us to reach the followingmain conclusions.

1. The presence of high tensile stresses in the liquid can cause light emission of sufficient intensity in the blue part of the color spectrum.

2. Light emission origins from and is localized in the zone of origin of cavitation caverns at the entrance edge of the diaphragm choke coil, where the absolute pressure is reduced, and not in the zone of closure of cavitation caverns, where the absolute pressure is much greater.

3. The intensity of light emission depends significantly on the absolute pressure in the zone of formation of vapor-gas cavities from cavitation nuclei, the flow rate determining the value of centripetal forces acting on the liquid particles located on the peripheral flow jets enveloping the sharp edge, as well as on the basis of mineral oil and the presence in it of a package of corrective additives.

4. The dependence of the light emission on the degree of sharpness of the input edge, necessitates the analysis of the dynamics of the process of envelope peripheral streams flow conditionally sharp entrance edge diaphragm choke coil.

5. It can be assumed that possible electrical processes in the flow of mineral oil in narrow slits can cause electrical erosion of the oil base and limiting the flow of the channel walls made of materials with a crystal structure, as well as "burning" of corrective additives. Electrical erosion of the channel walls made of amorphous material (organic glass) was not detected.

References

[1] Marinesco M and Trillat J 1933 Action des ultrasons sur les plaques photographiques*ComptesRendus de l'Académie des Sciences* (Paris) **196** pp 858-860

[2] Knapp R, Daily J and Hamit F *Cavitation* 687 p.

[3] Jarman P and Tailor K 1964 Light emission from cavitation water *British Journal of Applied Physics***15**(3) pp 321-322 DOI: 10.1088/0508-3443/15/3/412

IOP Conf. Series: Materials Science and Engineering **492** (2019) 012028 doi:10.1088/1757-899X/492/1/012028

[4] Jarman P 1960 Sonoluminescence: A discussion *Journ. Acoust. Soc. Am.* **32** (11) pp 1459-1462 DOI: 10.1121/1.1907940

[5] Peterson F 1966 *Light emission from hydrodynamic cavitation. Ph. D. Thes*is. Northwestern University

[6] Lomakin V O, Kuleshova M S and Kraeva E A 2015 *Fluid Flow in the Throttle Channel in the Presence of Cavitation*

[7] Xinbei L, Vladimir L and Songjing L 2018 *Performance and Flow Field Analysis of Flapper Deflection Servo Valve*

[8] Lomakin V O, Kuleshova M S and Bozh'eva S M 2016 *Numerical Modeling of Liquid Flow in a Pump Station*

[9] Lomakin V. O 2015 *Investigation of two-phase flow in axial-centrifugal impeller by hydrodynamic modeling methods*

[10] Lomakin V O, Chaburko P S and Kuleshova M S 2017 *Multi-criteria Optimization of the Flow of a Centrifugal Pump on Energy and Vibroacoustic Characteristics*

[11] MargulisM A 2000 Sonoluminescence*Physics-Uspekhi (Advances in Physical Sciences)* **170** (3) pp 263-287

[12] Koldamasov A I 1991 Plasma formation in cavitating dielectric liquid *Technical physics***61**(2) pp 188-190

[13] Gertsenstein S Ya and Monakhov A A 2009 Electrification and the glow of the liquid in the coaxial channel with dielectric walls *Izv. RAN Fluid and gas mechanics* (3) pp 114-119

[14] Kornilova A A, Vysotskii V I, Koldamasov A I, Hyun Ik Yang and Desyatov A V 2007 Generation of intense directional radiation during the fast motion of a liquid jet through a narrow dielectric channels *The Surface. X-ray, synchrotron and neutron studies* (3) pp 55-60

[15] Kornilova A A, Vysotskii V I, Sysoev N N and Desyatov A V 2009 Generation of X-rays at bubble cavitation of fast liquid jet in dielectric channels *The Surface. X-ray, synchrotron and neutron studies* (4) pp 17-26

[16] Bagrov V V, Desyatov A V, Kazantseva N N, Zaryankin A S, Kozlov N P Kornilov A A, Ksenofontov B S, Kubyshkin A P, Kuzhekin I P, Kuleshov N I, Nagel Y A and Cherkasov S G 2010 *Water: effects and technologies* (Moscow: OOO SRC "Inzhener", JSC "ONIKO-M") 488 p

[17] Crum C C and Reynolds G T 1985 Sonoluminescence produced by "stable" cavitation *J. Acoust. Soc. Am.***78** pp 137-139

[18] Prosperetti A 1997 A new mechanism for sonoluminescence*J. Acoust. Soc. Am.* **83** pp 502-514

[19] Margulis M A 1981 Study of electrical phenomena associated with cavitation, *Phis. Chemistry J.***55** pp 154-158

[20] MargulisM A1995 *Sonochemistry and Cavitation* (London: Gordon and Breach Publishers) p 543

IOP Conf. Series: Materials Science and Engineering **492** (2019) 012029 doi:10.1088/1757-899X/492/1/012029

Study of the dynamics of an electro-hydraulic servo drive physical model on FESTO learning system

D Kulakov[1,2,3] and B Kulakov[1,2]

[1]Bauman Moscow State Technical University, 5 Second Baumanskaya Street, Moscow, 105005, Russian Federation
[2]Peoples Friendship University of Russia, 3, Ordjonikidze str., Moscow 117923, Russian Federation

[3]E-mail: kulakov@bmstu.ru

Abstract: The task of conducting research for dynamics improvement of an electro-hydraulic servo drive by including compensators on FESTO hydraulic learningsystem with low permissible inertial loads on the drive was considered. It has been suggested to simulate large inertial loads, which increase the time constant of the loaded hydraulic cylinder by attaching additional volumes filled with working fluid to the cavities of the executive hydraulic cylinder of the electro-hydraulic servo drive. The efficiency and clarity of the synthesis of compensators for an electro-hydraulic servo drive achieved with the MATLAB/Simulink software environment integration into FESTO hydraulic learningsystem has been shown.

1. Introduction

The development of a mathematical model of an electro-hydraulic drive in the form applicable to controllersynthesis and thedetermination of its parameters is an integral part of the task of controllersynthesis forthe electro-hydraulic servo drives (EHSD) [1-4]. Since any mathematical model of

a plant has a certain degree of inaccuracy, it is necessary to check the quality of regulation on the real plant. However, in laboratory environment it is often impossible to conduct studies of hydraulic systems with high power and high inertial loads [5-10]. In this case, it becomes appropriate to conduct research of the controllers on the lessloaded physical model of the EHSD with the reduction of its mathematical model parameters to the analogous parameters of the mathematical model describing the controlled object from the viewpoint of the developed controller. This is especially important for educational institutions when it is necessary to learn the controller synthesis methods with experimental confirmation of the obtained results at thereal plants.

The aforementioned approach to the study of the control laws synthesis [11] and the study of the EHSD characteristics on its physical model, based on FESTO laboratory hydraulic learningsystem (figure 1), has been implemented in the laboratory of the "Hydromechanics, Hydromachines and Hydro-Pneumoautomatics" department of Bauman MSTU.

The hydraulic learning system includes MOOG high-speed hydraulic directly actuated servo valve (HSV) with a valve position feedback (figure 2), an inertial load mechanism with two hydraulic cylinders installed opposite to each other and with a linear potentiometric feedback sensor, electronic PID controllers and state feedback controllers, the personal computers.

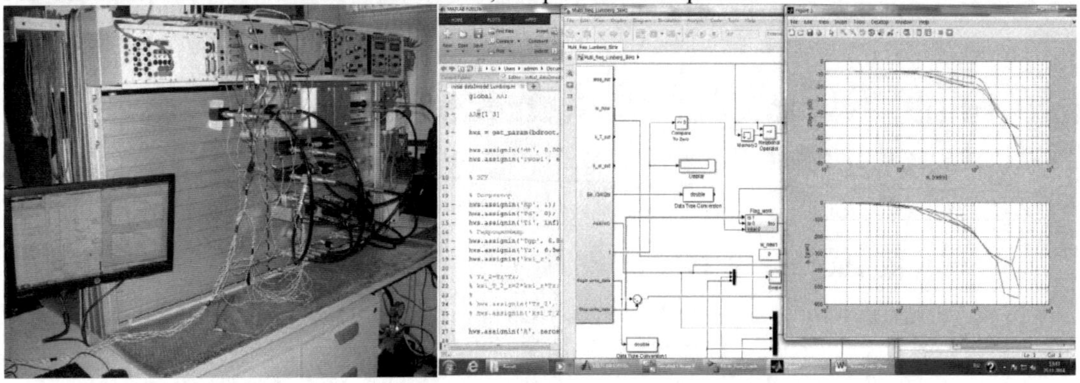

Figure 1. Research system based on the hydraulic learning system FESTO.

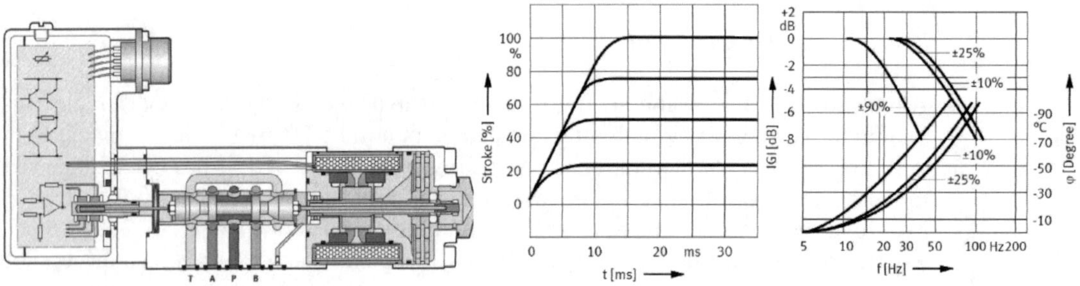

Figure 2. Hydraulic directly actuated servo valve (HSV) and its passport characteristics.

To enable the study of dynamic processes taking place in the EHSD and controller synthesis, the learning system was equipped with National Instruments PCI-6024E board, MATLAB/Simulink/RealTime Windows Target software, that allowto synthesize EHSD controllers and to develop control system programs that run on Windows in real time. The maximum sampling rate is 20 kHz.

To combine the process of controller synthesis with experimental studies of the EHSD and its elements, the authors have developed programs to control a research systemthat perform in an automated mode: the formation of typical step and harmonic input signals with automatic frequency variation in a given range; determination of the parameters of the EHSD mathematical model and its elements according to experimentally determined frequency characteristics; the EHSD control with synthesized compensators; experimental data registration and processing.

As a result of modernization, it becomes possible to conduct the entire chain of EHSD controller synthesis on one FESTO hydraulic learning system: development, implementation and research.

Works on the controller synthesis are performed in the following sequence:

1. Determination of the mathematical model parameters of the EHSD assembled on FESTO learning system according to experimentally determined transient processes or frequency characteristics of the hydraulic drive.

2. Synthesis of compensators and determination of their parameters using, in particular a LQR technique or numerical optimization methods in the MATLAB/Sisotool software package, Optimization Toolbox integrated in FESTO learning system.

3. Study of the EHSD work with synthesized controllers on the mathematical model in the MATLAB/Simulink software package integrated into FESTO learning system.

4. Study of the real work of the EHSD physical model with synthesized controllers with registration of experimental data in the MATLAB/Simulink/RealTime Windows Target software package.

2. Mathematical modeling of theEHSD physical model

A scheme of the studied EHSD physical model, assembled on the basis of FESTO hydraulic learning system, is shown in figure 3. Obtaining a linearized mathematical model of the EHSD is an important step in the controller synthesis [12, 13]. The mathematical description of the EHSD work, the linearized mathematical model and its areas of applicability are described in detail in [14, 15].

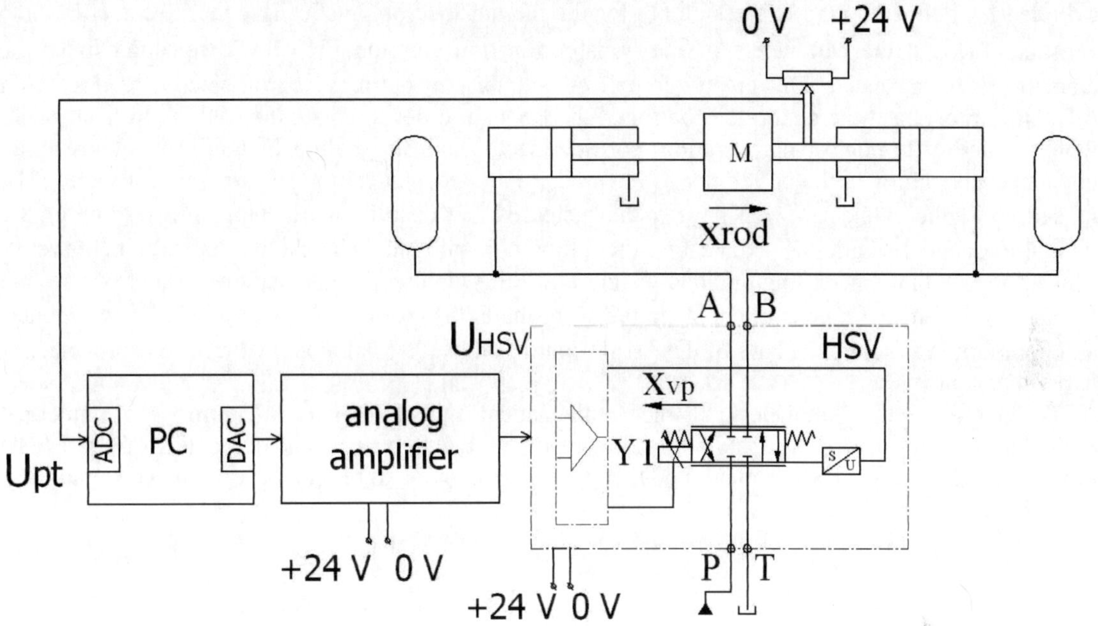

Figure 3. The scheme of the research system for studying the EHSD dynamics.

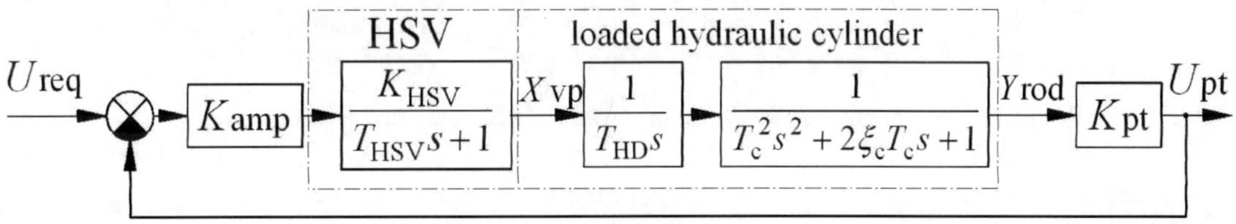

Figure 4. The EHSD linearized mathematical model block-diagram: K_{amp} – the electronic amplifier gain; K_{HSV} – the HSV gain; T_{HSV} – the HSV time constant; T_{HD} – the hydraulic drive time constant; T_c – the time constant of the loaded hydraulic cylinder; ξ_c – the hydraulic cylinder damping coefficient; K_{pt} – the potentiometric feedback sensor gain; Y_{rod} – the position of the hydraulic cylinder rod; U_{req} , U_{pt} – the input required signal and the rod position feedback sensor, respectively.

The EHSD linearized mathematical model can be represented by the block-diagram shown in figure 4 [14, 15]. The time constant of the loaded cylinder is determined by the formula:

$$T_c = \sqrt{\frac{mV_0}{2BF_p^2}}, \tag{1}$$

IOP Conf. Series: Materials Science and Engineering **492** (2019) 012029 doi:10.1088/1757-899X/492/1/012029

where
m isthe load mass moved by the EHSD;
V_0 is the volume of the cylinder cavity and the hydraulic lines connecting the hydraulic valve to this cavity;
B is the oil bulk modulus;
F_p is theworking area of the piston.

The mass of the load, which determines an inertial load of the EHSD assembled at FESTO hydraulic learning system, is 27.5 kg. It can be found that the time constant of the loaded hydraulic cylinder T_c, defined by the expression (1) for the inertial load, is practically equal to the HSV time constant T_{HSV}, if the parameters of the transfer functions for the EHSD with existing hydraulic elements were calculated. This greatly complicates the task of optimal control, since it requires, as a rule, an improvement the quality of control of the HSV. In order to create the EHSD physical model with the ability to control optimization, considering the existing EHSD hydraulic elements as an unchangeable part, with the possibility of linearizing its mathematical model, the time constant of the loaded hydraulic cylinder T_c (1) have to be increased. To do this, we increase the volume of the hydraulic cylinder cavities V_0, connecting them to additional tanks with the working fluid (figure 3). This allows, without increasing the load weight on FESTO hydraulic workstation, to approximate the EHSD physical model time constant T_c to the T_c of the EHSD with a large inertial load, and provides an opportunity to explore the laws of the EHSD control on its physical model with the existing electro-hydraulic elements.

To determine the parameters of the mathematical model shown in figure 4, a numerical optimization that ensures the maximum coincidence of the frequency characteristics of the EHSD physical model and its mathematical model (figure 5) at a given amplitude of a reference signal [16], is carried out.

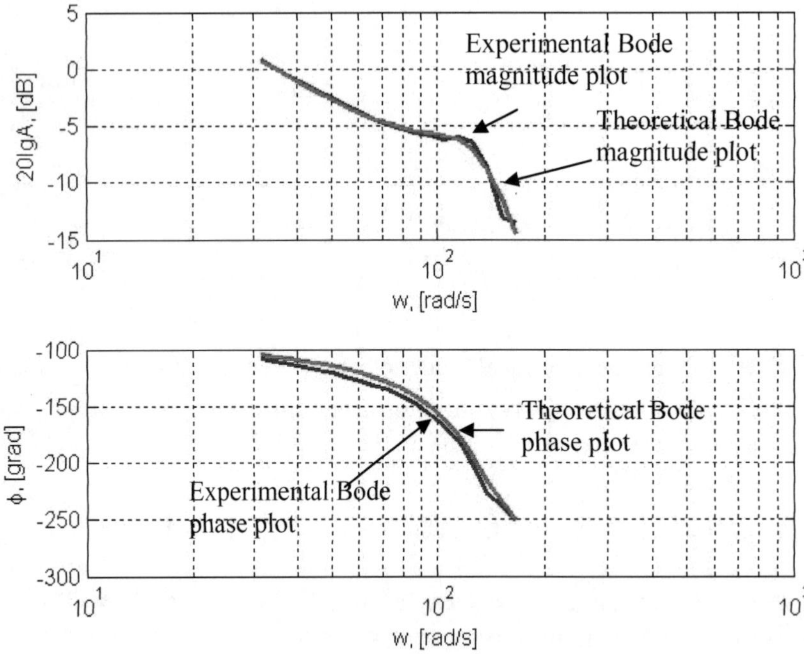

IOP Conf. Series: Materials Science and Engineering **492** (2019) 012029 doi:10.1088/1757-899X/492/1/012029

Figure 5. Frequency characteristics of the EHSD open loop, determined experimentally and calculated for a mathematical model with optimized parameters.

The frequency characteristics of the EHSD physical model are determined experimentally by the first harmonic of the measured EHSD response to the input harmonic signal. The processing of experimental data with the determination of Bode plot was carried out in accordance with the method given in [17]. The studies are carried out with an open loop plant for greater accuracy in determining the parameters of the mathematical model. The optimization of the mathematical model parameters was carried out by the lease squares method.

The parameters of the linear mathematical model (figure 4), which provide the best approximation of the frequency characteristics of the mathematical model to the experimentally determined frequency characteristics of the EHSD assembled on FESTO hydraulic workstation, are shown below:

$$K_{\text{amp}} = 20\,;\quad \frac{K_{\text{HSV}} \cdot K_{\text{pt}}}{T_{\text{HD}}} = 1.3763\,;\; T_{\text{HSV}} = 0.003\,;\; T_c = 0.0086\,;\; \xi_c = 0.2646\,.$$

3. Synthesis of the EHSDcontrollers on FESTO hydraulic learning system

Figure 6 shows the EHSD testing with a stepwise varying input signal, experimentally determined on FESTO hydraulic learning system, when operating without compensators and with compensators synthesized using the linearized mathematical model with the parameters defined above. To reduce the workstation oscillations during the EHSD operation (both in the experimental determination of frequency characteristics and in the study of transient processes), on the back of the workstation, a similar EHSD is assembled (figure 3) with the same parameters. The control signal is supplied to the HSV inputs of both drives from one computer in antiphase. Therefore, Fig. 6 shows the changes in the positions of the hydraulic cylinder rods of hydraulic actuators simultaneously working on two sides of the hydraulic workstation.

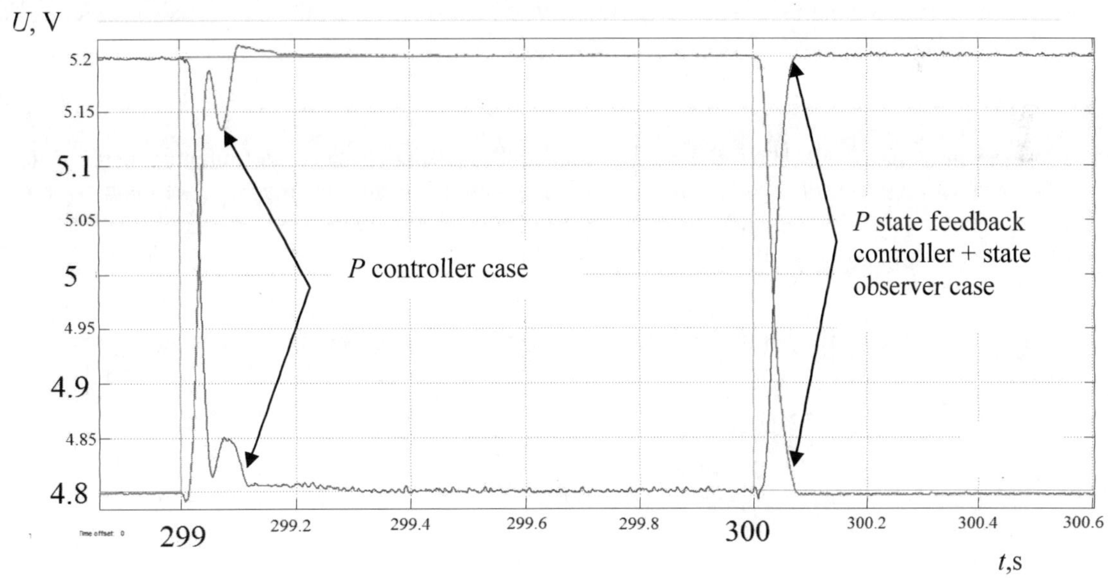

Figure 6.EHSD transients without compensators and with compensators, synthesized by the linearized mathematical model.

Synthesis of compensators in the EHSD control loop can be carried out using a variety of methods [3, 4,6, 8, 11, 18, 19], including the LQR method and numerical optimization methods using the

IOP Conf. Series: Materials Science and Engineering **492** (2019) 012029 doi:10.1088/1757-899X/492/1/012029

MATLAB/Sisotool software installed on the learning system. As an example, let us consider the improvement in the dynamics of the EHSD assembled on FESTO hydraulic workstation, using state components feedback. A block diagram of the EHSD with state feedback controller is shown in figure 7.

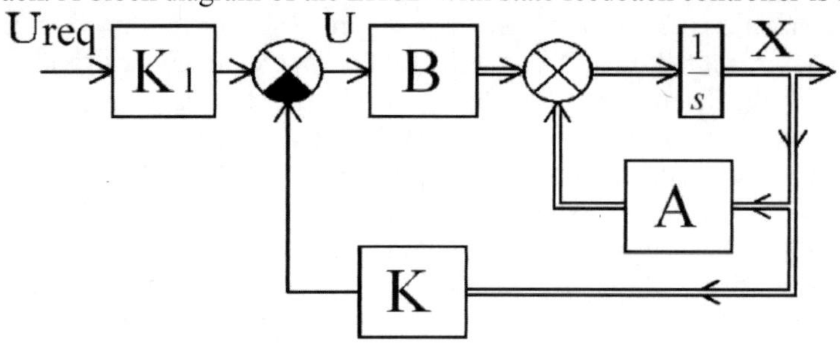

Figure 7. The EHSD block diagram with state feedback controller.

The notation conventions for figure 7:

A, B – matrices defining the parameters of the linear mathematical model of the plant, determined by the open loop of the EHSD block diagram shown in figure 4;

$$\dot{X} = \mathrm{A}X + \mathrm{B}u \quad (2)$$

X	–	plant state vector;
U	–	control signal applied to the input of the electronic amplifier;
K	–	state components feedback gain;
K_1	–	main feedback coefficient, the first element of the vector *K*.

It is necessary to track some external signal of the setting device, which is denoted $U_{\text{req}}(t)$. Since the system is tracking, the input signal $U_{\text{req}}(t)$ had to be multiplied by the main feedback factor K_1. Thus, the control signal is determined by the equation

$$u(t) = U_{\text{req}}(t) \cdot K_1 - K^T \cdot X(t) . (3)$$

The synthesis of a controller with state components feedback consists of defining the vector *K*. One of the ways to determine *K* is to use the MATLAB/Sisotool package. In this case, we choose a block diagram defining the location of the transfer functions of the EHSD unchangeable part and the compensators corresponding to equation (3), in Control Architecture window of MATLAB/Sisotool package (figure 8).

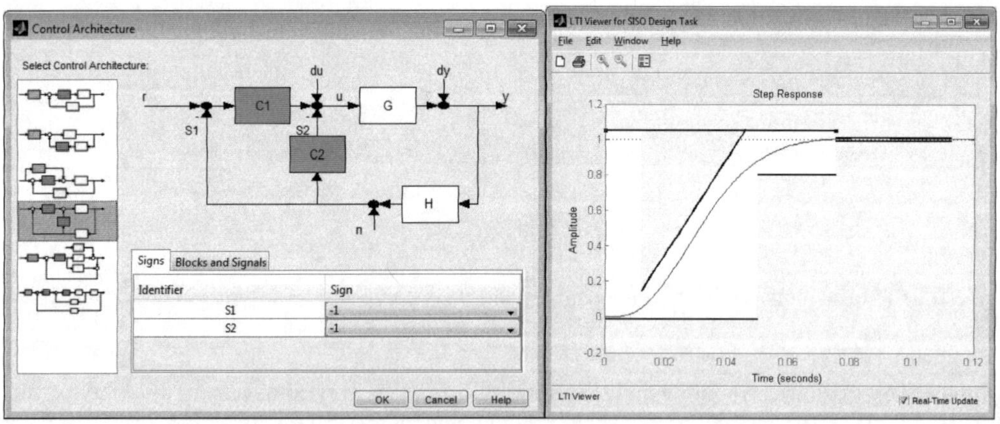

IOP Conf. Series: Materials Science and Engineering **492** (2019) 012029 doi:10.1088/1757-899X/492/1/012029

Figure 8.MATLAB/Sisotool software package windows.

After that, we introduce the defined above transfer function of the EHSD unchangeable part into MATLAB/Sisotool and determine the transfer functions of the compensators that need to be added to the EHSD loop. The software package determines the coefficients of the controller automatically if the required positions of the zeros and poles of the developed controller in the zero-pole portrait of the EHSD or the boundaries of the desired transition process in the corresponding package windows (Solodovnikov borders [13]), are set as shown in figure 8.

The EHSD transient processes without additional feedbacks and with state components feedback with the coefficients defined in the MATLAB/Sisotool package are shown in figure 9. Here and below, for comparison, both graphs of transient processes calculated numerically using the linearized mathematical model (dashed lines) and experimentally on EHSD physical model (solid lines) are presented. The transient processes of the EHSD without compensators (oscillatory processes in figure 9) and EHSD with calculated state components feedback gain (aperiodic processes in figure 9) are shown. The oblique and horizontal lines on the graph show the boundaries of the desired transition process, which were used to synthesize the compensators in MATLAB/Sisotool.

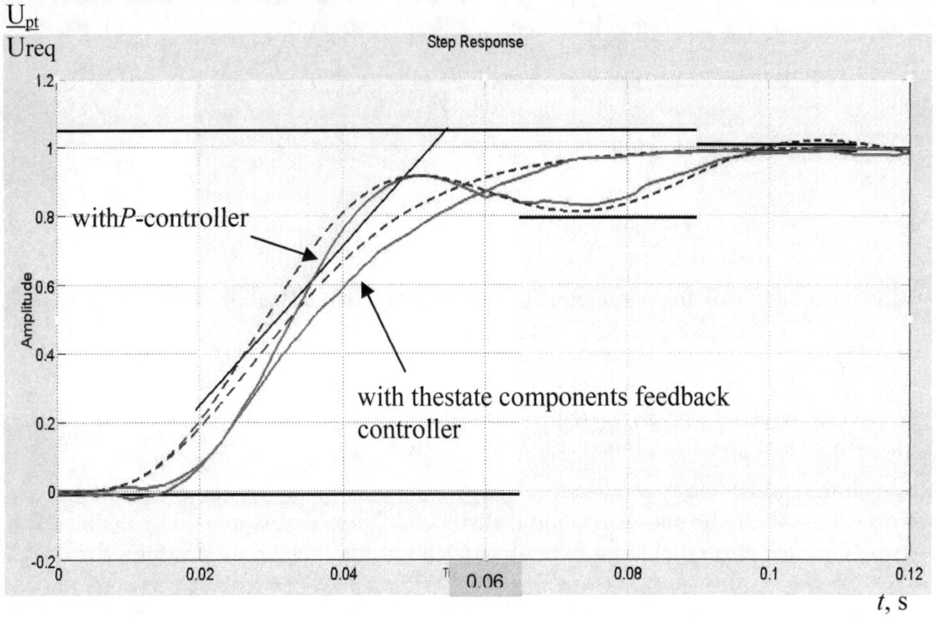

Figure 9. Theoretical (dashed) and experimental (continuous) transient processes of the EHSD on FESTO research system.

Another way to determine the state feedback gain K is to solve the LQR problem. According to the method described in [11], we introduce the additional requirement of finding the roots of the characteristic equation of a closed system inside a circle of radius ρ and with center at $(-\alpha, 0)$ with $\alpha > \rho \geq 0$.

If equation (2) is written in the Laplace form

$$sX(s) = AX(s) + BU(s) \quad (4)$$

IOP Conf. Series: Materials Science and Engineering **492** (2019) 012029 doi:10.1088/1757-899X/492/1/012029

and a new variable $\zeta = \dfrac{s+\alpha}{\rho}$ that converts the original circle with ρ radius in the s region into a unit circle on the ζ-plane located at the origin, is introduced, then the transformed state-space model has the form

$$\zeta X(\zeta) = \underbrace{\frac{1}{\rho}(\alpha \mathrm{I} + \mathrm{A})}_{Aq} X(\zeta) + \underbrace{\frac{1}{\rho}\mathrm{B}}_{Bq} U(\zeta). \quad (5)$$

In the shift operator region, it will be written

$$qX[k] = \mathrm{A}_q X[k] + \mathrm{B}_q u[k], \quad (6)$$

where

$q(f[k]) \overset{\Delta}{=} f[k+1]$ is the shift operator.

The stability of a discrete system requires the poles to have modules less than one, i.e. were inside the unit circle centered at the origin. Therefore, let's solve the task of constructing a discrete controller applied to (6) by solving the corresponding algebraic Riccati equation

$$\mathrm{A}^T\left(\mathrm{P} - \mathrm{PB}_q\left(R + \mathrm{B}_q^T\mathrm{PB}_q\right)^{-1}\mathrm{B}_q^T\mathrm{P}\right)\mathrm{A}_q + \mathrm{Q} - \mathrm{P} = 0,$$

and equations determining the coefficients of the K

$$K = \left(R + \mathrm{B}_q^T\mathrm{PB}_q\right)^{-1}\mathrm{B}_q^T\mathrm{PA}_q,$$

where

A, B are the matrices defining the parameters of the linear mathematical model of the plant (4);

Q, R are the weighting matrices of the cost function

$$J = \int_0^\infty X^T \mathrm{Q} X \mathrm{d}t + \int_0^\infty u^T R u \mathrm{d}t;$$

P is amatrix-solution to the Riccati equation.

Mentioned solutions can be obtained using the MATLAB function **dlqr**.

Figure 10 shows the theoretical and experimental transient processes of the EHSD with the state components feedback controller synthesized using LQR on FESTO research system for parameters $\rho = 50$, $\alpha = 120$; matrices Q, R are identity.

IOP Conf. Series: Materials Science and Engineering **492** (2019) 012029 doi:10.1088/1757-899X/492/1/012029

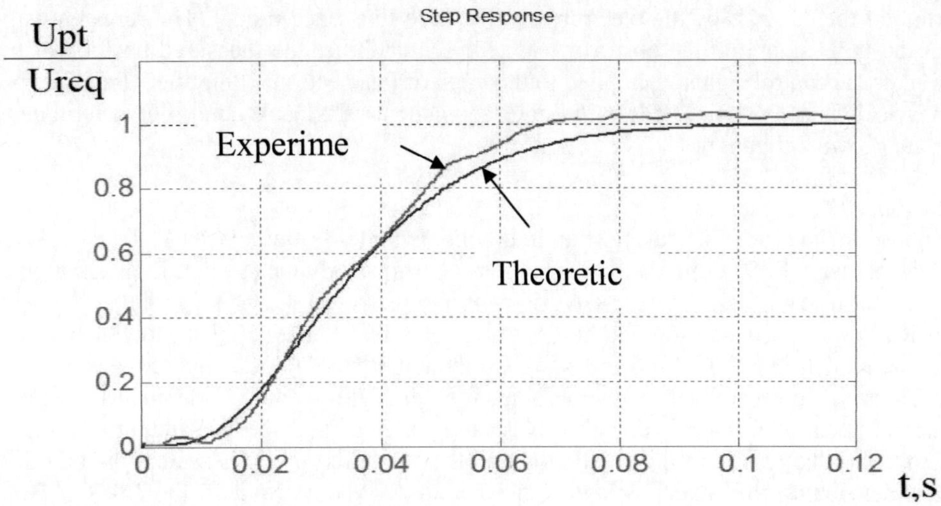

Figure 10. Theoretical and experimental transient processes of the EHSD with the controller synthesized using LQR on FESTO research system.

For the EHSD the state vector contains: position, speed and acceleration of the plunger, the position of the HSV valve. The rod position is determined by the potentiometric position sensor. The used proportional HSV has built-in feedback on the valve position with the electrical signal output indicating its current position. There are also FESTO electronic blocks, which determine the first and second derivatives of the main feedback signal and include them in the control loop. However, their effectiveness is limited by measurement noise in the control loop.

As a result of the fact that the EHSD is controlled by the MATLAB/Simulink/RealTime Windows Target package included in the EHSD control loop, it is possible to implement an observer that determines in real time the state components that are used in the synthesized controller but are not available for direct measurement. The block diagram of the EHSD with the developed observer [11, 19] is shown in figure 11.

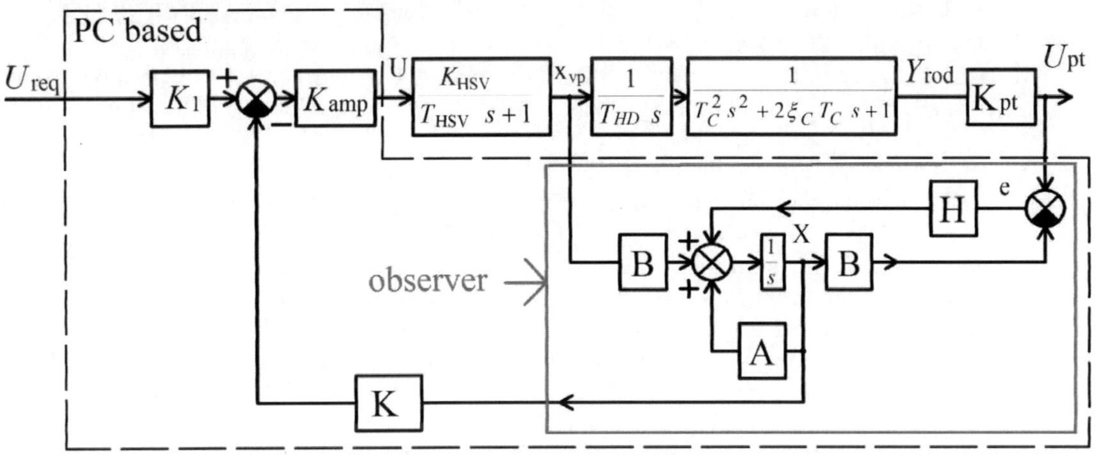

Figure 11.EHSD with the loaded hydraulic cylinder state observer and the state components feedback controller.

The dashed line highlights the part of the circuit implemented in the MATLAB/Simulink software package on a computer included in the EHSD control loop (the program runs in real time with 5 kHz sampling rate). The part, that is not highlighted, is the plant (without K_{amp}). The control object is represented on the diagram by the electrohydraulic drive linearized model. The research was carried out when the input signal on the observer was represented either the measured position of the spool (figure 11), or the control signal U, applied to the input of the electronic amplifier. Transient processes of the physical model of an EHSD with state components feedback controller determined by the observer are shown in figure 6.

4.Conclusions

The proposed method of increasing the hydraulic cylinder time constant greatly expands the possibilities of using FESTO hydraulic learning system for studying the EHSD dynamic properties. The integration of MATLAB software environment into the research system complements MATLAB with physical models of electro-hydraulic control objects that are assembled individually (similar to mathematical models in MATLAB/SimScape). On the other hand, FESTO hydraulic learning systems are provided with a powerful software environment for modeling, control and analysis, which makes the synthesis of control laws for hydraulic drives and their elements effective and intuitive. It becomes possible to verify the results of the simulation and the synthesis of compensators directly during the process of developing the modeling and control laws, "without leaving" the MATLAB software development environment. This modernization of FESTO hydraulic learning systems brings them to a level that allows conducting research of dynamic processes in the EHSD and its elements.

References

[1] Popov D N, Sosnovskiy N G and Siukhin M V 2017 Computer simulation of hydraulic systems with typical nonlinear characteristics *Nauka i Obrazovanie. Electronnoe nauchno-tehnicheskoe izdanie — Science and Education. Electronic scientific and technical journal* (7) pp 63–74 DOI: 10.7463/0717.0001136

[2] Kozlov O S and Skvortsov L M 2016 MVTU software package in scientific research and applied developments *Mathematical Models and Computer Simulations* **8** (4) pp 358–368 DOI: 10.1134/S2070048216040116

[3] Kulakov D B, Semenov S E, Kulakov B B, Shcherbachev P V and Tarasov O I 2015 Hydraulic Bipedal Robots Locomotion Mathematical Modeling *Procedia Engineering* **106** pp 62–70 DOI: 10.1016/j.proeng.2015.06.009

[4] Skarpetis M G and Koumboulis F N 2013 Robust PID controller for electro-hydraulic actuators *IEEE International Conference on Emerging Technologies and Factory Automation*, ETFA DOI: 10.1109/ETFA.2013.6648165 DOI: 10.1109/ETFA.2013.6648165

[5] Shumilov I S 2015 The steering hydraulic drives with centralized or independent power supplies for the main aircrafts *Proceedings of 2015 International Conference on Fluid Power and Mechatronics* FPM 2015 pp 596–602 DOI: 10.1109/FPM.2015.7337186

[6] Adeoye A O M, Aderoba A A and Oladapo B I 2017 Simulated Design of a Flow Control Valve for Stroke Speed Adjustment of Hydraulic Power of Robotic Lifting Device *Procedia Engineering* **173** pp 1499–1506 DOI: 10.1016/j.proeng.2016.12.230

[7] Konstantinov S V, Red'ko S A and Ermakov S A 2006 *Electrogidravlicheskiye ruleviye privody system upravleniya poletom manevrennih samoletov* [Electro-hydraulic steering drives for flight control systems of maneuverable aircraft] (Moscow: Yanus-K Publ.) 315 p (in Russian)

IOP Conf. Series: Materials Science and Engineering **492** (2019) 012029 doi:10.1088/1757-899X/492/1/012029

[8] Ovsyannikova E and Gouskov A M 2016 The analysis of the ventricle assist device controlled rotor dynamics. *Proceedings of the Vibroengineering: 22nd International Conference, Dynamics of Strongly Nonlinear Systems* (Moscow, Russia, October 4–7, 2016) **8** pp 169–173

[9] ObolenskiiIu G, Ermakov S A and Sukhorukov R V 2011 *Vvedenie v proektirovanie system aviatsionnykh rulevykh privodov* [Introduction to the design of systems of aircraft steering gear] (Moscow: the State Unitary Enterprise "OkruzhnayagazetaYuZAO" Publ.) 344 p (in Russian)

[10] Redko P G, Ermakov S A, Selivanov A M et al. 2008 Development Concept of Future Aircraft Steering Gears *Obshcherossiyskiy nauchno-tekhnicheskiy zhurnal "Polyot" — All-Russian Scientific-Technical Journal "Polyot"* (*"Flight"*) (1) pp 50–60 (in Russian)

[11] Goodwin G C, Graebe S F and Salgado M E 2001 *Control system design. Upper Saddle River* (Prentice-Hall) 908 p

[12] Pupkov K A, KuiCh N 2006 Relay-linear control law in a system of a homing guidance. *Problem i mashinostroeniya i avtomatizazii— Engineering and Automation Problems* (4) pp 55–58 (in Russian)

[13] Solodovnikov V V, Plotnikov V N and Yakovlev A V 1993 *Teoriya avtomaticheskogo upravleniya tekhnicheskimi sistemami* [The theory of automatic control of technical systems] (Moscow: Bauman MSTU Publ.) 492 p (in Russian)

[14] Popov D N 1987 *Dinamika i regulirovanie gidro- I pnevmosistem* [Dynamics and regulation of hydro- and pneumatic systems] (Moscow: Mashinostroenie Publ.) 464 p (in Russian)

[15] Popov D N 2001 *Mekhanikagidro- i pnevmoprivodov*[Hydraulic- and pneumatic actuators mechanic] (Moscow, Bauman MSTU Publ.) 320 p (in Russian)

[16] Malishev V N, Popov D N and Sosnovskiy N G 2007 Identification of autonomous electrohydraulic servo drive. *Nauka i Obrazovanie. Electronnoe nauchno-tehnicheskoe izdanie— Science and Education. Electronic scientific and technical journal* (6) (in Russian)

[17] Ivanov V A, Chemodanov B K and Medvedev V S 1971 *Matematicheskiye osnovy teorii avtomaticheskogo regulirovaniya* [Mathematical Basis of the Theory of Automatic Control]. (Moscow:Vysshaya Shkola Publ.) 807 p (in Russian)

[18] Ren H-P and Wang X 2018 PLC based variable structure control of hydraulic position servo system considering controller saturation *Proceedings of the 13th IEEE Conference on Industrial Electronics and Applications* ICIEA 2018 pp 135–140 DOI: 10.1109/ICIEA.2018.8397703

[19] Pupkov K A, Egupov N D, et al. 2004 *Metody klassicheskoy i sovremennoy teorii avtomaticheskogo upravleniya v 5 tomakh. Tom 1. Matematicheskiye modeli, dinamicheskiye harakteristiki i analiz system avtomaticheskogo upravleniya* [Methods of Classic and Modern Control Theory in 5 volumes Vol. 1 Description and Analysis of Automatic Control Systems] (Moscow: Bauman MSTU Publ.) 655 p (in Russian)

IOP Conf. Series: Materials Science and Engineering **492** (2019) 012030 doi:10.1088/1757-899X/492/1/012030

Analysis of hydrodynamic properties of a gas phase in water by hydroacoustic method

S Gavrilev[1] and M Ivanov[1,2]

[1]Bauman Moscow State Technical University, 5 Second Baumanskaya Street, Moscow, 105005, Russian Federation

[2]E-mail: mivanov2005@mail.ru

Abstract. The paper deals with the existing methods for determining the hydrodynamic properties of the gas phase in a liquid medium. Most modern sonar methods belong to the class of active methods. In a number of hydrodynamic systems, their use is not desirable; therefore, passive methods are used. These methods are still not well developed. In this work, the authors show the experimental studies of the noise of air bubbles in water. In conclusion, the paper points that the noise emitted by bubbles is due to three processes at once: exfoliation, floating and collapse.

1. Introduction

At the present time, the number of industrial technologies based on the use of hydromechanical systems containing substances in different phase states, primarily gas and liquid, is growing. The dynamics of the processes occurring in such systems have been the subject of many studies for many years [1–3]. Much attention in such works is paid to the dispersed composition of the gas phase in a liquid medium, since This parameter is key in the design and calculation of most hydromechanical systems [4, 5].

Methods for determining the dispersed composition of the gas phase in a liquid medium can be divided into three types: optical, photometric and acoustic. Optical and photometric methods are most common [6–9], but they have one major drawback: they are applicable only in optically transparent media, i.e. they are not suitable for measuring the size distribution of air bubbles in turbid liquid media. Therefore, acoustic methods are considered more versatile.

Most modern acoustic methods for determining the distribution of bubbles in size are based on measurements of the attenuation and velocity of an acoustic wave passing through a layer of a bubble cloud. These methods are called active and are well described in [10–12]. But this approach has one minor drawback. An acoustic wave generated to irradiate a bubble cloud can affect bubbles. This phenomenon can be seen in the results of [13]. The authors of this work conducted a series of experiments, while changing the intensity of the emitted acoustic wave and in each of the experiments obtained different results. Thus, in a number of cases, where the maintenance of regimes with a certain bubble size is especially important, the use of active methods is unacceptable. Therefore, it is preferable to use passive methods that do not affect the dynamics of the studied processes.

Passive methods are based on the fact that air bubbles in water are sources of acoustic signals. They emit an acoustic signal due to the variable pressure of the gas inside the bubble.

In 1933, Minauert showed that under adiabatic conditions (the heat exchange between the bubble gas and the liquid is insignificant) the frequency of the acoustic signal f, a gas emitted by a bubble depends on its size as follows:

$$f = \frac{1}{R}\left(\frac{3\gamma P}{(2\pi)^2 \rho}\right)^{\frac{1}{2}}, (1)$$

where
f – sound wave frequency emitted by a bubble;
P – absolute liquid pressure;
γ – gas specific heat coefficient;
ρ – liquid density;
R – bubble radius.

This expression is valid for spherical bubbles. The shape of the bubbles usually changes during the ascent. Depending on the size, they may take the form of a sphere, an oblate spheroid of a spherical segment, or a mushroom cap. In [14], a criterion is presented, in accordance with which bubbles can be considered approximately spherical, if the condition:

$$R < \delta_\sigma, (2)$$

where
R – bubble radius;
δ_σ – capillary constant.

According to this characteristic, bubbles in water are shaped as spheres, if their radius is $R < 2.7$ mm.

The most complete description of bubble acoustics was given by Leighton (1994). According to Leighton and Walton, the spectrum of the sound emitted by the bubbles can be used to determine their size.

The essence of passive methods is that the noise emitted by bubbles is measured, and the size of the bubbles is calculated from equation 1. Later this idea was developed in [15–20]. However, it is worth noting that these works do not take into account the fact that air bubbles in water emit noise not only during pulsations during ascent, but also when detached from the aerator and collapsing at the boarder “water-air”.

2. Statement of problem

As mentioned before, the noise emitted by a bubble can be divided into 3 stages:

1) exfoliation noise;
2) floating noise;
3) collapse noise.

Since in most cases we are dealing with a cloud of bubbles, rather than single bubbles, using the passive acoustic methods for determining the size of bubbles creates the problem of mixing the noise of all three stages. Due to the fact that the bubbles are continuously generated in a sufficiently large number, it is impossible to separate these stages in time. It is important to note that to determine the size of the bubbles; we need to select exactly the floating noise.

3. Experimental procedure

To separate the noise of the bubbles into components, an experimental setup was assembled. Its scheme is shown in figure 1. The installation consists of a vertically arranged pipe (2) made of organic glass. The diameter of the pipe is 54 mm and the length is 1000 mm. Inside the tube, at the very bottom there is a porous fine-bubble aerator (3). Next to, it is a hydrophone (4) of type 8103 from Bruel&Kjaer. Another such hydrophone (5) moves freely along the entire height of the pipe. Both hydrophones are connected to the Pulse LAN-XI multichannel analyzer. At the upper end of the pipe is fixed tank (1). A hole with a diameter of 15 mm is drilled at the bottom of this tank.

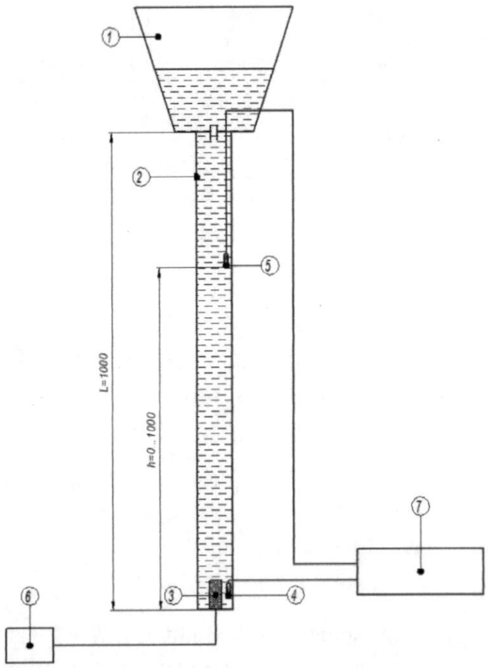

Figure 1. Schematic diagram of the experimental setup: 1 – collapse noise separation tank; 2 – vertical tube; 3 – fine-bubble porous aerator; 4, 5 – hydrophones type 8103 Bruel&Kjaer; 6 – air compressor; 7 – Pulse LAN-XI multichannel analyzer.

The experiments were carried out as follows. With the help of a compressor (6), air was supplied to the pipe through a fine-bubble aerator. The bubbles came off the aerator and floated up through the pipe. The collapse of bubbles occurred at the "water-air" boundary in the reservoir (1). The bottom of this tank serves as a cover for the upper end of the pipe, and it isolates the collapsing noise.

During the entire experiment, a constant flow of air through the aerator was maintained; therefore, it can be argued that the formation of bubbles was steady. The noise emitted by the bubbles was recorded from the hydrophones (4) and (5) connected to the multi-channel analyzer (7). The position of the hydrophone (5) has changed. The height of its installation varied from 0 to 1000 mm with a step of 100 mm. Next, the recorded signals were processed with Fast Fourier Transform (FFT) and the spectra of noise emitted by bubbles were plotted.

4. Results

The results of the experiments are presented in figure 2. The graphs show the noise spectra measured by a hydrophone (5). It moved along the height of the pipe from 0 to 1000 mm in increments of 100 mm. It was assumed that in this way, the contribution of the noise of sticking bubbles will be less and less, due to the increasing removal of the hydrophone from the aerator.

After analyzing the results of the experiments, it can be noted that when the hydrophone is removed from the aerator, the noise level decreases. You should also pay attention to the fact that the spectrum of noise, measured directly at the aerator, has a high frequency component, up to about 3.5 kHz. Further, this component gradually disappears (see graphs at $h = 100...400$ mm).

In the plots with $h > 400$ mm, the spectrum can be described as noise up to 2 kHz. That it is the noise of oscillations of bubbles during the ascent and the noise from 2 to 3.5 kHz, which can be seen in the graphs at $h < 400$ mm, is due to the sticking of bubbles.

IOP Conf. Series: Materials Science and Engineering **492** (2019) 012030 doi:10.1088/1757-899X/492/1/012030

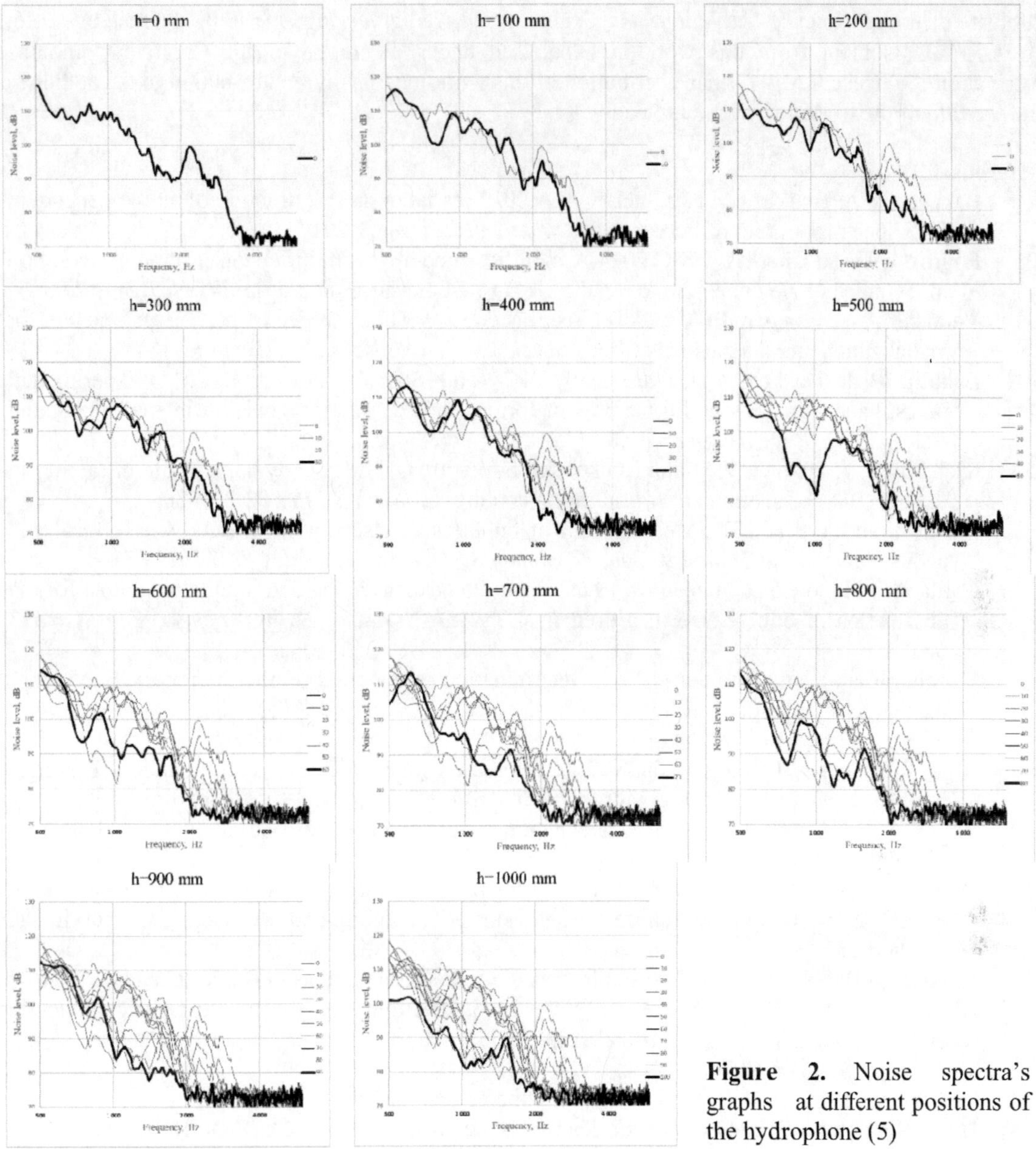

Figure 2. Noise spectra's graphs at different positions of the hydrophone (5)

5. Conclusions

According to the results of the conducted experiments we can concluded that the noise during the detachment of bubbles from the aerator is more broadband than the noise during ascent. In our case, it turned out that, depending on the height of the hydrophone, the noise spectrum is up to 3.5 kHz or up to 2 kHz.

Thus, clearly showing that the noise emitted by bubbles is due not only to fluctuations during ascent, but also to noise when detaching from the aerator, we conclude that the existing passive acoustic methods for determining the gas phase in water require substantial modifications.

IOP Conf. Series: Materials Science and Engineering **492** (2019) 012030 doi:10.1088/1757-899X/492/1/012030

References

[1] Babenko V A, Bunkin N F and Sychev A A 2017 Variation in the structure of a time-dependent SRS spectrum in microfiltered water *Quantum Electron.* **47** 901–5
[2] Bagrov V V and Cherkasov S G 2014 A modified model for simplified description of evolution of a single bubble upon increase in the pressure of the surrounding fluid *High Temp.* **52** 93–7
[3] Varaksin A Y, Protasov M V and Teplitskii Y S 2014 About choice of particle parameters for visualization and diagnostics of free concentrated air vortices *High Temp.* **52** 554–9
[4] Bakhtin B I, Ivashov A I, Kuznetsov A V and Skorokhodov A S 2014 Experimental investigation of the specific features of formation of cavitation zones in intense ultrasound fields *J. Eng. Phys. Thermophys.* **87** 672–85
[5] Kichatov B V, Polyaev V M and Runkovskii S V 2001 Gas bubble dynamics with instability in the operation of a system of transpiration cooling *Heat Transf. Res.* **32** 273–80
[6] Zhang Y and Sun H 2012 Measurement of bubble size distribution in liquids by optical and acoustical methods *Proc. Int. Conf. Commun. Syst. Netw. Technol. CSNT 2012* 671–4
[7] Gaillard T, Honorez C, Jumeau M, Elias F and Drenckhan W 2015 A simple technique for the automation of bubble size measurements *Colloids Surfaces A Physicochem. Eng. Asp.* **473** 68–74
[8] Mazahernasab R and Ahmadi R 2016 Determination of bubble size distribution in a laboratory mechanical flotation cell by a laser diffraction technique *Physicochem. Probl. Miner. Process.* **52** 690–702
[9] Argo T F, Wilson P S and Palan V 2008 Measurement of the resonance frequency of single bubbles using a laser Doppler vibrometer *J. Acoust. Soc. Am.* **123** EL121-EL125
[10] Wu X J and Chahine G L 2010 Development of an acoustic instrument for bubble size distribution measurement *J. Hydrodyn.* **22** 325–31
[11] Christensen M and Thomassen P 2014 Experimental Determination of Bubble Size Distribution in a Water Column by Interferometric Particle Imaging and Telecentric Direct Image Method 121
[12] Xue J 2004 Bubble Velocity, Size and Interfacial Area Measurements in Bubble Columns 1–210
[13] Al-Masry W A, Ali E M and Aqeel Y M 2005 Determination of Bubble Characteristics in Bubble Columns Using Statistical Analysis of Acoustic Sound Measurements *Chem. Eng. Res. Des.* **83** 1196–207
[14] Volkov P K 1996 Fluid dynamics with gas bubbles *Fluid mechanics* 75–88
[15] Husin S and Mba D 2010 Correlation between Acoustic Emission (AE) and Bubble Dynamics **II** 0–5
[16] Manasseh R, LaFontaine R F, Davy J, Shepherd I and Zhu Y-G 2001 Passive acoustic bubble sizing in sparged systems *Exp. Fluids* **30** 672–82
[17] Leighton T 1994 Acoustic bubble detection I: The detection of stable gas bodies
[18] Devaud M, Hocquet T, Bacri J, Leroy V, Devaud M, Hocquet T, Bacri J and Leroy V 2007 The Minnaert bubble : a new approach To cite this version : HAL Id : hal-00145867
[19] Chen L, Wood S, Moore S and Nguyen B 2012 Acoustic Emission of Bubbly Flow and Its Size Distribution Spectrum *Proc. Acoust.* 1–6
[20] Greene C, Wilson P S and Coffin R B 2011 Laboratory measurements on gas hydrates and bubbly liquids using active and passive low-frequency acoustic techniques 045001

IOP Conf. Series: Materials Science and Engineering **492** (2019) 012031 doi:10.1088/1757-899X/492/1/012031

Modification of polymeric membranes with unipolar corona discharge to intensify the separation of oil-in-water emulsions

I Shaikhiev[1], V Dryahlov[1], M Galikhanov[1], S Sverguzova[2] and M Ivanov[3,4]

[1] Kazan National Research Technological University, Kazan, Russian Federation
[2] Belgorod State Technological University named after V.G. Shukhov, Belgorod, Russian Federation
[3] Bauman Moscow State Technical University, 5 Second Baumanskaya Street, Moscow, 105005, Russian Federation

[4]E-mail: mivanov2005@mail.ru

Abstract. The data on the unipolar corona treatment of polymer membranes made of polyethersulfone and polyacrylonitrile and their use for the separation of a 20% model oil-in-water emulsion based on an "I-20A" industrial oil stabilized by a "Kosintol-242" surfactant are given.It is shown that corona treatment of polymer membranes contributes to an increase in membrane productivity and the efficiency of separation of water-oil emulsion.It has been determined that corona discharge treatment leads to a decrease in the degree of crystallinity of the membranes, roughness due to oxidation of the surface with ozone produced during corona processing and the edge wetting angle, which indicates an increase in hydrophilic characteristics.

1. Introduction

In connection with the intensive development of industrial production, it is formed a greater amount of wastewater (WW), which containing various impurities in its composition. A special place in the problem under consideration is occupied by WWs containing petroleum and petroleum products. Crude oil, as well as its numerous refined products, flow in large quantities into storm sewage, industrial wastewater. At the same time, they can partially or completely dissolve in water, form emulsions or form a slick on the water surface.

Various methods are used for cleaning WW containing water-insoluble petroleum products in the form of corresponding emulsions – flotation [1-3], coagulation [4, 5], adsorption [6-8], etc. In recent years worldwide for purifying WW including those containing petroleum products, membrane methods are being intensively developed. As a rule, ultrafiltration processes using polymer membranes are used to remove petroleum products.

In order to improve the efficiency and selectivity of the release of pollutants from aqueous environments, the membranes are subjected to modification, which is carried out by treatment: using chemical reagents [9-11], flame, plasma, corona discharge, laser radiation and high-energy radiation.

Recently, the industrial scale widely used processing of polymeric materials using corona discharge, mainly to change their adhesive characteristics. It is also shown that the corona discharge may be used for processing polymer membranes, however, it is indicated that corona discharge treatment leads to hydrophilization of the surface and an increase in the rate of emulsion separation through the membrane. At the same time, it should be noted that a variable corona discharge is mainly used abroad. In contrast, a constant (unipolar) corona discharge can not only modify the surface of dielectric materials, but also give them electret properties, i.e. the ability for a long time to maintain a polarized state after the removal of external influences. This circumstance leads to the polarization of this dielectric, which then creates a quasi-permanent electric field in the surrounding space [9]. This circumstance helps to increase the performance of polymer membranes when cleaning emulsions of the type "oil in water".

2. Methods

In light of the above, we carried out experiments on the separation of 20% water-oil emulsion (WOE) using polyethersulfone (PES) and polyacrylonitrile (PAN) membranes with a mass of cut-off particles 10, 30, 50 and 10, 25, 60 kDa, respectively. The emulsion was prepared by mixing 200 cm^3 of industrial oil of the "I-20A" brand from 20 cm^3 of surfactant of the "Kosintol-242" brand and 780 cm^3 of distilled water until an aggregate-stable disperse system was formed. The chemical oxygen consumption (COC) of the resulting emulsion was 1,66550 mg O / dm^3. Using the "Malvern ZetasizerNano ZS" brand nanoparticle analyzer, the particle size of the dispersed phase of the emulsion,

179 nm, was determined using the dynamic light scattering method, which corresponds to the pore size range of selected ultrafiltration PES and PAN membranes. The separation process was carried out in a membrane module with a "cross-flow", a filtration area of 47 cm^2 at a temperature of 20-25 °C and a pressure of 2 bar. Corona processing of the membranes was carried out at a voltage of U = 5, 10 or 15 kV of negative polarity and exposure time τ = 0.5, 1.0 or 1.5 minutes.

3. The results

Analysis of the performance of the initial and modified PES membranes showed an increase in the flow of WOE with an increase in the mass of cut-off membrane particles, which is quite natural. At the same time, the largest increase in the indicator under consideration from 10 to 20 $dm^3/(m^2 \cdot h)$ for modified filter elements compared to the initial ones is observed for membranes with a mass of 50 kDa cut-off particles. For membranes with a mass of cut-off particles of 10 and 30 kDa, the increase in productivity of separation of WOE after corona processing occurs from 2 to 4 $dm^3/(m^2 \cdot h)$ and from 12 to 15 $dm^3/(m^2 \cdot h)$. COC values are presented in table 1.

Table 1.Values of COC WOE before and after separation of the original and corona-treated PES membranes with a mass of cut-off particles of 10, 30 and 50 kDa.

Mass of cut-off particles of PES membrane, kDa	Corona discharge voltage U_a, kV	COC values, mgO / dm^3		
		Corona time τ,min		
		0.5	1.0	1.5
10	5	13202	11316	12731
	10	10373	8487	11316
	15	28290	18860	11788
	The filtrate after separation of the emulsion by the original membrane	9902		
30	5	3920	784	3920
	10	2352	1960	3136
	15	2352	784	1568
	The filtrate after separation of the emulsion by the original membrane	7056		
50	5	34215	6077	13554
	10	23520	9985	14525

	15	6195	6382	7273
	The filtrate after separation of the emulsion by the original membrane	19010		
The original emulsion		166550		

According to the results of table 1, it is obvious that the best effect of corona processing was noted, as in the case with the performance parameter for membranes with a larger mass of cut-off particles — 30 and 50 kDa. At the same time, the efficiency for the indicated membranes as a result of corona processing increased from 95.8 to 99.5% (τ = 1 min and U_a = 5 and 15 kV) and from 88.6 % (initial PES membrane 30 kDa) up to 96.3% (U_a = 10 kV and τ = 0.5 min), respectively.

Unlike PES membranes for PAN membranes with the smallest pore size (10 kDa), the increase in productivity as a result of corona processing was not revealed and is ≈15 $dm^3/(m^2{\cdot}h)$, at the same time for membranes with a mass of cut-off particles 25 and 60 kDa marked increase in the parameter under consideration after exposure to corona discharge with 10 to 20 $dm^3/(m^2{\cdot}h)$ and from 20 to 35 $dm^3/(m^2{\cdot}h)$, respectively. Values of COC of the filtrates are given in table 2.

Table 2. The values of COC WOE before and after separation of the original and corona-treated PAN membranes with a mass of cut-off particles 10, 25 and 60 kDa.

Mass of cut-off particles of PAN membrane, kDa	Corona discharge voltage U_a, kV	COC values, mgO / dm^3		
		Corona time τ,min		
		0,5	1,0	1,5
10	5	1850	4040	1400
	10	1190	3830	2720
	15	1280	1620	1430
	The filtrate after separation of the emulsion by the original membrane	1900		
25	5	10002	7803	7639
	10	6578	5642	6810
	15	6812	6339	7462
	The filtrate after separation of the emulsion by the original membrane	8520		
60	5	31591	19126	34023
	10	20112	24783	44326
	15	43989	44045	45760
	The filtrate after separation of the emulsion by the original membrane	61046		
The original emulsion		166550		

Analysis of the data in table 2 shows that the greatest effect of the corona discharge in terms of COC, namely more than 3 times the decrease in the values of the indicator is observed for membranes with a mass of 60 kDa cut-off particles, in which case the lowest corona rending effect on performance is noted.

The surface and structural characteristics of the original and most effective coronaned membranes were determined by IR spectroscopy of internal reflection, sessile drop, atomic force microscopy, and diffractometry. It should be noted that in the infrared spectra of corona-treated membranes in most cases there is a decrease in the intensity of the absorption bands compared to those of the original sample, which is a consequence of the partial destruction of the polymer material of the membrane.

The above is confirmed by a decrease in the degree of crystallinity of the polymer membrane matrix of a corona-treated membrane compared to the initial one with Y = 0.14 to Y = 0.12 (from 14 to 12%), as determined by X-ray analysis using a “RigakuUltima IV” diffractometer. Using the device

"Kruss DSA 20E" revealed that as a result of exposure to a unipolar corona discharge is increased wettability as evidenced by a decrease in contact angle of a corona-treated membrane compared to the original $\alpha = 59°$ to $48°$ for PAN membranes, in particular.

Using a probe microscope "MultiMode V", images and histograms of the surface topography of the original (untreated) and corona-discharged PAN membranes are presented, shown in figure 1. Hereinafter, the surface height at a given point relative to the base plane is taken into account as the main topographic parameter, which in this case represents the plane in contact with the lowest point of the surface under consideration.

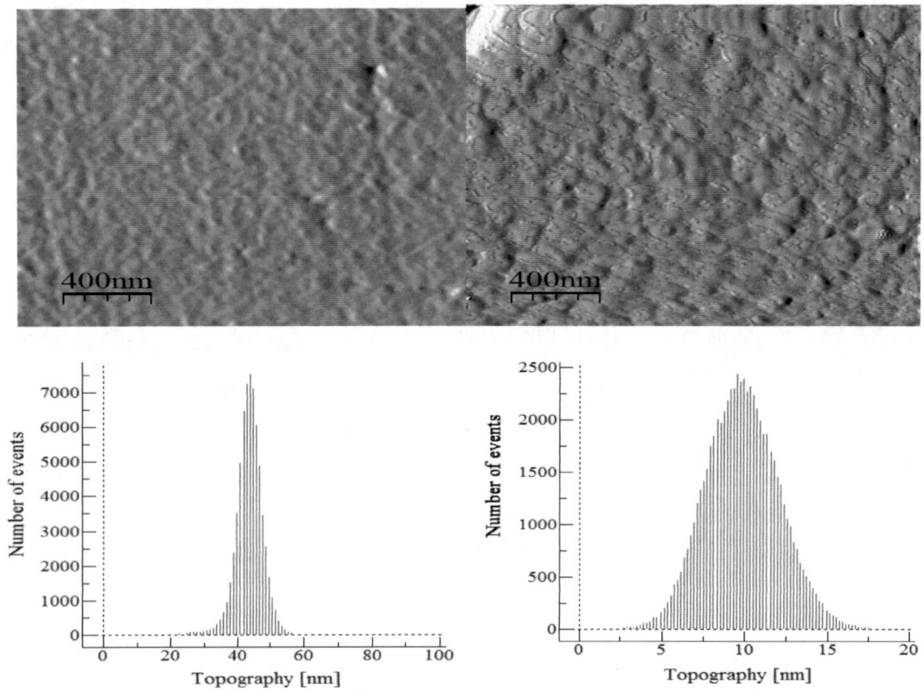

(a) (b)

Figure 1. Surface images with corresponding histograms of PAN membranes:a) source membrane;b) 10 kDa, $U_a = 10$ kV, $\tau = 0.5$ min

On the basis of the presented images of the surfaces of the original and corona-treated PAN membranes, a reduction in roughness was revealed. This circumstance can be explained from the following positions. It is known that under the action of a corona discharge, ozone is formed [12], which contributes to the oxidation of protruding parts of the membrane surface. Its action is based on the destruction of the crystallites of the polymer, while the reaction products are mainly dicarboxylic acids.

Thus, it has been shown that the use of a corona discharge, in order to intensify the separation of WOEs, is advisable not only for PAN and PES membranes, but also suggests the possibility of such a modification for other polymeric materials.

On the basis of a comparative analysis of the results of our own research [13–17] and foreign authors [18–20], it was found that as a result of the effect of a corona discharge on the surface of polymer membranes, their performance characteristics change - performance and efficiency. The latter depend on changes in the surface properties of the membranes, the most significant of which are their thickness, roughness, hydrophilicity, porosity and pore size and the presence of charge, which, in turn, depend on the corona discharge parameters — the exposure time and power (voltage) of the corona. In this regard, the main task of research on the modification of polymer membranes by corona discharge is to identify the dependencies of membrane properties on corona parameters, followed by the determination of targeted membrane treatment modes, which increase their performance in a specific field of science and technology.

References

[1] Ksenofontov B S, Kozodaev A S, Taranov R A, Ivanov M V, Petrova E V, Vinogradov M S andBalina A A 2013 *Ecology and Industry of Russia* **12** 4–8
[2] Ksenofontov B S and Ivanov M V 2013 *Water Science and Technology* **68** (4) 807–812
[3] Ksenofontov B S and Ivanov M V 2014 *Water Practice and Technology* **9**(3) 392–397
[4] Martínez P C F, Jiménez C, Sáez C and Rodrigo M A 2008 *Journal of Hazardous Materials* 151 44–51
[5] Zouboulis A I and Avranas A 2000 *Colloids and Surfaces A: Physicochemical and Engineering Aspects* 172 153–161
[6] Cambiella A, Ortea E, Ríos G, Benito J M, Pazos C and Coca J 2006 *Journal of Hazardous Materials* 131 195–199
[7] Adebajo M O, Frost R L, Kloprogge J T, Carmody O and Kokot S 2003 *Journal of Porous Materials* 10(3) 159–170
[8] Deschamps G, Caruel H, Borredon M, Albasi C, Riba J, Bonnin C and VignolesCh 2003 *Environ. Sci. Technol.*, **37**(21) 5034–5039
[9] Fazullin D D, Mavrin G V, Shaikhiev I G and Nizameev I R 2018 *Petroleum Chemistry* **58**(2) 145–151
[10] Fazullin D D, Mavrin G V and Shaikhiev I G 2017 *Petroleum Chemistry* **57**(2) 165–171
[11] Fazullin D D, Mavrin G V, Haritonova E A and Shaikhiev I G 2016 *Petroleum Chemistry* **56**(5) 454–458
[12] Sessler G1987 *Electrets* Berlin Springer 453
[13] Dryakhlov V O, Nikitina M Yu, Shaikhiev I G, Galikhanov M F, Shaikhiev T I and Bonev B S 2015 *Surface Engineering and Applied Electrochemistry* **51**(4) 406–411
[14] Dryakhlov V O, Shaikhiev T I, Shaikhiev I G, Zagidullina I A and Bonev B S 2015 *Bulgarian Chemical Communications* B **47** 109–115
[15] Shaikhiev I G, Galikhanov M F, Dryakhlov V O, Alekseeva M Yu and Shaikhiev T I 2016 *Chemical and Petroleum Engineering* **52** 352–356
[16] Dryakhlov V O, Shaihiev T I, Shaihiev I G, Galihanov M F and Bonev B S 2014 *Core Programming Area at the 2014 AIChE Annual Meeting "Separations Division 2014"* Atlanta, United States **2** 644–651
[17] Alekseeva M Y, Dryakhlov V O, Galikhanov M F, Shaikhiev I G and Nizameev I R 2018 *Petroleum Chemistry* **58**(2) 152–156
[18] Sadeghi I, Aroujalian A, Raisi A and Dabir B 2013 *Journal of Membrane Science* **430** 24–36
[19] Dorai R and Kushner M J 2003 *J. Phys. D: Appl. Phys.* **36** 666
[20] Cramariuc R, Tudorache A, Popa M E, Branduse E, Nisiparu L, Mitelut A, Turtoi M O and Fotescu L 2008 *J. Phys.: Conf. Ser.* **142** 012062

IOP Conf. Series: Materials Science and Engineering **492** (2019) 012032 doi:10.1088/1757-899X/492/1/012032

Obtaining of dependence of sustained angular rotor speed of centrifugal pump with hydrostatic bearings

S Korsakova[1] and A Protopopov[1,2]

[1]Bauman Moscow State Technical University, 5 Second Baumanskaya Street, Moscow, 105005, Russian Federation

[2]E-mail: proforg6@yandex.ru

Abstract. Manned and cargo vehicles in space industry often use low-capacity centrifugal pumps with direct-current motors with hydrostatic bearings. Such pumps have plenty of benefits, but their main drawback is instability of characteristics. It is caused by wide spread of rotor frequency which influences other parameters. This article describes mathematical model of rotor start-up developed and solved in order to determine sustained rotation frequency. Dependence between angular rotation speed of rotor shaft and all basic construction parameters of pump was determined by Runge-Kutta numerical method with varied pitch.

1. Introduction

Operational scheme of low-capacity centrifugal pump is shown on figure1.

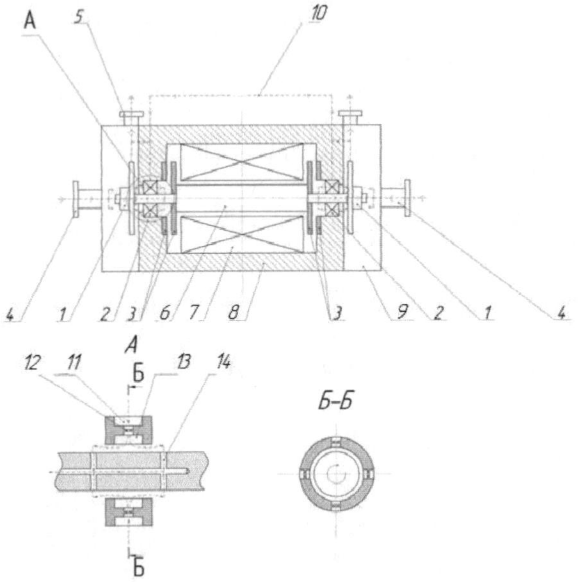

Figure 1. Structure of low-capacity centrifugal pump with opposite scheme and hydrostatic bearings: 1– Impeller; 2 – Hydrostatic bearing; 3 – Pair of footstep bearing; 4 – Inlet connection; 5 – Outlet connection; 6 – Motor rotor; 7 – Motor stator; 8 – Motor body; 9 – Body of one of pumps of electric pump unit; 10 – Directions of working liquid are shown by dash lines; 11– Downstream side; 12 – Throttle; 13 – Pocket; 14 – Hole system in the shaft, providing working liquid return to the impeller

IOP Conf. Series: Materials Science and Engineering **492** (2019) 012032 doi:10.1088/1757-899X/492/1/012032

Operating principle of the pump is as following: working liquid is directed to the input of the impeller, from its output it is directed to diverter. Liquid comes to throttles of hydrostatic bearings through channels in pump's body. It is important to note that problems concerning design of centrifugal pump rotors are well described in [1-6].

Research concerning methods of hydrodynamic modelling are of most interest [7-18].

However, research concerning rotor's dynamics is described scantily.

2. Mathematical model

Force diagram is shown on figure 2.

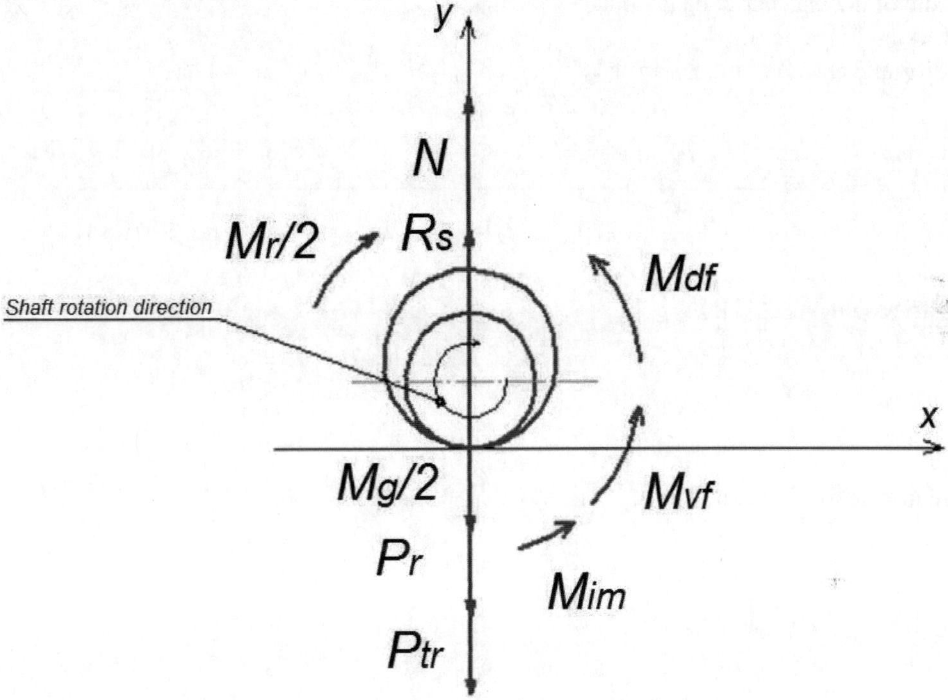

Figure 2. Forces and moments applied to the shaft: *forces* Mg – rotor weight; Pr – radial force emerging in impeller; N – normal force of support reaction; Rs – static reaction of hydrostatic bearing; *moments* Mim – impeller moment; Mdf – dry friction moment; Mvf – viscous friction moment; Mr – rotor moment.

They are known from pump shaft equation [1]

$$\begin{cases} M \cdot \dfrac{d^2 y}{dt^2} = 2 \cdot Rs(t) + 2 \cdot N(t) - 2 \cdot Pr(t) - M \cdot g - 2 \cdot Ptr(t) \\ J \cdot \dfrac{d\omega}{dt} = Mr(t) - 2 \cdot Mim(t) - 2 \cdot Mst(t) - 2 \cdot Mvf(t) \end{cases} \tag{1}$$

where

M — rotor mass;

J — rotor inertia about its axis;

ω — angular rotational speed of pump shaft;
t— time from the moment of pump start-up;
g — gravity acceleration.

Let's write down members of the system (1). Radial force Pr(t) applied to the impeller [1]:

$$Pr(t) = 0{,}1 \cdot \rho \cdot D_2^3 \cdot b_2 \cdot \eta \cdot \omega(t)^2 \tag{2}$$

where
H(t) — pump head;
D_2 — diameter of output from the impeller;
b_2—width of the impeller at the output;
ρ — working liquid density.

Hydrostatic reaction of the bearing [1]:

$$Rc = l \cdot d \cdot \left[\frac{p_н \cdot \left(\mu_c \cdot \frac{\pi \cdot d_c^2}{4} \right)^2}{\left(\mu_c \cdot \frac{\pi \cdot d_c^2}{4} \right)^2 + \left(\mu_m \cdot 2 \cdot f_2(y) \right)^2} - \frac{p_н \cdot \left(\mu_c \cdot \frac{\pi \cdot d_c^2}{4} \right)^2}{\left(\mu_c \cdot \frac{\pi \cdot d_c^2}{4} \right)^2 + \left(\mu_m \cdot 2 \cdot f_1(y) \right)^2} \right]. \tag{3}$$

Dry friction moment [1]:

$$M_{df} = \begin{cases} \frac{\left[2 \cdot Pr + M \cdot g - 2 \cdot Rs\right] \cdot \kappa_m \cdot d}{4} & for \;\; y = 0 \\ 0 & for \;\; y > 0 \end{cases} \tag{4}$$

Viscous friction moment [1]:

$$Mvf(y) = \left(\beta_1 + \beta_2 \cdot (y - \delta)^6 \right) \cdot \omega, \tag{5}$$

where

$$\beta_1 = 6{,}28 \cdot 10^{-5} N \cdot m$$

$$\beta_2 = 7{,}15 \cdot 10^{19} N \cdot m^{-5}$$

Impeller moment [1]:

$$M_{im}(t) = \frac{\omega(t) \cdot \pi \cdot \mu \cdot R_2^4}{a} + \rho \cdot Q \cdot R_2^2 \cdot \omega \tag{6}$$

Rotor moment [1]:

$$M_r(t) = K - K_1 \cdot \omega(t), \tag{7}$$

where
K and K_1 — coefficients of moment-mechanical characteristic of the motor.
Viscous friction moment [1]:

$$P_{vf} = 0{,}4 \cdot V \tag{8}$$

Let's write down equation system (1) as follows [1]:

$$\begin{cases} \frac{dV}{dt} = \frac{2 \cdot \rho \cdot \omega^2 \cdot R_2^2 \cdot \eta \cdot l \cdot d - 0{,}2 \cdot \rho \cdot D_2^3 \cdot b_2 \cdot \eta \cdot \omega^2 - M \cdot g}{M} \\ \frac{dy}{dt} = V \\ \frac{d\omega}{dt} = \frac{Mr(t) - 2 \cdot Mim(t) - 2 \cdot Mdf(t) - 2 \cdot Mvf(t)}{J} \\ V(0) = 0 \\ y(0) = 0 \\ \omega(0) = 0 \end{cases} \quad (9)$$

The system (9) is solved in Mathcad, using Runge – Kutta method of 4th order.
The following formula is obtained:

$$\omega_{sus} = A \cdot (-462{,}56 \cdot lnK1 + 1211.4) \cdot (119.05 \cdot K + 174.95) \cdot (-10.498 \cdot a + 569.09) \times \\ \times (-13.434 \cdot D1 + 587.15) \cdot (-2.7616 \cdot D2^2 - 1.1665 \cdot D2 + 568.78) \times \\ \times (-0.0335\rho^3 + 0.4509\rho^2 - 15.304\rho + 589.69),$$

where
A – scale coefficient describing influence of other construction parameters.

Thus, dependence between sustained angular rotor speed and construction parameters of the pump is obtained. It is important to note that construction parameters are given in non-dimensional form and brought to the following values:

$K_1 = 0.0015$ N·m·sec
$K = 0.9$ N·m
$a = 0.001$ m
$D_2 = 0.05$ m
$D_1 = D2/5$ m
$\rho = 1000$ kg/m^3

3. Conclusion

Obtained model allows to determine sustained angular rotor speed of centrifugal pump with hydrostatic bearings. Such model can be useful in different industry segments, in particular, in space systems.

References

[1] Borovin G.K. et al. 2016 Dinamika rotorov maloraskhodnykh tsentrobezhnykh nasosov s gidrostaticheskimi podshipnikami i privodom ot elektrodvigateley pstoyannogo toka *Preprinty IPM imeni M V Keldysha* **142** p 24
[2] Cherkasskiy V M 1984 *Nasosy, ventilyatory, kompressory* Moscow Energoatomizdat Publ 416 p
[3] Lomakin A A 1966 *Tsentrobezhnye i osevye nasosy* Moscow Mashinostroyeniye Publ 354 p
[4] Lomakin V O, Artemov A V and Petrov A I 2012 Opredeleniye vliyaniya osnovnykh geometricheskikh parametrov otvoda nasosa HM 10000-210 na ego kharakteristiki *Nauka i obrazovaniye: electronnoye nauchno-tekhicheskoye izdaniye*(8) p 5

[5] Lomakin V O, Petrov A I and Shcherbachyov P S 2012 Razrabotka bokovogo poluspiral'nogo podvoda s uvelichennym momentum skorosti na vkhode v rabochee koleso *Izvestiya vysshikh uchebnykh zavedeniy. Mashinostroyeniye* (S) pp 3–5

[6] Petrov A I, Martynov N D, Pokrovskiy P A, Pashchenko V I, Ustyuzhanin P Yu, Korolyov P V and Artemov A V 2010 Opyt razrabotki stenda dlya ispytaniy krupnykh tsentrobezhnykh nasosov *Nauka i obrazovaniye: electronnoye nauchno-tekhicheskoye izdaniye* (11) p 2

[7] Lomakin V O, Petrov A I and Kuleshova M S 2014 Issledovaniye dvukhfaznogo techeniya v ose tsentrobezhnom kolese metodami gidrodinamicheskogo modelirovaniya *Nauka i obrazovaniye: electronnoye nauchno-tekhicheskoye izdaniye* (9) pp 45–64

[8] Lomakin V O and PetrovA I 2012 Verifikatsiya resul'tatov rascheta v pakete gidrodinamicheskogo modelirovaniya STAR-CCM+ protochnoy chaste tsentrobezhnogo nasosa AX 50-32-200 *Izvestiya vysshikh uchebnykh zavedeniy. Mashinostroyeniye*(S) p 6

[9] Guskov A M, Lomakin V O, Banin E P and Kuleshova M S 2017 Minimization of Hemolysis and Improvement of the Hydrodynamic Efficiency of a Circulatory Support Pump by Optimizing the PumpFlowpath*Biomedical Engineering* **4** pp229–233

[10] Lomakin V O, Chaburko P S and Kuleshova M S 2017 Multi-criteria Optimization of the Flow of a Centrifugal Pump on Energy and Vibroacoustic Characteristics *Procedia Engineering* **176** pp 476–482

[11] Gouskov A M, Lomakin V O, Banin E P and Kuleshova M S 2016 Assessment of Hemolysis in a Ventricular Assist Axial Flow Blood Pump *Biomedical Engineering* 4 pp 12–15

[12] Lomakin V O, Kuleshova M S and Bozh'eva S M 2016 Numerical Modeling of Liquid Flow in a Pump Station*Power Technology and Engineering* **5** pp 324–327

[13] Lomakin V O 2015 *Proceedings of 2015 International Conference on Fluid Power and Mechatronics*

[14] Lomakin V O, Kuleshova M S and Kraeva E A 2015 Fluid Flow in the Throttle Channel in the Presence of Cavitation *Procedia Engineering* **106** pp 27–35

[15] Pugachev P V, Svoboda D G and Zharkovsky A A 2016 *Calculation and design of blade hydraulic machines. Calculation of viscous flow in blade hydraulic machines using the ANSYS CFX package* St. Petersburg Publishing house of Polytechnic University 120 p

[16] Petrov A I 2016 Technique of continuous obtaining of characteristics of a vane pump for variable temperature and viscosity of a working fluid during tests in a thermal pressure chamber *Engineering Bulletin Electronic Journal* (10) Available at: http://technomag.edu.ru/doc/850931.html

[17] Petrov A I and Isaev N Yu 2017 Gidrodinamicheskoye modelirovaniye raboty tsentrobezhnogo nasosa v zone otritsatel'nykh podach *Gidravlika* (3) pp 91–101

[18] Petrov A I and Isaev N Yu 2017 Issledovaniye raboty lopastnogo nasosa v zone otritsatel'nykh podach metodami gidrodinamicheskogo modelirovaniya *Nauchnoye obozreniye* (13) pp 75–80

IOP Conf. Series: Materials Science and Engineering **492** (2019) 012033 doi:10.1088/1757-899X/492/1/012033

Multistage model of flotation process for water purification

B Ksenofontov

Bauman Moscow State Technical University, 5 Second Baumanskaya Street, Moscow, 105005, Russian Federation

E-mail: ksenofontov@bmstu.ru

Abstract. The paperdeals with the features of multistage model's application of the flotation process that was devoted by the author. In his paper, the author described the applicability of ion flotation for wastewater treatment using the flotation machines. The author showed the approach associated with the use of a multi-stage process model is effective for determining the floating time. The author gives the theoretical and experimental data of the flotation process and their comparison. The paper shows that the proposed model reasonable describes the process of flotation water purification, in particular from metal ions. In this case, the kinetics of the formation of flotation complexes and the concentration of metal ions in the foam layer are also quite understandable.

1. Introduction

Modeling of various technological processes allows not only to more deeply study the phenomena underlying these processes, but also to increase their efficiency [1–9]. The development of modeling is also important for water purification technology, in particular using flotation [10–20]. Previously used models of the flotation process did not consider the particular gas bubble flotation complex as the main object, which did not allow an objective presentation of the elementary flotation act and the process as a whole.

2. Basic proposals and ideas

B.S. Ksenofontov developed a multistep flotation model, taking into account the main features of the process, including the formation of a flotation complex [18]. The main stages of the flotation process according to this model are showт in the figure. 1.

B.S. Ksenofontov determined the values of the flotation process's constants for the constant K1 [18].

The constant K_1 in a variety of ways flotation:

$$K_1 = \frac{1}{t_0} P_{CN}$$

where

P_{CN} – probability of flotation complex formation bubble – a particle over a period of time t_0.

IOP Conf. Series: Materials Science and Engineering **492** (2019) 012033 doi:10.1088/1757-899X/492/1/012033

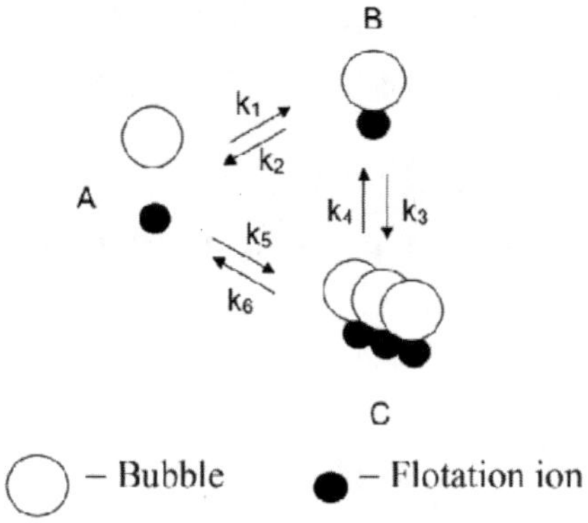

Figure 1. Diagram of multistage flotation model:
A, B, C – system condition, Ki – kinetic constant.

It is generally accepted that during the flotation process of clarifying fine suspension or wastewater treatment, the constant K_1 can be determined from the ratio:

$$K_1 = \frac{1.5qE}{k_0 \bar{D}}$$

where
q – speed of barbotage;
E – the capture efficiency of particles by a floating gas bubble during flotation, which characterizes the probability of a bubble contacting with a particle (a dimensionless quantity);
$\bar{D}$ – average bubble diameter in the flotation cell;
k_0 – bubblepolydispersity factor. The most of all the value of this constant is taken 10-3.

The probability of destruction of the formed flotation complexes is characterized by a constant, which can be approximately determined by the formula

$$K_2 = A\, C_{\phi}\, GaM^2\, C^{-1}_{пп}$$

where
A – nondimensional coefficient;
C_{ϕ} – particulate bubble flotation concentration;
Ga – velocity gradient in the aeration zone, determined by the ratio of the difference of the speeds to the difference of the distance between the points in question;
M – ratio of particle diameter to bubble diameter;
Cпп – bubble concentration in the under bubble column.

Displacements of flotation complexes particulate bubble from a liquid to a bubble column are characterized by a constant

$$K_3 = \frac{\upsilon_{под}}{h}$$

where
$\upsilon_{под}$ – lifting speed of the flotation complex;
h – distance from the zone of suspended water to the bubble column (depth of the flotobox).

The constant K_4, characterizing the loss of flotation complexes from the bubble column, provided it is immediately removed, is determined by the following approximate formula

$$K_4 = F\, Gп\, Cп\, d^3 cp$$

where
F – constant of proportionality;
Gп – rate of shear in under bubble column;
Cп – concentration of bubbles in the bubble column;
dcp – bubbles average diameter in bubble column.

The constant K_5, determining the precipitation of solid particles from the foam layer to the aeration zone:

$$K_5 = \frac{\upsilon_{oc}}{h}$$

where
υ_{oc} – sedimentation rate of solid particles falling out of the bubble column (as a rule, can be calculated using Stokes formula).

The probability of transition of particles of a solid phase from a liquid to a bubble column is characterized by a constant:

$$K_6 = \psi \frac{\partial}{\partial x}\left\{\frac{1}{2\sqrt{\pi\psi t}}\left[\exp\left(-\frac{(x-h)^2}{4\psi t}\right) - \exp\left(-\frac{(x+h)^2}{4\psi t}\right)\right]\right\}$$

where
t – time;
x – current distance from the boundary of the bubble column;
ψ – diffusion coefficient of solid particles in a liquid.

As an example, it seems appropriate to consider a multistage model of ion flotation, which can be shown as a sequence of the following system states (figure 2):

-condition A – collegial and collector ions and gas bubbles exist autonomously;

- condition B – the formation of a sublate as a result of the interaction of the collector and the colligend;
- condition C – flotation complex gas bubble collector;
- condition D – the formation of the flotation complex ion collegiate-collector-gas bubble;
- condition E – the formation of a bubble column containing ions of a colligend and a collector and gas bubbles;
- condition F – formation of foam containing colligend ions and a collector without gas bubbles (sublate concentrate).

The mathematical description of the flotation process that showed in figure 2 can be described by the following system of equations:

$$\begin{cases} \frac{dC_A}{dt} = -k_1C_A + k_2C_B - k_3C_A + k_4C_C - k_{15}C_A + k_{16}C_F; \\ \frac{dC_B}{dt} = k_1C_A - k_2C_B - k_5C_B + k_6C_D - k_{13}C_B + k_{14}C_F; \\ \frac{dC_C}{dt} = k_3C_A - k_4C_C - k_7C_C + k_8C_D; \\ \frac{dC_D}{dt} = k_5C_B - k_6C_D + k_7C_C - k_8C_D - k_9C_D + k_{10}C_E; \\ \frac{dC_E}{dt} = k_9C_D - k_{10}C_E - k_{11}C_E + k_{12}C_F; \\ \frac{dC_F}{dt} = k_{11}C_E - k_{12}C_F + k_{13}C_B - k_{14}C_F + k_{15}C_A - k_{16}C_F. \end{cases} \quad (1)$$

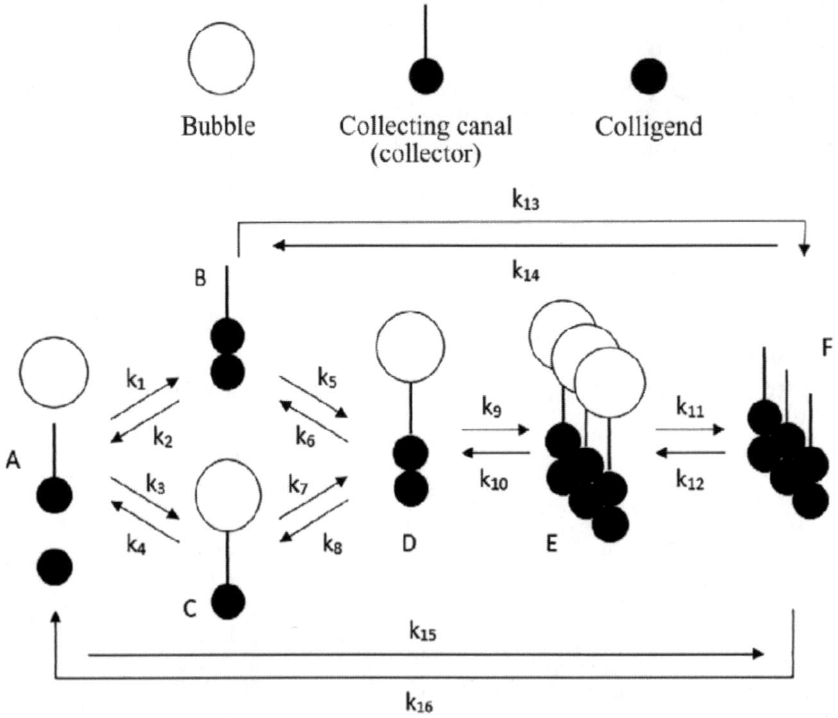

Figure 2. Diagram of multistage model of ion flotation.

The proposed system must satisfy at least two conditions, namely, at the initial moment of time; the concentration of the colligend at the first stage is equal to the initial concentration in the solution and at any moment of time the sum of the concentrations of the colligend at all stages is equal to its initial concentration.

3. Calculations and discussion of results

Usually the proposed system is solved with using the numerical methods. For practical cases, as shown by the author's calculations, it is possible to use simplified calculations according to the scheme shown in figure 3.

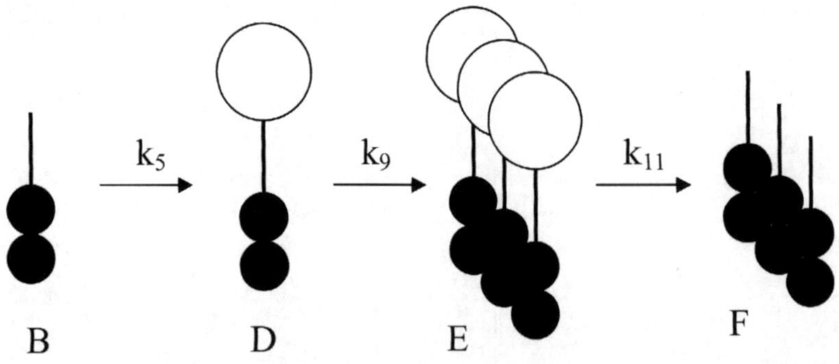

Figure 3. Simplified diagram of the ion flotation process in a flotation machine with a conditioning chamber.

Considering that the collector, which in some cases can be used, activated sludge [21-22], fully reacted with a colligend in the mixing reactor, and then the system of equations (1) will take the form:

$$\begin{cases} \dfrac{dC_B}{dt} = -k_5 C_B; \\ \dfrac{dC_D}{dt} = k_5 C_B - k_9 C_D; \\ \dfrac{dC_E}{dt} = k_9 C_D - k_{11} C_E; \\ \dfrac{dC_F}{dt} = k_{11} C_E. \end{cases} \tag{2}$$

Solving the system of equations (2), we take the following initial data on the example of the extraction of nickel from waste water:

$C_0 = 0.9$ mg/l – concentration of the extracted substance in the initial solution;

$k_5 = 0.005$ c^{-1};

$k_9 = 0.15$ c^{-1};

$k_{11} = 0.001$ c^{-1}.

to$t = 0$ $C_B = C_0$; $C_D = C_E = C_F = 0$.

The efficiency of extraction of individual metals from wastewater, obtained by calculation and determined experimentally, is showed in table 1. Comparison of calculated and experimental values shows a small discrepancy, not exceeding about 7%, which allows you to use the calculated data to assess the effectiveness of wastewater treatment, including from metals, using ion flotation.

Table 1. Extraction of individual metals from waste water using ion flotation.

No	Metal	Concentration of metal in waste water,Mg/l	Flotation time,min	Estimated cleaning efficiency,%	Experimentally determined cleaning efficiency, %
1	Chrome (generic)	2.2	15.5	97.7	91.4
2	Pb (plumb)	4.4	15.5	96.9	89.6
3	Ni (nickel)	3.5	15.5	98.4	92.7
4	W (wolfram)	2.9	15.5	97.8	93.3
5	Co (cobalt)	4.1	15.5	96.5	89.6

It should also be noted that this possibility allows to determine the concentration of the extracted substance at each of the considered stages of the process at any time point without conducting expensive experiments, which is especially important when designing a sewage treatment system. In addition, it is possible to find a time-limiting stage, affecting the flow of which can reduce the total time of the process of extracting contaminants, including metal ions. In this case, the intensification of the flotation process can be achieved by using reagents.

4. Conclusion

Analysis of the solutions obtained and calculation examples show that the proposed model rather accurately describes the process of flotation water purification, in particular from metal ions. At the same time, the kinetics of the formation of flotation complexes and the concentration of metal ions in the foam layer are also quite understandable.

References

[1] Ferri R B 2006 *Theoretical and empirical differentiations of phases in the modelling process* ZDM **38**(2) 86–95

[2] Galbraith P and Stillman G 2006 *A framework for identifying student blockages during transitions in the modelling process* ZDM **38**(2) 143–162

[3] Mostafa N, Syed H M, Igor S and Andrew G 2009 A study of melt flow analysis of an ABS-Iron composite in fused deposition modelling process *Tsinghua Science & Technology* **14** 29–37

[4] Dimitriyenko Yu I, Kashkarov A I 2002 A final and element method for calculation of effective characteristics of the spatial reinforced composites *Herald of the Bauman Moscow State Technical University. Series Natural Sciences* (2) 95–108

[5] Dimitriyenko Yu I, Gubareva E A and Sborschikov S V 2014 The asymptotic theory constructive ортотропных plates with two-periodic structure *Mathematical Modeling and Computational Methods* (1) 36–57

[6] Dimitriyenko Yu I, Gubareva E A and Yakovlev D O 2014 Asymptotic theory of viscoelasticity of multilayered thin composite plates *Science and education. Electron. journal* (1) 359–382 DOI: 10.7463/1014.0730105

[7] Dimitriyenko Yu I, Yurin Yu V and Gubareva E A 2014 Asymptotic theory of thermocreep of multilayered thin plates *Mathematical Modeling and Computational Methods* (4) 36–57

[8] Algazin O D and Kopayev A V 2017 Filtration of liquid in a non-uniform layer with the filtration coefficient changing under the square law *Engineering Journal: Science and Innovation* issue 6 DOI: 10.18698/2308-6033-2017-6-1624

[9] Wills B A and Finch J 2015 *Wills' mineral processing technology: an introduction to the practical aspects of ore treatment and mineral recovery* Butterworth-Heinemann

[10] Grammatika M and Zimmerman W B 2001 Microhydrodynamics of flotation processes in the sea surface layer *Dynamics of Atmospheres and Oceans* **34**(2-4) 327–348

[11] Van der Westhuizen A P and Deglon D A 2007 Evaluation of solids suspension in a pilot-scale mechanical flotation cell: The critical impeller speed *Minerals Engineering* **20**(3) 233–240

[12] Szpyrkowicz L 2005 Hydrodynamic effects on the performance of electro-coagulation/electro-flotation for the removal of dyes from textile wastewater *Industrial & Engineering Chemistry research* **44**(20) 7844–7853

[13] Yianatos J B 2007 Fluid flow and kinetic modelling in flotation related processes: columns and mechanically agitated cells—a review *Chemical Engineering Research and Design* **85**(12) 1591–1603

[14] Rodrigues R T and Rubio J 2007 DAF–dissolved air flotation: Potential applications in the mining and mineral processing industry *International Journal of Mineral Processing* **82**(1) 1–13

[15] Miettinen T, Ralston J and Fornasiero D 2010 The limits of fine particle flotation *Minerals Engineering* **23**(5) 420–437

[16] Grattoni C, Moosai R and Dawe R A 2003 Photographic observations showing spreading and non-spreading of oil on gas bubbles of relevance to gas flotation for oily wastewater cleanup *Colloids and Surfaces A: Physicochemical and Engineering Aspects* **214**(1-3) 151–155

[17] Santander M, Rodrigues R T and Rubio J 2011 Modified jet flotation in oil (petroleum) emulsion/water separations *Colloids and Surfaces A: Physicochemical and Engineering Aspects* **375**(1-3) 237–244

IOP Conf. Series: Materials Science and Engineering **492** (2019) 012033 doi:10.1088/1757-899X/492/1/012033

[18] Ksenofontov B S 2015 *Ochistka stochnyh vod: kinetika flotacii i flotokombajny* Moscow ID "FORUM": INFRA-M Publs 256 p
[19] Ksenofontov B S and Titov K V 2017 *Ochistka stochnyh vod: komp'yuternye tekhnologii v reshenii zadach flotacii* Moscow ID "FORUM": INFRA-M Publs 240 p
[20] Ksenofontov B S and Titov K V 2017 Reshenie kraevyh zadach ochistki vody flotaciej *Ehkologiya promyshlennogo proizvodstva* (4) 31–39
[21] Ksenofontov B S and Goncharenko E E 2016 Intensification of Purification of Surface Sewage by use a Bioflocculant*Herald of the Bauman Moscow State Technical University, Series Natural Sciences***3**(66)
[22] Ksenofontov B S 2018 Engineering problems of dehydration of rainfall of sewage *Ecology and Industry of Russia***22**(9) 4–7

IOP Conf. Series: Materials Science and Engineering **492** (2019) 012034 doi:10.1088/1757-899X/492/1/012034

Compromise Resource-Efficiency Curve for a Centrifugal Pump

A Protopopov[1,2] and C Jakovich[1]

[1] Bauman Moscow State Technical University, 5 Second Baumanskaya Street, Moscow, 105005, Russian Federation

[2]E-mail: proforg6@yandex.ru

Abstract. This article discusses centrifugal pumps with half-open impellers. One of the two most important parameters of a centrifugal pump is its resource and efficiency. However, often the race for efficiency leads to such constructive solutions that help reduce the pump life and vice versa. The solution of such problems is attributed to multicriteria optimization, one of the methods of which is to build a trade-off curve. In this paper, this curve is constructed using the application of such methods as the LP-Tau search method. The variable parameters in this article are axial clearance and rotor speed.

1. Introduction

The main characteristics of centrifugal pumps, as well as many other mechanisms, include the resource and efficiency. Often, these are two competing parameters: increasing the first, the second has to sacrifice. There is a need to find a compromise solution [1-7], when at the selected value of one parameter, the second is maximum.

Such a task is completely non-trivial, given that many parameters have an impact on these characteristics. To concretize the solution to this problem we will consider the example of a low flow centrifugal pump with a switching valve, which is used in the thermal control system on the ISS.

The resource and efficiency of such a pump, among other parameters, are significantly influenced by the axial clearance [8-12] between the impeller and the housing of the outlet and the frequency of rotation of the pump rotor. Determining the final dependence causes some difficulties, since both criteria are weighty and it is not possible to determine weight coefficients for them. Therefore, it was decided to use LP-Tau search, since this method does not imply an exact definition of the objective function.

It should be noted that the currently existing methods of finding a compromise between resource and efficiency [13-18] are oriented towards small rotational speeds of the pump shaft, which creates large errors in the calculation of the high-speed low-flow pump from them.

2. Getting a compromise curve

The LP-Tau search generates points in a quasi-random manner in the specified interval for two parameters. For the above pump, the axial clearance “a” can vary from 0.5 to 1.5 mm, the rotor speed “n” from 3000 to 8000 rpm (figure 1). Then the field of the generated operating points will be a rectangle.

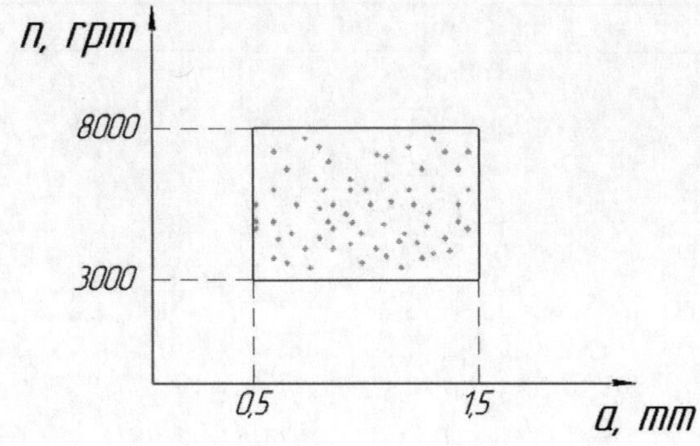

Figure 1. Point generation area.

The calculated points obtained using LP-Tau search are tabulated (table 1).

Table 1. LP-Tau search data points.

Number of point	*a*, mm	*n*, rpm
1	1	5500
2	0.75	6750
3	1.25	4250
4	0.625	6125
5	1.125	3625
6	0.875	4875
7	1.375	7375
8	0.5625	7687.5
9	1.0625	5187.5
10	0.8125	3937.5
11	1.3125	6437.5
12	0.6875	4562.5
13	1.1875	7062.5
14	0.9375	5812.5
15	1.4375	3312.5
16	0.53125	5656.25
17	1.03125	3156.25
18	0.78125	4406.25
19	1.28125	6906.25
20	0.65625	3781.25
21	1.15625	6281.25
22	0.90625	7531.25
23	1.40625	5031.25
24	0.59375	5343.75
25	1.09375	7843.75
26	0.84375	6593.75
27	1.34375	4093.75
28	0.71875	7218.75
29	1.21875	4718.75

Number of point	*a*, mm	*n*, rpm
30	0.96875	3468.75
31	1.46875	5968.75
32	0.515625	6984.375
33	1.015625	4484.375
34	0.765625	3234.375
35	1.265625	5734.375
36	0.640625	5109.375
37	1.140625	7609.375
38	0.890625	6359.375
39	1.390625	3859.375
40	0.578125	4171.875
41	1.078125	6671.875
42	0.828125	7921.875
43	1.328125	5421.875
44	0.703125	6046.875
45	1.203125	3546.875
46	0.953125	4796.875
47	1.453125	7296.875
48	0.56875	4328.125
49	1.046875	6828.125
50	0.796875	5578.125
51	1.296875	3078.125
52	0.671875	7453.125
53	1.171875	4953.125
54	0.921875	3703.125
55	1.421875	6203.125
56	0.609375	6515.625
57	1.109375	4015.625
58	0.859375	5265.625
59	1.359375	7765.625
60	0.734375	3390.625
61	1.234375	5890.625
62	0.984375	7140.625
63	1.484375	4640.625
64	0.5078125	6320.125

Each point corresponds to certain values of the resource and efficiency, which can be obtained from the relations in [3]. After calculations, it is possible to construct the distribution of all calculated points in the coordinates $T(\eta)$ (figure 2).

For this distribution, we build a compromise curve for the following reasons: for the same value of one parameter, the point with the highest value of the second is considered to be the “winning”. From this it is clear that the compromise curve must pass through these “winning” points (figure 3).

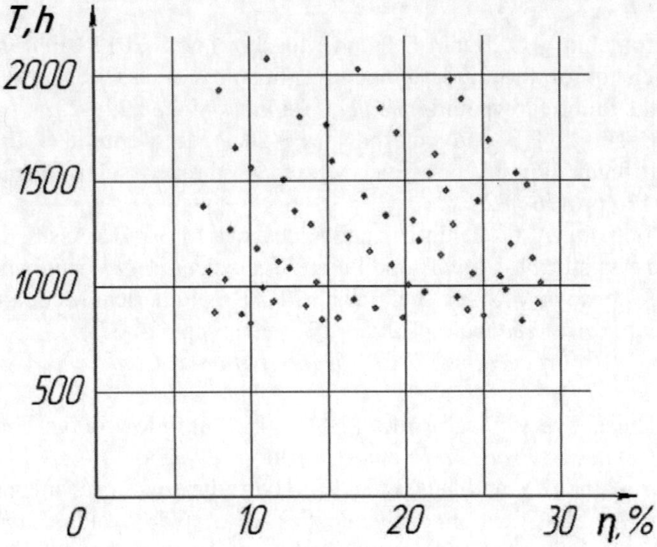

Figure 2. Distribution of calculated points.

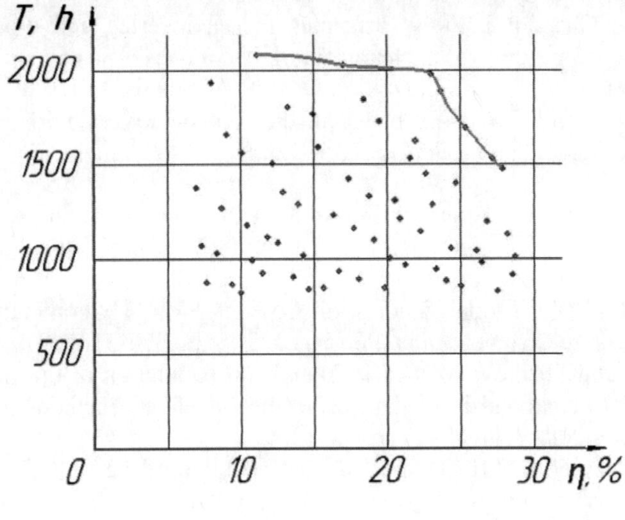

Figure 3. Compromise curve.

3. Conclusion

The use of LP-Tau sequence for finding a compromise between resource and efficiency when varying the rotor speed and axial clearance by an effective method. This is clearly seen from the obtained compromise curve - when considering 64 points of the LP-tau sequence, 7 of them entered the Pareto set and formed a compromise curve. Thus, the above method can be recommended when calculating high-speed centrifugal pumps.

References

[1] Guskov A M, Lomakin V O, Banin E P and Kuleshova M S 2017 Minimization of Hemolysis and Improvement of the Hydrodynamic Efficiency of a Circulatory Support Pump by Optimizing the Pump Flowpath*Biomedical Engineering***4** pp 229-233

[2] Lomakin V O, Chaburko P S and Kuleshova M S 2017 Multi-criteria Optimization of the Flow of a Centrifugal Pump on Energy and Vibroacoustic Characteristics *Procedia Engineering***176** pp 476-482

[3] Gouskov A M, Lomakin V O, Banin E P and Kuleshova M S 2016 Assessment of Hemolysis in a Ventricular Assist Axial Flow Blood Pump *Biomedical Engineering***4** pp 12-15

[4] Lomakin V O, Kuleshova M S and Bozh'eva S M 2016 Numerical Modeling of Liquid Flow in a Pump Station *Power Technology and Engineering***5** pp 324-327

[5] Lomakin V O 2015 *Proceedings of 2015 International Conference on Fluid Power and Mechatronics*

[6] Lomakin V O, Kuleshova M S and Kraeva E A 2015 Fluid Flow in the Throttle Channel in the Presence of Cavitation *Procedia Engineering***106** pp 27-35

[7] Min'kov L L, Krokhina A V and Dueckc J 2011 Hydrodynamic mechanisms of the influence of injection on the classification characteristics of a hydrocyclone*Journal of Engineering Physics and Thermophysics***84** (4) pp 807-819

[8] Isaev S A, Kornev N V, Leontiev A I and Hassel E 2010 Influence of the Reynolds number and the spherical dimple depth on turbulent heat transfer and hydraulic loss in a narrow channel *International Journal of Heat and Mass Transfer***53** (1-3) pp 178-197

[9] Ionaitis R R and Chain P L 2009 Principles for engineering active-passive hydrodynamic control-and-safety systems for nuclear reactors *Atomic Energy***106** (3) pp 175-184

[10] Pelevin F V, Avraamov N I, Ir'yanov N Y, Orlin S A, Lozovetskii V V and Ponomarev A V 2018 Intensification of Heat Exchange in the Regenerative Cooling System of a Liquid-Propellant Rocket Engine *Journal of Engineering Physics and Thermophysics***91** (3) pp 601-610

[11] Aref'ev K Yu, Voronetskii A V and Il'Chenko M A 2013 Dynamic characteristics of a resonant gas-dynamic system for ignition of a fuel mixture *Combustion, Explosion and Shock Waves***49** (6) pp 657-661

[12] Krapivtsev V G 2017 Model Studies of Coolant Flow Hydrodynamics in VVER-1000 In-Reactor Pressure Channel *Atomic Energy* 122 (5) pp 304-310

[13] Marchevskii I K and Puzikova V V 2016 Numerical simulation of the flow around two fixed circular airfoils positioned in tandem using the LS-STAG method *Journal of Machinery Manufacture and Reliability***45** (2) pp 130-136

[14] Petrov A I and Isaev N Y 2017 Hydrodynamic modeling of the centrifugal pump operation in the zone of negative innings *Hydraulics* (3) pp 91-101

[15] Petrov A I and Isaev N Y 2017 Study of the operation of a vane pump in the zone of negative innings by methods of hydrodynamic modeling *Scientific Review* (13) pp 75-80

[16] Lomakin V O, Artemov A V and Petrov A I 2012 Determination of the influence of the main geometrical parameters of the pump outlet HM 10000-210 on its characteristics *Science and education: electronic scientific and technical publication* (8) p 5

[17] Lomakin V O, Petrov A I and Scherbachev P S 2012 Development of a side helix supply with an increased moment of speed at the impeller inlet. *Proceedings of Higher Educational Institutions. Machine Building* (S) pp 3-5

[18] Petrov A I, Martynov N D, Pokrovsky P A, Pashchenko V I, Ustyuzhanin P Y, Korolev P V and Artemov A V 2010 Experience in developing a test bench for large centrifugal pumps *Science and education: electronic scientific and technical publication* (11) p. 2

[19] Lomakin V O, Petrov A I and Kuleshova M S 2014 Investigation of a two-phase flow in an axial centrifugal wheel by methods of hydrodynamic modeling *Science and education: electronic scientific and technical publication* (9) pp 45-64

[20] Lomakin V O and Petrov A I 2012 Verification of the calculation results in the STAR-CCM+ hydrodynamic simulation package of the flow-through part of the centrifugal pump AX 50-32-200 *Proceedings of Higher Educational Institutions. Machine Building* (C) p 6

IOP Conf. Series: Materials Science and Engineering **492** (2019) 012035 doi:10.1088/1757-899X/492/1/012035

Obtaining the head characteristic of a Low Flow Centrifugal Pump by numerical methods

A Shablovskiy[1,2] and E Kutovoy[1]

[1]Bauman Moscow State Technical University, 5 Second Baumanskaya Street, Moscow, 105005, Russian Federation

[2]E-mail: shabl@bmstu.ru

Abstract. During the study of various problems related with centrifugal pumps, it would be extremely convenient to have a head characteristic (head-capacity curve) approximated by any mathematical function. In so doing, in the field of small pump output flow experiments show that its characteristic is linear. In order to verify that this characteristic is truly linear, a mathematical experiment was carried out using the methods of hydrodynamic modeling and its approximation was carried out using the least square method and the maximum deviation from it was calculated.

1. Introduction

To solve a number of scientific and technical problems [1-5], it is interesting to obtain a head-capacity curve of a low flow centrifugal pump as an analytical function. It is supposed [6-12] that in the field of small pump output flow the head-capacity curve is linear as friction losses are proportional to the square of the head and in the field of small pump output flow the square of the head can be neglected.

It should be emphasized that the problems related to the study of head-capacity curve of centrifugal pumps are highlighted in the literature [13-18], but the problem of the approximation of the head-capacity curves by the analytical function has not been adequately debated.

2. Mathematical model

For a solving this problem was developed a 3D model of a low flow centrifugal pump and a number of points of the head-capacity curve of this pump were calculated by numerical methods using the STAR-CCM + software package. The results of this calculation are presented in table 1:

Table 1. Estimated points of head-capacity curve of experimental low flow centrifugal pump.

No	H, m	Q, l/min
1	5.1	0
2	5.02	0.25
3	4.94	0.5
4	4.8	1.0
5	4.75	1.5
6	4.66	1.75
7	4.59	2.0

Therefore, we have the following estimated head-capacity curve characteristic of a low flow centrifugal pump. Figure 1 illustrates that the characteristics of a low flow centrifugal pump are approximately linear, which accords well enough with practice.

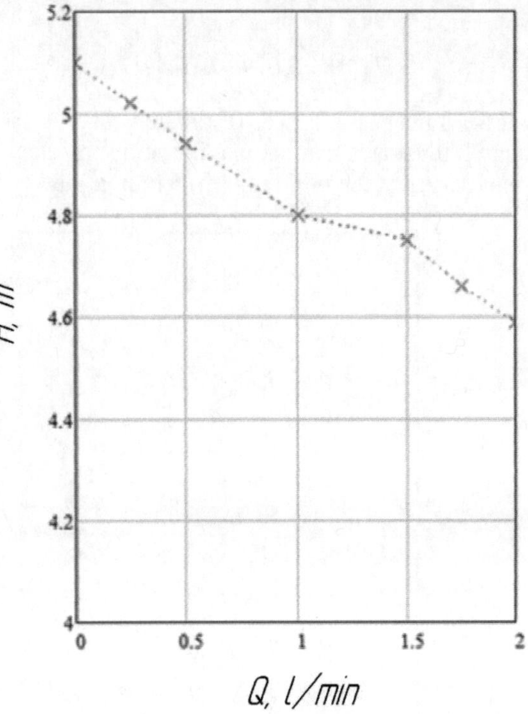

Figure 1. Estimated head-capacity curve characteristic of a low flow centrifugal pump.

However, it is interesting to check how linear the characteristic of a low flow centrifugal pump is. For this purpose, the method of least squares is used. We will carry out the approximation of the estimated head-capacity curve of the following function:

$$y = a + b \cdot x \tag{1}$$

As is known [18], in the case of a function form (1), the method of least squares for finding the coefficients a and b gives the following formulas:

$$\begin{cases} b = \dfrac{\sum_{t=1}^{n} x_t \cdot y_t - \sum_{t=1}^{n} \dfrac{x_t}{n} \cdot \sum_{t=1}^{n} \dfrac{y_t}{n}}{\sum_{t=1}^{n} \dfrac{x_t^2}{n} - \left(\sum_{t=1}^{n} \dfrac{x_t}{n} \right)^2} & (2) \\ a = \sum_{t=1}^{n} \dfrac{y_t}{n} - b \cdot \sum_{t=1}^{n} \dfrac{x_t}{n} & (3) \end{cases}$$

where
x_t – is the t-th (current) value of pump output flow in table 1;
y_t – is the t-th (current) value of head in table 1;
n – is the number of points, in our case $n = 7$.

Substituting the values from table 1 into the system of equations (2)-(3), for our particular case we have

$$\begin{cases} b = -0,241(m \cdot min)/l \\ a = 5,079\,m \end{cases}$$

Substitute the coefficients a and b into formula (1) and obtain the following approximation function:

$$H(Q) = 5,079 - 0,241 \cdot Q \tag{4}$$

Therefore, obtain following head-capacity curves (estimated curve is shown by dots, approximation curve is shown by line). Figure 2 illustrates that the maximum discrepancy between the approximation curve and the target points is observed at the point (4.8 m; 1 l / min) and is 0.037 m.

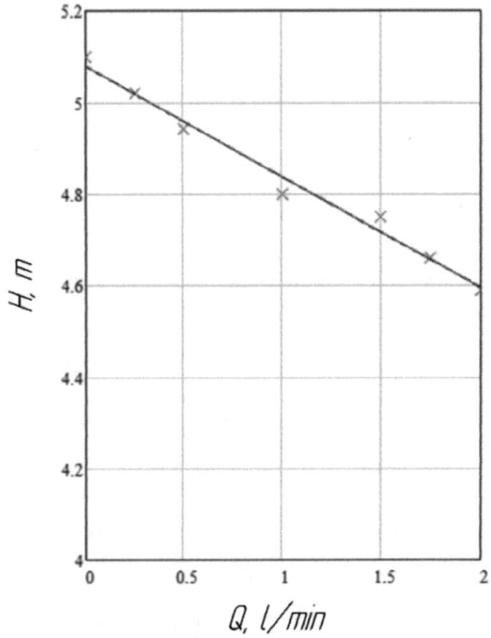

Figure 2. Estimated head-capacity curve of a low flow centrifugal pump.

Define the maximum admissible error of this discrepancy:

$$\sigma = \frac{4,837 - 4,8}{4,837} = 0,76\%$$

3. Conclusions

Therefore, it can be concluded that in the field of small pump output flow, the head-capacity curve of the low flow centrifugal pump is linear with an accuracy of less than 1%. This result can be used in scientific and practical activities for the mathematical modeling of low flow centrifugal pumps.

IOP Conf. Series: Materials Science and Engineering **492** (2019) 012035 doi:10.1088/1757-899X/492/1/012035

References

[1] Lomakin V O, Petrov A I and Kuleshova M S 2014 Investigation of a two-phase flow in an axial centrifugal wheel by methods of hydrodynamic modeling *Science and education: electronic scientific edition of Bauman MSTU* (9) pp 45–64 (in Russian)

[2] Guskov A M, Lomakin V O, Banin E P and Kuleshova M S 2017 Minimization of Hemolysis and Improvement of the Hydrodynamic Efficiency of a Circulatory Support Pump by Optimizing the Pump Flow path *Biomedical Engineering* 4 pp 229–233

[3] Lomakin V O, Chaburko P S, Kuleshova M S 2017 Multi-criteria Optimization of the Flow of a Centrifugal Pump on Energy and Vibroacoustic Characteristics *Procedia Engineering***176** pp 476–482

[4] Gouskov A M, Lomakin V O, Banin E P and Kuleshova M S 2016 Assessment of Hemolysis in a Ventricular Assist Axial Flow Blood Pump *Biomedical Engineering***4** pp 12–15

[5] Lomakin V O, Kuleshova M S and Bozh'eva S M 2016 Numerical Modeling of Liquid Flow in a Pump Station *Power Technology and Engineering***5** pp 324–327

[6] Lomakin V O 2015 *Proceedings of 2015 International Conference on Fluid Power and Mechatronics*

[7] Min'kov L L, Krokhina A V and Dueckc J 2011 Hydrodynamic mechanisms of the influence of injection on the classification characteristics of a hydrocyclone*Journal of Engineering Physics and Thermophysics***84** (4) pp 807–819

[8] Isaev S A, Kornev N V, Leontiev A I and Hassel E 2010 Influence of the Reynolds number and the spherical dimple depth on turbulent heat transfer and hydraulic loss in a narrow channel *International Journal of Heat and Mass Transfer***53** (1-3) pp 178–197

[9] Marchevskii I K and Puzikova V V 2016 Numerical simulation of the flow around two fixed circular airfoils positioned in tandem using the LS-STAG method *Journal of Machinery Manufacture and Reliability***45**

[10] Ionaitis R R and Chain P L 2009 Principles for engineering active-passive hydrodynamic control-and-safety systems for nuclear reactors *Atomic Energy***106** (3) pp 175–184

[11] Pelevin F V, Avraamov N I, Ir'yanov N Y, Orlin S A, Lozovetskii V V and Ponomarev A V 2018 Intensification of Heat Exchange in the Regenerative Cooling System of a Liquid-Propellant Rocket Engine *Journal of Engineering Physics and Thermophysics***91** (3) pp 601–610

[12] Aref'ev K Yu, Voronetskii A V, Il'Chenko M A 2013 Dynamic characteristics of a resonant gas-dynamic system for ignition of a fuel mixture *Combustion, Explosion and Shock Waves* **49** (6) pp 657–661

[13] Krapivtsev V G 2017 Model Studies of Coolant Flow Hydrodynamics in VVER-1000 In-Reactor Pressure Channel *Atomic Energy***122** (5) pp 304–310 (2), pp. 130-136.

[14] Lomakin V O, Artemov A V and Petrov A I 2012 Determination of the influence of the main geometrical parameters of the pump outlet NM 10000-210 on its characteristics *Science and education: electronic scientific edition of Bauman MSTU* no 8 p 5 (in Russian)

[15] Lomakin V O, Petrov A I and Shcherbachev P S 2012 Development of a casing with an increased velocity moment at the impeller inlet. *Proceedings of Higher Educational Institutions. Machine Building* (S) pp 3–5 (in Russian).

[16] Petrov A I, Martynov N D, Pokrovsky P A, Pashchenko V I, Ustyuzhanin P Yu, Korolev P V, Artyemov A V 2010 Experience in developing a testing workbench for large centrifugal pumps *Science and education: electronic scientific edition of Bauman MST*U, (11) 2 p (in Russian)

[17] Lomakin V O, Kuleshova M S and Kraeva E A 2015 Fluid Flow in the Throttle Channel in the Presence of Cavitation *Procedia Engineering***106** pp 27–35 (in Russian)

[18] Linnik Yu V 1962 *The method of least squares and the basics of mathematical-statistical theory of processing observations* 2nd ed Moscow (in Russian)

IOP Conf. Series: Materials Science and Engineering **492** (2019) 012036 doi:10.1088/1757-899X/492/1/012036

Test bench flow straightener design investigation and optimization with computational fluid dynamics methods

A Petrov[1,2], N Isaev[1] and M Kuleshova[1]

[1]Bauman Moscow State Technical University, 5 Second Baumanskaya Street, Moscow, 105005, Russian Federation

[2]E-mail: alex_i_petrov@mail.ru

Abstract: The article contains the application of computational fluid dynamics methods to optimization of flow straightener, which is a part of test bench for centrifugal pumps manufactured with additive technologies. Problems faced during testing of large-scale prototypes are shown, e.g. unsteady cavitation processes occurring in case there is no straightener, the ways to solve such problems are suggested. The design of the optimized flow straightener is described, and its effectiveness is proved by calculations carried in STAR-CCM+ software.

1. Introduction

For the last 10 years the E-10 department of BMSTU is actively working on optimization of centrifugal pumps wet parts using the methods of hydrodynamic modeling [1], as well as experimental verification of the results obtained based on testing of large-scale prototypes, manufactured with additive technologies. A 3D model and a photograph of such a test bench are shown in figure 1.

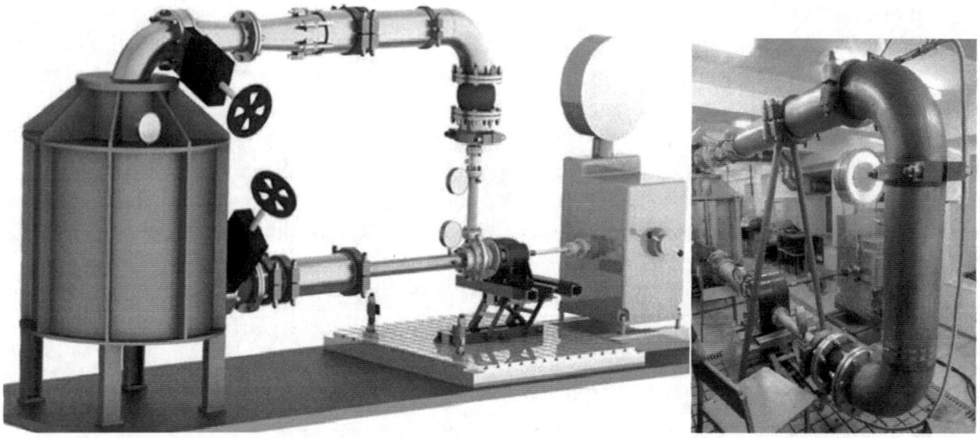

Figure 1. The test bench used to obtain characteristics of pump prototypes.

During testing several problems were faced due to relatively small size of the tank installed. In particular, when testing the fuel pump prototype with inlet flow of 400 m^3/h, unsteady effects (pressure low-frequency pulsations) as a result of control valve cavitation were observed in the tank

IOP Conf. Series: Materials Science and Engineering **492** (2019) 012036 doi:10.1088/1757-899X/492/1/012036

and suction line. These phenomena led to the impossibility of obtaining a stable experimental characteristic of the prototype, so it was important to study the flow in the test bench tank and optimize its design to eliminate the influence of cavitation phenomena on the pump characteristic.

In order to solve this problem methods of computational fluid dynamics in application to multiphase flows were used, as a result an optimized test bench and tank designs were obtained, and the tests were fully completed.

2. Methods

As it is known from years of experience [2–6], for correct testing of centrifugal pumps, it is necessary to achieve stabilization of the fluid flow at the pump inlet. This can be achieved in case the tank installed has a large size (it is recommended to use the tank with volume providing not less than 2 minutes of mass flow of the largest pump tested), several tanks of smaller volume installed consequently can also be used or flow straightener can be introduced to the tank. There are 3 main requirements for the flow straightener performance:

1. Full damping of the velocity head in the tank at the outlet of the discharge pipe;
2. Removal of air bubbles and cavitation pockets from the fluid, entrained in the flow from the discharged pipeline;
3. Creating minimal resistance of the flow in the tank.

It is quite difficult to develop a straightener capable of solving these problems in a tank of minimal volume purely empirically, therefore, there is a need to investigate and optimize the design of such a straightener with hydrodynamic modeling methods [7].

The difficulty in formulating the problem of hydrodynamic modeling lies in the fact that to correctly calculate the flow in the tank and the straightener, it is necessary to take into account both the two-phase nature of the working fluid (taking into account cavitation pockets and air cavities formed at the outlet of the control valve in the discharge pipeline) and the tank free surface of the fluid through which the exchange of matter between the fluid and the air pad in the tank happens.

As initial conditions for modeling, the initial gas phase concentration in the fluid at the outlet of the control valve (during the tests it can vary, but for convenience of comparative analysis of the calculation results a constant value of 5% volumetric gas concentration was taken), flow through the tank (speed at the outlet and the total pressure at the inlet) were set [8].

Criteria for assessing the quality of the straightener performance are based on the percentage of the gas phase at the outlet of the tank, i.e. at the inlet to the pump suction piping, and its ratio to the percentage of the gas phase at the tank inlet (estimated by the introduced value of the gas concentration reduction ratio calculated separately for the average and peak gas concentration values in the suction pipe), the uniform distribution of velocities in the suction pipe (described by the standard deviation of speed) and hydraulic losses from the entrance to the tank to the exit from the tank (described by the coefficient of the total hydraulic losses in the straightener).

The method of numerical simulation is based on solving discrete analogs of the basic hydrodynamic equations. [9–11]. In case of incompressible fluid (ρ = const):

Mass conservation equation (continuity equation)

$$\frac{\partial \bar{u}_j}{\partial x_j} = 0,$$

where

$\bar{u}_j$ – averaged velocity value in projection on axis j (j = 1, 2, 3).

Momentum equation (Reynolds averaged):

$$\rho\left[\frac{\partial U_i}{\partial t} + U_i\frac{\partial U_i}{\partial x_j}\right] = -\frac{\partial P}{\partial x_i} + \frac{\partial}{\partial x_i}\left[\bar{T}_{ij}^{(v)} - \rho\langle u_i u_j\rangle\right],$$

where
U, P – averaged velocity and pressure;
$\bar{T}_{ij}^{(v)} = 2\mu\bar{S}_{ij}$ – viscous stress tensor for incompressible fluid,
$\bar{S}_{ij} = \frac{1}{2}\left[\frac{\partial \tilde{u}_i}{\partial x_j} + \frac{\partial \tilde{u}_j}{\partial x_i}\right]$ – strain velocity tensor;
$\rho\langle u_i u_j\rangle$ – Reynolds stresses.

The introduction of Reynolds averaged Navier—Stokes equations makes the system of equations nonclosed, since additional unknown Reynolds stresses appear. To solve this system, a semiempirical k–ω SST model of turbulence was used, which introduces the necessary additional equations: the transfer equations for the kinetic energy of turbulence and the relative dissipation rate of this energy:

$$\frac{\partial k}{\partial t} + U_j \frac{\partial k}{\partial x_j} = P_k - \beta k\omega + \frac{\partial}{\partial x_j}\left[\left(\upsilon + \sigma_k \upsilon_T\right)\cdot\frac{\partial k}{\partial x_j}\right],$$

$$\frac{\partial \omega}{\partial t} + U_j \frac{\partial \omega}{\partial x_j} = \alpha \cdot S^2 - \beta \cdot \omega^2 + \frac{\partial}{\partial x_j}\left[\left(\upsilon + \sigma_\omega \upsilon_T\right)\cdot\frac{\partial \omega}{\partial x_j}\right] + 2\cdot(1 - F_l)\cdot\sigma_{\omega 2}\cdot\frac{1}{\omega}\cdot\frac{\partial k}{\partial x_j}\cdot\frac{\partial \omega}{\partial x_j}.$$

To simulate gas bubbles entering the tank, a two-phase model was used. An approach known as VOF (Volume of Fluid) [12] was adopted. In the VOF method, a multiphase medium is considered as a single fluid, which properties vary according to the volume fraction of each of the phases present:

$$\alpha_i = \frac{V_i}{V},$$

where
V_i – each phase volume fraction;
V – cell volume.

The equation for the mass concentration of each of the phases:

$$\frac{\partial(\alpha_i \rho_i)}{\partial t} + \nabla\cdot(\alpha_i \rho_i V) = 0.$$

The minimum bubble size and the minimum gas concentration in any calculation cell are strictly limited to very small, but non-zero, values. The time step was chosen based on the characteristic flow rates and cell size. The selected time step is 1e-3s. The number of internal iterations at each time step is 10. To obtain discrete analogues of the original continuous equations STAR-CCM + software uses the control volume method described in [13]

The wet part of the tank was discretized into a set of finite cells (figure 2), for each of which discrete analogs of continuous equations were composed. The combination of all discrete analogs forms a closed system of algebraic equations.

The number of cells is 580 000. Cell base size equals 12 mm.

Reduction ratio of the mean gas concentration:

$$\bar{g} = \frac{\sigma_{out}}{\sigma_{in}} = \frac{\frac{\sum_{i=0}^{n}\sigma_i}{n}}{\sigma_{in}},$$

where

σ_{in} – gas concentration at the tank inlet (a constant value of 5% volumetric gas concentration was taken);

σ_{out} – gas concentration at the tank outlet;
σ_i – gas concentration at the i-time step;
n – number of time steps.

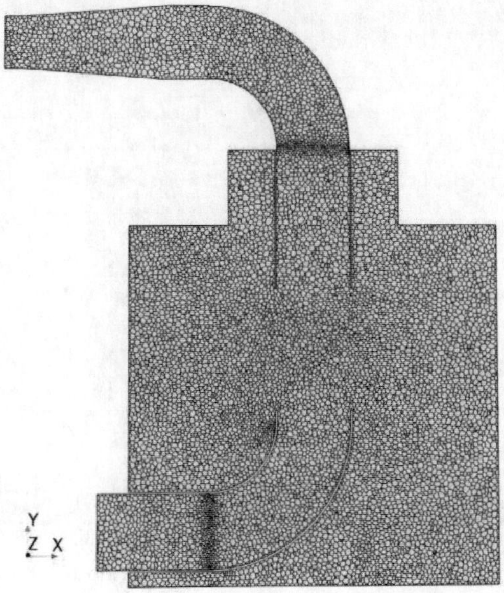

Figure 2. The mesh of the tank without the straightener.

Reduction ratio of the mean gas concentration:

$$g_{max} = \frac{\sigma_{max}}{\sigma_{in}}$$

where
σ_{max} – maximum gas concentration at the tank outlet.
Total hydraulic losses in the tank:

$$\sum h_t = \frac{P_{out}}{\rho \cdot g} - \frac{P_{in}}{\rho \cdot g},$$

where
P_{in}, P_{out} – pressure value at inlet and outlet of the tank respectively.

3. Results

Consequently, three configurations of tank and the straightener geometries were calculated – the tank without the straightener (initial design with worse test results), the tank with type 1 straightener and the tank with the optimized type 2 straightener design.

Figure 3 shows velocity distribution in the initial tank design, figure 4 – volume fraction of phase distributions in this tank.

Figure 5 describes the change of gas concentration in the suction line representing the unsteady behavior of 2-phase flow (gas concentration at the inlet of the tank is constant).

IOP Conf. Series: Materials Science and Engineering **492** (2019) 012036 doi:10.1088/1757-899X/492/1/012036

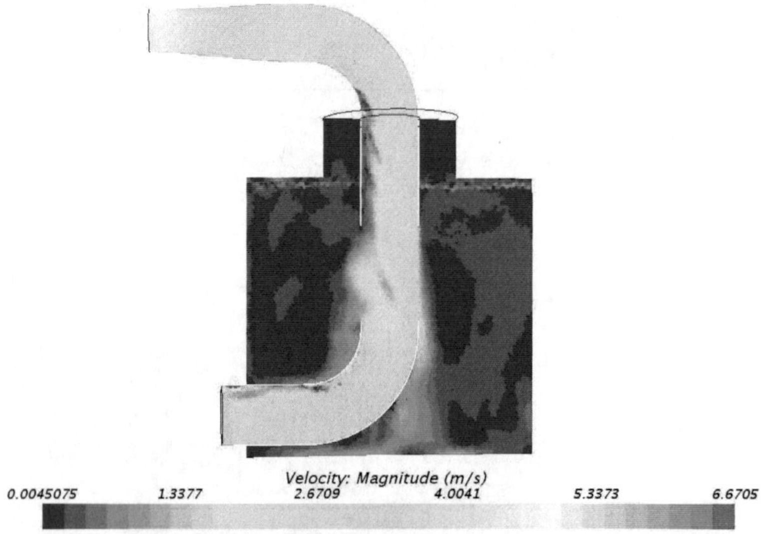

Figure 3. Velocity distribution in the tank without the straightener.

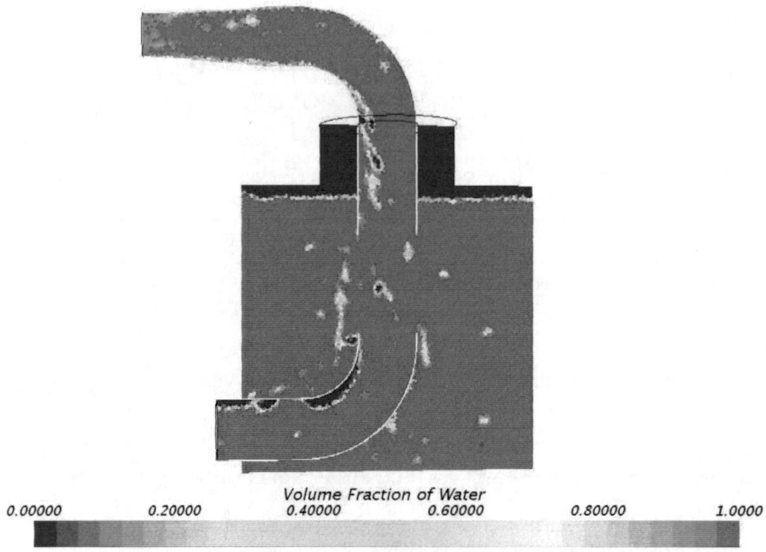

Figure 4. Volume fraction of phases in the tank.

IOP Conf. Series: Materials Science and Engineering **492** (2019) 012036 doi:10.1088/1757-899X/492/1/012036

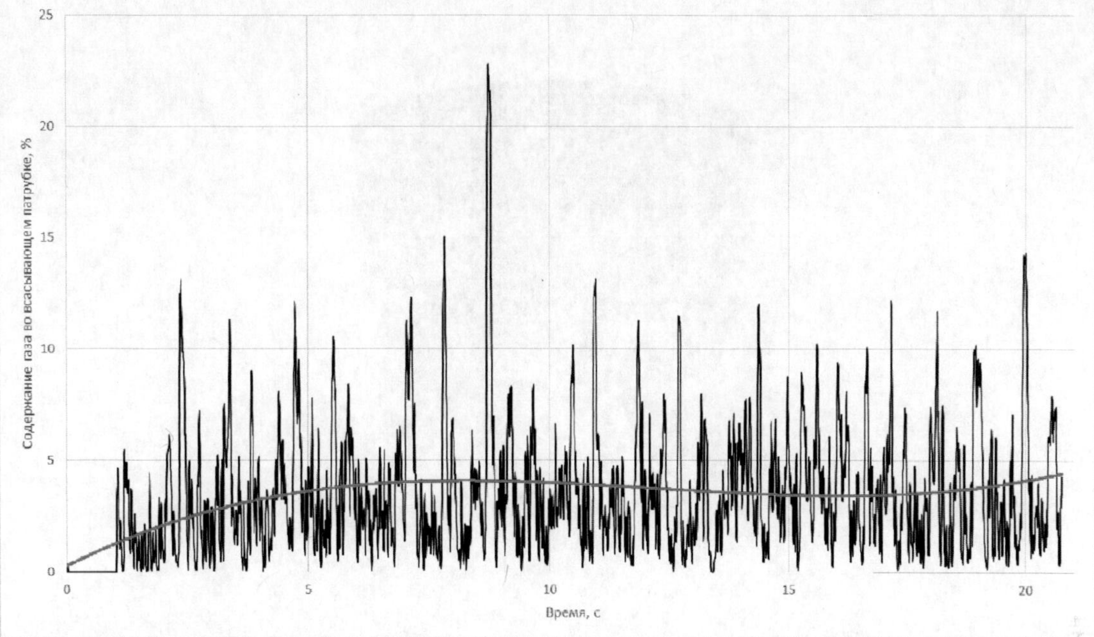

Figure 5. The change in gas concentration at the outlet of the tank depending on time, red line – averaged values.

According to the results shown above (figure 4), there is almost no gas separation obtained in the initial tank design (specific gas concentration at the outlet decreases on 25% maximum, table 1). At the same time as a result of flow separation on the turn there are pulsating cavity pockets at the inlet of suction line, which lead to unsteady pump operation during the tests. As practice of such tests in a number of enterprises has shown, such a picture is not an exception and is often found in the test benches with tanks of insufficient volume and without a flow straightener.

Table 1. Results of calculations carried.

No.	Straightener type	Reduction ratio of the mean gas concentration/ Reduction ratio of the peak gas concentration	Total losses in the straightener, m
1	No straightener	0.74/2.4	1.09
2	Straightener, type 1	0.09/0.31	3.27
3	Straightener, type 2	0.14/0.42	2.44

One of the possible ways to resolve such unsteady processes is tank pressurization that eliminates cavitation behind control valve. But the test bench design does not always allow to include such a pressurization, moreover in some cases it may lead to additional fluid aeration and gaseous cavitation [14–17].

Based on initial tank design the flow straightener geometry was developed, its 3D model is shown on figure 6.

However, simulation of the flow in such a straightener has shown that the flow, even in case of small velocities, leads to intensive fluctuations of the free surface (figure 7, 8). It does not affect gas concentration, apart from cases when there is an additional hydrostatic head in the tank (additional fluid aeration), but negatively affects the value of energy losses in the tank [18].

The values of mean gas concentration in the tank with a straightener has decreased 8.2 times, peak values – 7.9 times (table 1). Nevertheless, type 1 straightener is not appropriate in terms of values of losses in the tank and fluctuations of the free surface there [19–20].

Figure 6. The flow straightener.

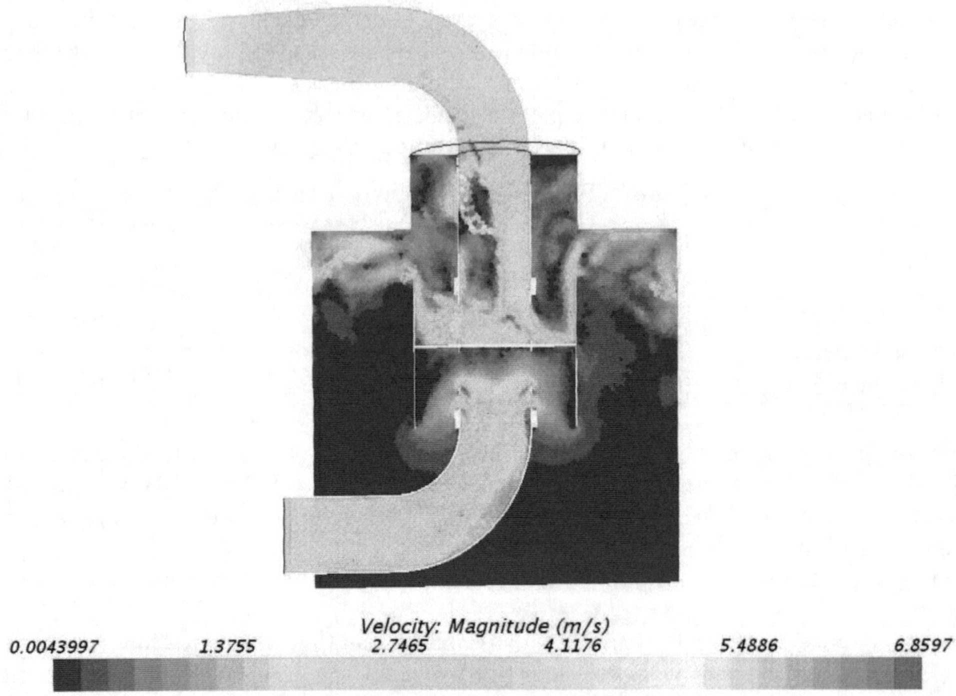

Figure 7. Velocity distribution in the tank with the straightener, type 1.

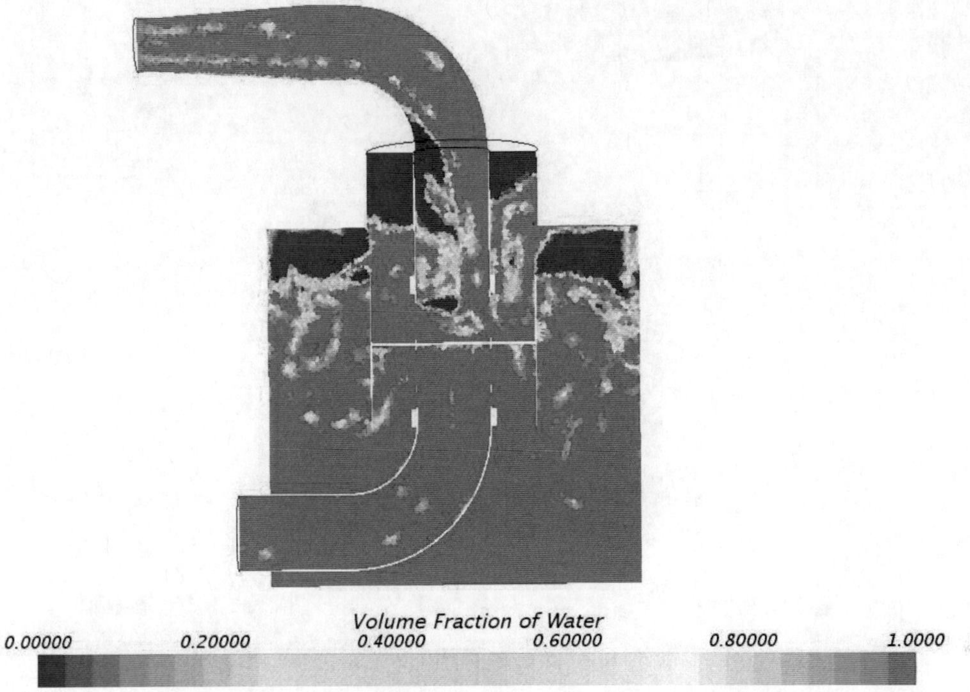

Figure 8. Volume fraction of phases in the tank with the straightener, type 1.

Type 2 of straightener does not include upper straightener pipe ring in contrast to type 1 design. It led to decreased value of hydraulic losses and free surface fluctuations, but relative gas concentration at the suction line was increased insignificantly (from 0.09 to 0.14). Calculation results are show on figures 8 and 9 and in table 1.

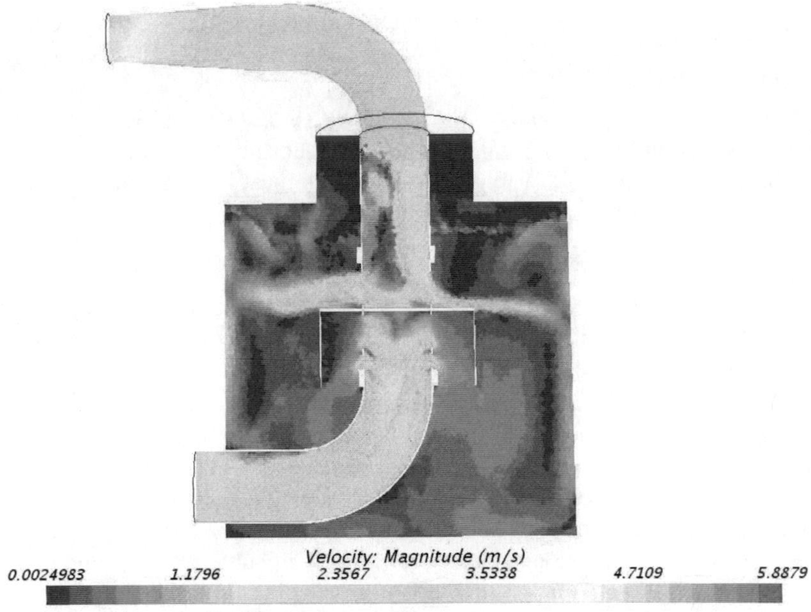

Figure 9. Velocity distribution in the tank with the straightener, type 2.

IOP Conf. Series: Materials Science and Engineering **492** (2019) 012036 doi:10.1088/1757-899X/492/1/012036

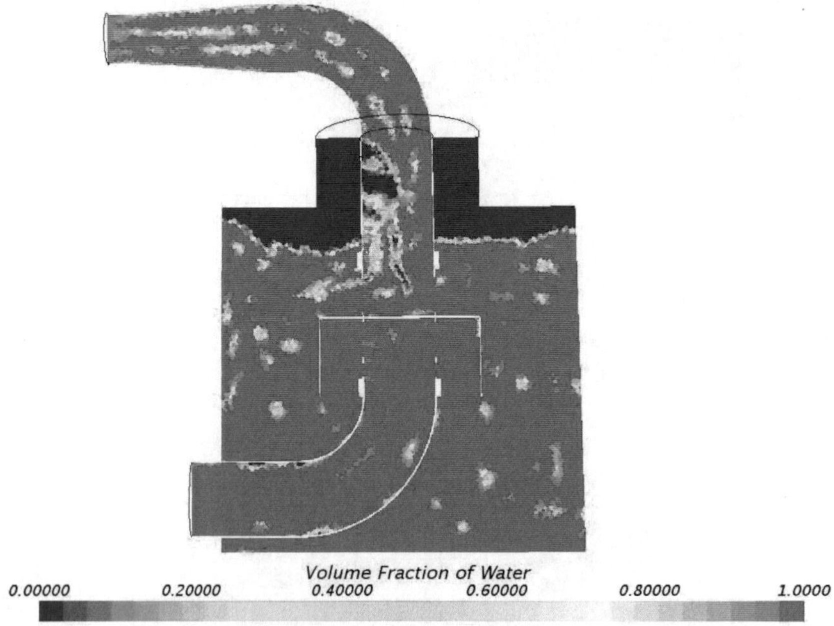

Figure 10. Volume fraction of phases in the tank with the straightener, type 2.

The results of all simulations are presented in table 1.

4. Conclusions and discussion

1. Hydrodynamic simulations performed for the flow in the tank with/without the straightener allowed to identify the source of unsteady effects during pump prototypes tests;
2. Optimized design of the tank-straightener was developed, allowing to eliminate unsteady processes in the test bench without increasing tank volume;
3. The effectiveness of proposed designs was shown during pump prototypes tests on the modified test bench.

References

[1] Lomakin V O, Chaburko P S and Kuleshova M S 2017 Multi-criteria Optimization of the Flow of a Centrifugal Pump on Energy and Vibroacoustic Characteristics *Procedia Engineering***176** pp 476-482

[2] Guskov A M, Lomakin V O, Banin E Pand Kuleshova M S 2017 Minimization of Hemolysis and Improvement of the Hydrodynamic Efficiency of a Circulatory Support Pump by Optimizing the Pump Flowpath *Biomedical Engineering* **4** pp 229-233

[3] Gouskov A M, Lomakin V O, Banin E P and Kuleshova M S 2016 Assessment of Hemolysis in a Ventricular Assist Axial Flow Blood Pump *Biomedical Engineering* **4** pp 12-15

[4] Lomakin V O, Kuleshova M S and Bozh'eva S M 2016 Numerical Modeling of Liquid Flow in a Pump Station *Power Technology and Engineering***5** pp 324-327

[5] Lomakin V O 2015 *Proceedings of 2015 International Conference on Fluid Power and Mechatronics*

IOP Conf. Series: Materials Science and Engineering **492** (2019) 012036 doi:10.1088/1757-899X/492/1/012036

[6] Lomakin V O, Kuleshova M S and Kraeva E A 2015 Fluid Flow in the Throttle Channel in the Presence of Cavitation *Procedia Engineering* **106** pp 27-35
[7] Shargatov V A, Gorkunov S V and Il'ichev A T 2019 Dynamics of front-like water evaporation phase transition interfaces *Communications in Nonlinear Science and Numerical Simulation* **67** pp 223-236
[8] Kraposhin M V, Banholzer M, Pfitzner M and Marchevsky I K 2018 A hybrid pressure-based solver for nonideal single-phase fluid flows at all speeds *International Journal for Numerical Methods in Fluids* **88** (2) pp 79-99
[9] Pelevin F V, Avraamov N I, Ir'yanov N Y, Orlin S A, Lozovetskii V V and Ponomarev A V 2018 Intensification of Heat Exchange in the Regenerative Cooling System of a Liquid-Propellant Rocket Engine *Journal of Engineering Physics and Thermophysics* **91** (3) pp 601-610
[10] Bondarenko V L, Valyakina A V, Borisenko A V, Trotsenko A V and Valyakin V N 2018 Vapor-Liquid Equilibrium of the Ethylene–Butane Mixture *Chemical and Petroleum Engineering* **53** (11-12) pp 778-787
[11] Arefyev K Y, Prokhorov A N and Saveliev A S 2018 Study of the breakup of liquid droplets in the vortex wake behind pylon at high airspeeds *Thermophysics and Aeromechanics* **25** (1) pp 55-66
[12] Yakush S E, Iskhakov A S, Melikhov V I and Melikhov O I 2018 Pressure Waves due to Rapid Evaporation of Water Droplet in Liquid Lead Coolant *Science and Technology of Nuclear Installations* Article no 3087051
[13] Vladimirov I Y, Korchagin N N and Savin A S 2018 Effects of the Influence of a Wave on an Underwater Obstacle by a Stratified Flow of Finite Depth *Rep. Earth Sciences* **478** (1) pp 132-136
[14] Krapivtsev V.G. 2017 Model Studies of Coolant Flow Hydrodynamics in VVER-1000 In-Reactor Pressure Channel *Atomic Energy* **122** (5) pp 304-310
[15] Marchevskii I K and Puzikova V V 2017 Numerical simulation of the flow around two circular airfoils positioned across the stream using the LS-STAG method *Journal of Machinery Manufacture and Reliability* **46** (2) pp 114-119
[16] Arefyev K Y, Voronetsky A V, Suchkov S A and Ilchenko M A 2017 Computational and experimental study of the two-phase mixing in gas-dynamic ignition system *Thermophysics and Aeromechanics* **24** (2) pp 225-237
[17] Leont'ev A I, Kuzma-Kichta Y A and Popov I A 2017 Heat and mass transfer and hydrodynamics in swirling flows (review) *Thermal Engineering* **64** (2) pp 111-126
[18] Devisilov V A and Sharai E Y 2016 Numerical study of the flow structure in a hydrodynamic filter *Theoretical Foundations of Chemical Engineering* **50** (2) pp 209-216
[19] Isaev S A, Schelchkov A V, Leontiev A I, Baranov P A and Gulcova M E 2016 Numerical simulation of the turbulent air flow in the narrow channel with a heated wall and a spherical dimple placed on it for vortex heat transfer enhancement depending on the dimple depth *International Journal of Heat and Mass Transfer* **94** pp 426-448
[20] Leontiev A I, Milman O O and Fedorov V A 1992 Thermohydraulic vapor generation processes in vertical tubes of low-pressure natural convection systems *Experimental Thermal and Fluid Science* **5** (1) pp 1-4

IOP Conf. Series: Materials Science and Engineering **492** (2019) 012037 doi:10.1088/1757-899X/492/1/012037

Numerical prediction of the gas content effect on the cavitation characteristics of the pump using the simplified Rayleigh–Plesset equation

V Lomakin[1,2] and O Bibik[1]

[1]Bauman Moscow State Technical University, 5 Second Baumanskaya Street, Moscow, 105005, Russian Federation

[2]E-mail: lomakin@bmstu.ru

Abstract. The paper presents the ongoing research on cavitation modelling so far which includes validation studies of the gas content effect on the cavitation characteristics of the pump. Cavitation flow characteristics are modeled by performing Computational Fluid Dynamics (CFD) simulations using Schnerr–Sauer cavitation model implementing empirical coefficients. As a result, it was found that the inclusion of the third phase into the cavitation model significantly affects the cavitation characteristics of the pump and the NPSH.

1. Introduction

Cavitation is a hydrodynamic phenomenon consisting of intensive formation and subsequent collapse of gas bubbles in a moving liquid (Figure 1). Cavitation affects the characteristics of the pump. Developed cavitation makes the head, pressure and efficiency of the pump decreasing, and then leads to a complete breakdown of its operation.

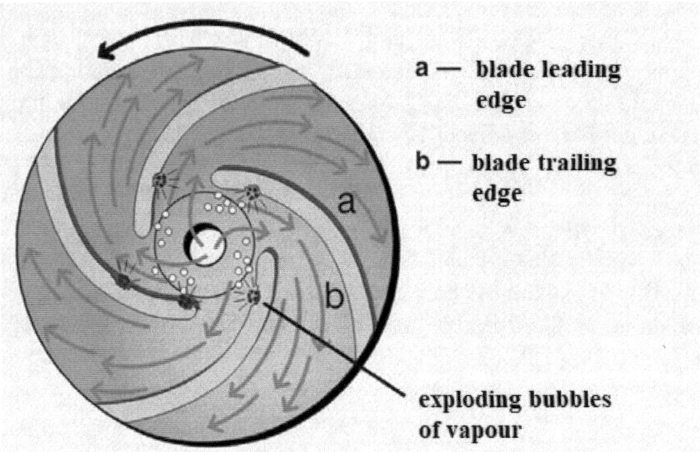

Figure 1.The cavitation in the pump.

IOP Conf. Series: Materials Science and Engineering **492** (2019) 012037 doi:10.1088/1757-899X/492/1/012037

Nowadays the problem of improving the cavitation qualities of the pump equipment is very urgent because cavitation and cavitation erosion damage the most critical pumping equipment: energy pumps, special purpose pumps, etc. Numerous researches are devoted to study the process of cavitation [1].

For example, in article [2], the cavitation model was tested experimentally for a pump with a radial impeller. The volume of fluid (VOF) model was chosen as a cavitation model for the calculation of multiphase flow. To study the cavitation characteristics at the inlet boundary, the pressure was decreased until the failure occurred. After the experiment, the results were compared (Figure 2).

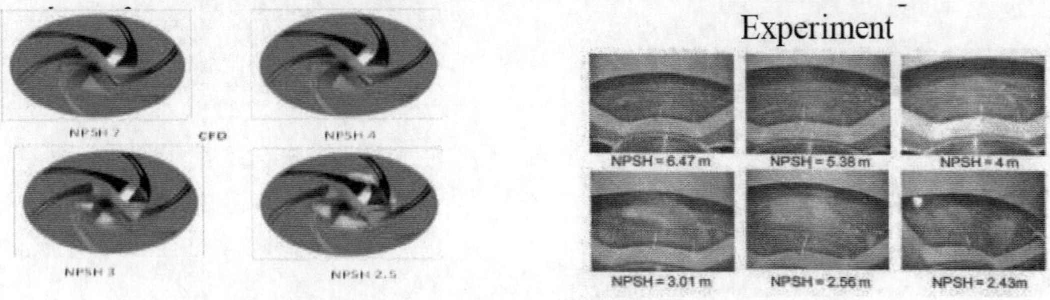

Figure 2.Comparison of numerical simulation and test results.

In study [3] the model was compared with experimental data for different flow rates. According to the graphs in the Figure 3, the difference between experimental data and numerical simulation results is considerable.

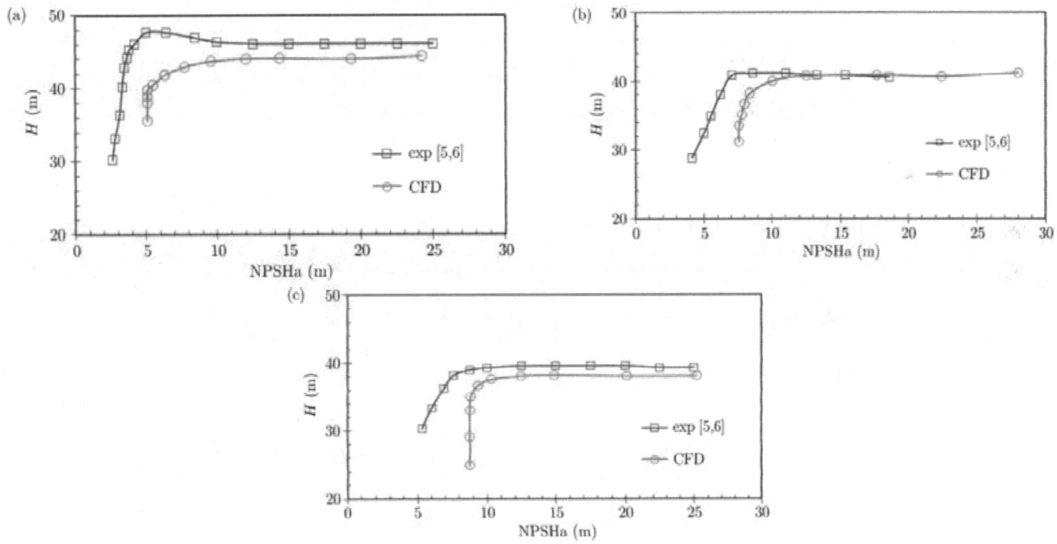

Figure 3. Comparison of numerical simulation and test results for various flow rates:
a) $Q = 168\ m^3 / h$; b) $Q = 210\ m^3 / h$; c) $Q = 226.8\ m^3 / h$

The experimental studying of cavitation can be carried out in various ways. One of the most common used is ultrasonic. For example, this method can be used to study the resistance of different materials to cavitation [4, 5]. Whereas the cavitation tests of the pump are carried out on special stands with the subsequent plotting of cavitation characteristics. [6]

With the invention of computational fluid dynamics methods it became possible to calculate the partial cavitation characteristics of pumps by numerical methods, but this approach is collocated with many difficulties. The main problem is the complex physical nature of the phenomena, therefore, it is

IOP Conf. Series: Materials Science and Engineering **492** (2019) 012037 doi:10.1088/1757-899X/492/1/012037

necessary to take into account a large number of factors in the mathematical model. During the simulation of cavitation in [7], a simplified Rayleigh–Plesset equation was used, which led to applying of the additional empirical coefficients (Figure 4).

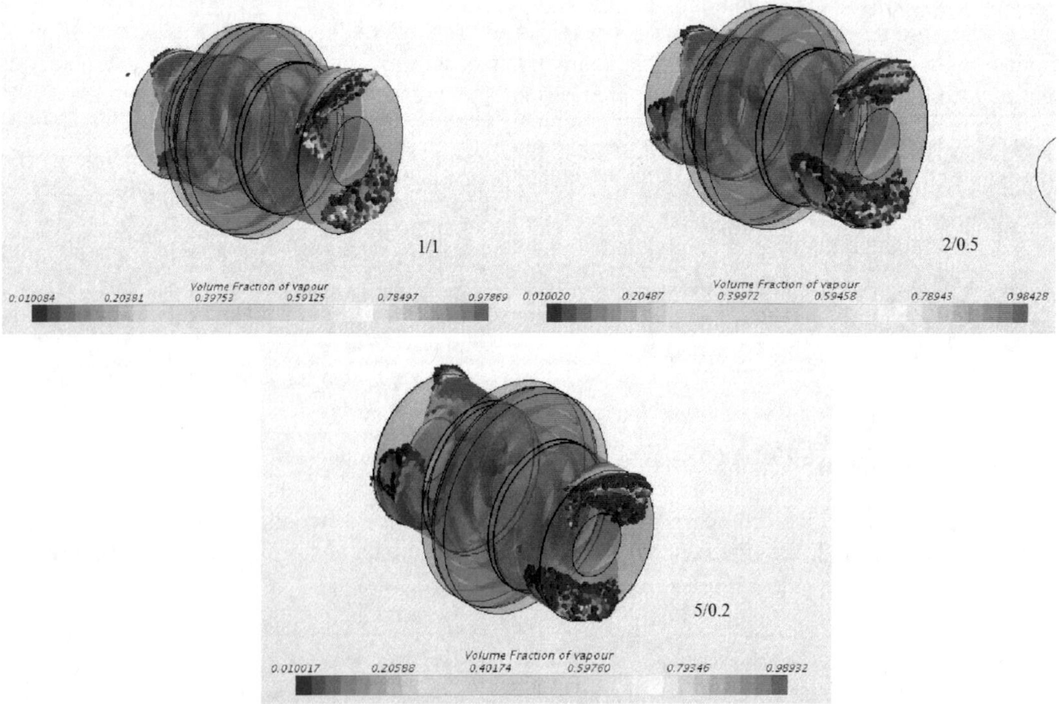

Figure 4.Volume fraction of vapour at the NPSH 2 m.

As a result, the cavitation model including certain empirical coefficients showed a very good agreement with experimental data (Figure 5).

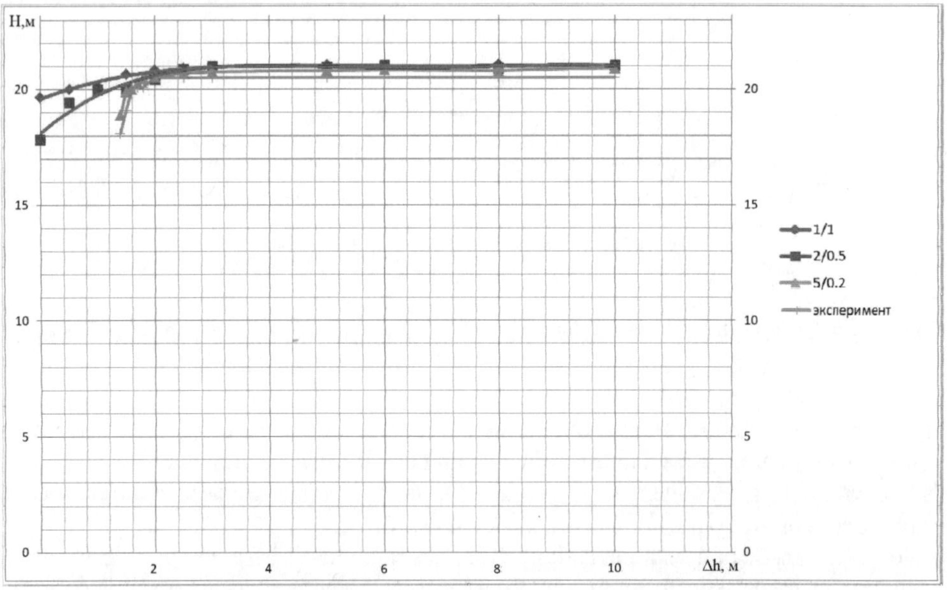

Figure 5.Cavitation characteristics of the pump for different coefficients.

IOP Conf. Series: Materials Science and Engineering **492** (2019) 012037 doi:10.1088/1757-899X/492/1/012037

Actually, the use of a simplified Rayleigh–Plesset model leads to the necessity of correction this equation by introducing empirical coefficients. Probably, it is due to the fact that this model does not take into account many factors, such as the presence of undissolved air in the liquid, the surface tension of the gas bubble, acceleration and others.

In the current article it was supposed that the dissolved air has a greater influence on cavitation than other factors. Therefore, the aim of the study was to find out whether gas content has an influence on cavitation characteristics obtained by numeric calculations or does not have any.

2. Mathematical models

Experiments were carried out with the pump which 3D model and volume mesh are represented in Figure 6.

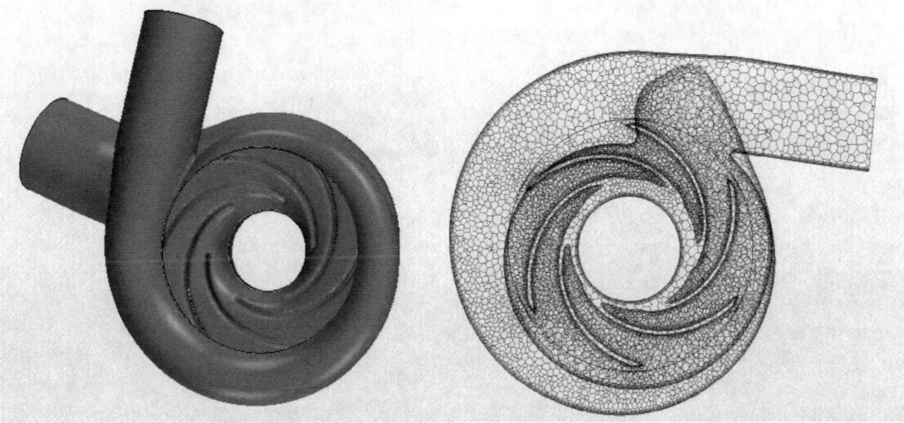

Figure 6. 3D model (left) and volume mesh (right) of the considering pump.

In order to numerically predict cavitation CFD simulation was used [8,9]. The effectiveness of such method was validated by numerous researches and publications [10,11]. Flow field was simulated by the standard k–ω two-equation turbulence model and multiphase flow model [12,13].

Regarding the process of cavitation, it was performed by Schnerr–Sauer [14] model which is widely used in numerous cavitating flow simulations in literature [15,16]. This model is based on the Rayleigh–Plesset equation [17,18,19,20].

$$\frac{p_v - p_\infty}{p_l} = R\frac{d^2R}{dt^2} + \frac{3}{2}\left(\frac{dR}{dt}\right)^2 + \frac{2\sigma}{\rho_l R} \quad (1)$$

Neglecting acceleration, surface tension and viscosity, a simplified form is obtained:

$$\frac{dR}{dt} = \sqrt{\frac{2}{3}\frac{|p_v - p|}{p_l}} \quad (2)$$

The boundary conditions adopted in this calculation are the total pressure at the input and speed at the output. An important point is that such boundary conditions are preferable to set in the calculation of the cavitational phenomena, as in this case, the inlet pressure is a calculated value and pulses during the solution. So, the inlet pressure is a fixed value in the considered problem. By decreasing (changing) reference pressure the intensity, distribution area and cavitation characteristics were obtained.

After finding the suitable method to simulate cavitation process, besides water and vapour, the third phase consisting of undissolved air was included to study the gas content effect on the cavitation characteristics.

IOP Conf. Series: Materials Science and Engineering **492** (2019) 012037 doi:10.1088/1757-899X/492/1/012037

3. Numerical simulation and flow field analysis

Figure 7 shows the difference between two-phase and three-phase simulation of cavitation.

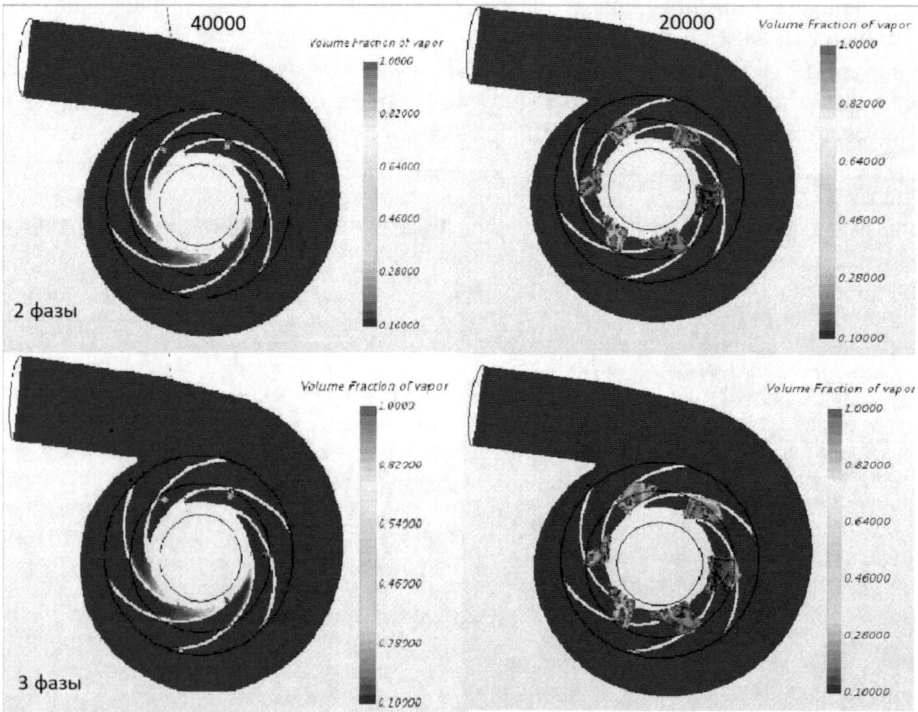

Figure 7.Volume fraction of vapour at pressure 40 kPa, 20 kPa.

Several models with different gas content were considered in the course of the study: with 1,3 and 5% of undissolved air in liquid. The numerical simulation results are represented in the Figure 8, the change in the intensity of cavitation with increasing gas content is shown.

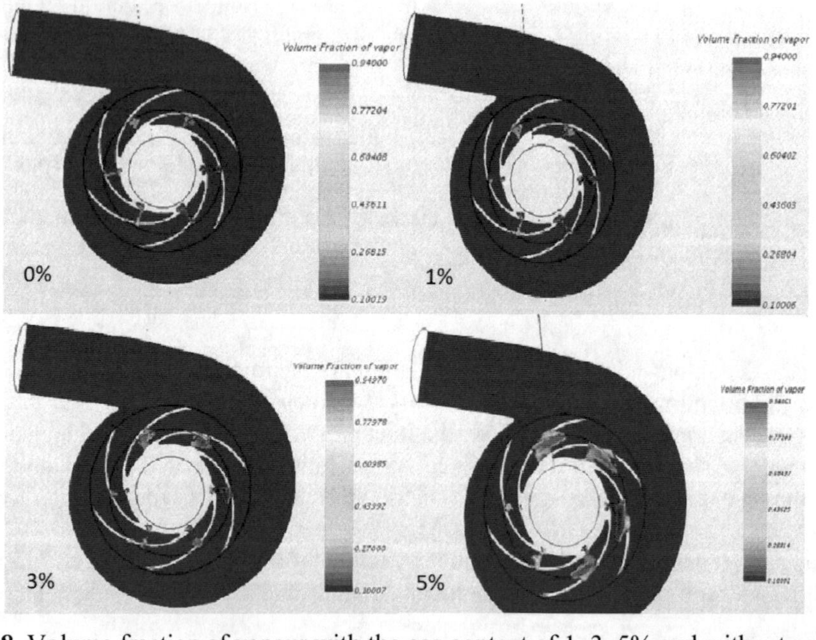

Figure 8. Volume fraction of vapour with the gas content of 1, 3, 5% and without gas at all.

IOP Conf. Series: Materials Science and Engineering **492** (2019) 012037 doi:10.1088/1757-899X/492/1/012037

Cavitation flow characteristics for various gas content are shown in the Figure 9.

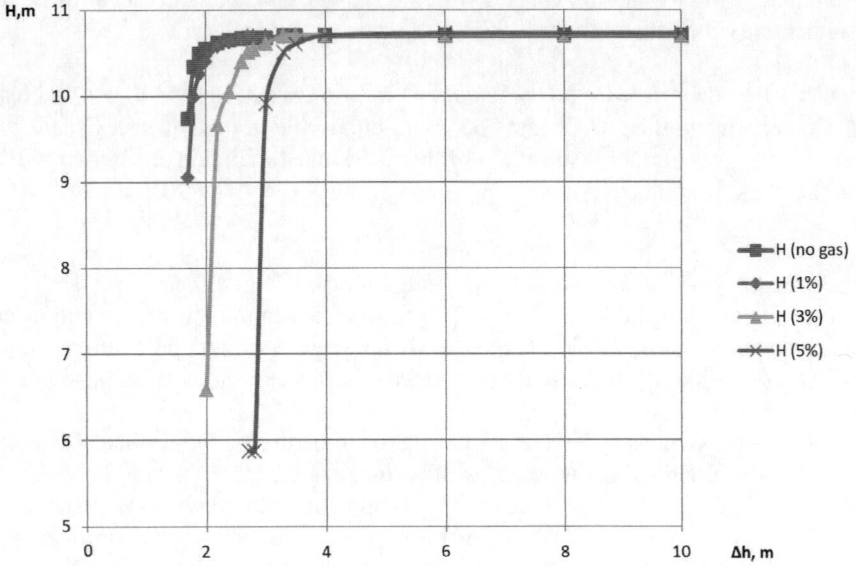

Figure 9. Cavitation characteristics of the pump for various gas content.

Based on above the results of numerical simulation, Table 1 is made.

Table 1. The obtained results of NPSH

Dissolved air content, %	NPSH, m
0	1,8
1	2
3	2,6
5	3,2

According to table 1, NPSH of 5% model is much higher than without gas at all. Inspite of the fact that such amount of gas is rather small, the inclusion of the third phase into the cavitation model significantly affects the cavitation characteristics of the pump and the NPSH. The same difference between NPSH results is obtained by using a simplified Rayleigh–Plesset model without insertion of the empirical coefficients. What is more, such amount of undissolved air is always contained in the liquid during the pump tests. Thus, the characteristics obtained during such tests will incorporate the gas content.

4. Conclusions

Despite numerous research efforts cavitation modelling is rather a difficult task and it is necessary to model it much more accurately. The standard model for numeric simulation of cavitation is Schnerr–Sauer one, which is based on the simplified Rayleigh–Plesset equation. Unfortunately, this simplification doesn't allow to take into account the surface tension of the liquid, acceleration and, especially, the undissolved air according to the results of this study.

In order to find out whether the model of pump with undissolved air differs from the standard Schnerr–Sauer one, the third phase was inserted. The current paper mainly analyzes the difference of cavitation development between two-phase and three-phase flow on the condition of the same pressure boundary conditions at both ends and other settings. From the above results of flow field analysis, it is shown that the intensity of cavitation increases with the rising of gas content. As a result, undissolved air in liquid significantly influences on the NPSH of the pump.

References

[1] Rodrigues Cunha M A 2013 *Cavitation modeling of a centrifugal pump impeller*

[2] Somashekar D and Purushothama H R 2012 Numerical simulation of cavitation inception on radial flow pump *Journal of mechanical and civil engineering* 5 pp 21–26

[3] Li W-G 2014 Validating full cavitation model with an experimental centrifugal pump *Task Quarterly* 1 pp 81-100

[4] Kovalev A A and Kuznetsov N N 2017 Cavitation Fracture of the Typical Materials Used in Hydraulic Machines and Units *Russian Metallurgy (Metally)* 13 pp 1202–1206

[5] Bakhtin B I, Ivashov A I, Kuznetsov A V and Skorokhodov A S 2014 Experimental investigation of the specific features of formation of cavitation zones in intense ultrasound fields *Journal of Engineering Physics and Thermophysics* **87** (3) pp 672–685

[6] Kochevsky A N, Kozlov S N, Aye K M, Schelyaev A Y and Konshin V N 2005 Simulation of flow inside an axial-flow pump with adjustable guide vanes *Proceedings of 2005 ASME Fluids Engineering Division Summer Meeting, FEDSM 2005* pp 1570–1577

[7] Lomakin V O and Bibik O Y 2018 The experimental confirmation of the right use of empirical rates in the simplified Rayleigh-Plesset's model in case of calculating the cavitation characteristics of the centrifugal pump *Pumps. Turbines. Systems* **2**(27) pp 47–52

[8] Bui V T, Kalugin V T, Lapygin V I and Khlupnov A I 2017 Numerical study of the influence of flow blockage on the aerodynamic coefficients of models in low-speed wind tunnels *Thermophysics and Aeromechanics* **24** (6) pp 857–866

[9] Yan D, Kovacevic A, Tang Q, Rane S and Zhang W 2017 Numerical modelling of twin-screw pumps based on computational fluid dynamics *Proceedings of the Institution of Mechanical Engineers, Part C: Journal of Mechanical Engineering Science* **231** (24) pp 4617–4634

[10] Panaiotti S S, Soldatov V A, Rohatgi U S, Chumachenko B N and Timuchev S F 2007 CFD study of prospective 1st stage centrifugal impeller design *Proceedings of the 5th Joint ASME/JSME Fluids Engineering Summer Conference, FEDSM 2007, 2 FORA (PART B)* pp 861–866

[11] Li Y, Walker A D and Irving J 2019 Improved modeling capabilities of the airflow within turbine case cooling systems using smart porous media *Journal of Engineering for Gas Turbines and Power*, **141** (5), article no 051003.

[12] Dhande D Y and Pande D W 2018 Multiphase flow analysis of hydrodynamic journal bearing using CFD coupled Fluid Structure Interaction considering cavitation *Journal of King Saud University — Engineering Sciences* **30** (4) pp 345–354

[13] Yan D, Kovacevic A, Tang Q, Rane S Numerical investigation of cavitation in twin-screw pumps 2018 *Proceedings of the Institution of Mechanical Engineers, Part C: Journal of Mechanical Engineering Science* **232** (20) pp 3733–3750

[14] Reboud J-L, Stuts B, Coutier O 1998 Two-phase flow structure of cavitation: Experiment and modelling of unsteady effects *Third International Symposium on Cavitation*

[15] Usta O and Korkut E A 2018 Study for cavitating flow analysis using DES model *Ocean Engineering* **160** pp 397–411

[16] Sun S, Wu K, Huang Y, Guo P and Luo X 2017 Numerical simulation on flow field and cavitation in scroll hydraulic pump *Journal of Drainage and Irrigation Machinery Engineering* **35** (2) pp 100–105

[17] Yilmaz N, Atlar M and Khorasanchi M 2019 An improved Mesh Adaption and Refinement approach to Cavitation Simulation (MARCS) of propellers *Ocean Engineering* **171** pp 139–150

[18] Fuster D A 2018 Review of Models for Bubble Clusters in Cavitating Flows *Flow, Turbulence and Combustion* Article in Press

[19] Liu X, Hu Q, Shi G and Zhao Q 2018 Cavitation characteristics of multiphase pump at low flow rate *Journal of Drainage and Irrigation Machinery Engineering* **36** (1) pp 15–20

[20] Ye W X, Yu W P, Huang R F and Luo X W 2018 The analysis on the cavitation performance of a steam turbine lubricating oil pump with consideration of the fluid temperature effect *IOP Conference Series: Earth and Environmental Science* **163** (1) article no 012069

IOP Conf. Series: Materials Science and Engineering **492** (2019) 012038 doi:10.1088/1757-899X/492/1/012038

Simulation of non-stationary loads on the rotor of a pumping unit with an assembly pipeline

T Valiev[1,2] and A Petrov[1,3]

[1] Bauman Moscow State Technical University, 5 Second Baumanskaya Street, Moscow, 105005, Russian Federation

[2] E-mail: timurbmstu@gmail.com
[3] E-mail: alexeypetrov@bmstu.ru

Abstract. The article describes the method of calculating unsteady loads on the rotors of oil main pumps, based on the use of CFD-modeling methods. The limiting values of pressures in the pump, as well as the magnitude of the instantaneous and average values of axial and radial forces on the pump rotor depending on the intensity of transients in the pipeline system are determined. It is shown that peak loads are 20 times higher than the standard values of the forces acting on the pump rotor, which should be taken into account when designing systems and pumps for them.

1. Introduction

Modern methods of hydrodynamic modeling in centrifugal pumps, as a rule, are used to optimize the flow parts of pumps in order to increase their energy efficiency. However, one area has not been sufficiently studied by CFD modeling. These are non-stationary phenomena occurring in the pumps and the pipelines attached to them during transient processes in the hydraulic system. Especially such transients are important in systems with long pipelines, for example, in trunk pipelines [1-7].

As a result of a water hammer phenomenon occurrence with a sharp closure of valves, there appear, as the pumping experience shows, very significant axial and radial loads on the rotor of the hydraulic machine, briefly exceeding the standard by an order of magnitude or more. Such loads can lead to the destruction of the bearings and rotors of the pumps, and their magnitude should be taken into account when designing pumps and systems [8-16].

This article attempts to simulate the interaction of a pump and a system based on CFD-modeling methods for calculating a water hammer phenomenon in a pipe.

2. Methods

The STAR CCM+ software product was used in this work. It specializes in hydrodynamic modeling. In the case of calculating the flow of compressible fluid inside the pump, the following equations were applied.

The continuity equation (mass conservation equation):

$$\frac{\partial \rho}{\partial t}+\frac{\partial\left(\rho u_j\right)}{\partial x_j}=0, (1)$$

IOP Conf. Series: Materials Science and Engineering **492** (2019) 012038 doi:10.1088/1757-899X/492/1/012038

where
u_j is the average value of the velocity of the fluid in the projection on the axis j.

Navier—Stokes equations averaged over Reynolds (equations of conservation of momentum of motion averaged over time):

$$\frac{\partial}{\partial t}(\rho u_j)+\frac{\partial}{\partial x_j}(\rho u_i u_j)=-\frac{\partial p}{\partial x_i}+\frac{\partial}{\partial x_i}\left[T_{ij}-\rho\overline{\langle u_i' u_j'\rangle}\right], (2)$$

where
$T_{ij}=2\mu s_{ij}-\frac{2}{3}\frac{\partial u_j}{\partial x_j}$ is the viscous stress tensor,
μ is the dynamic fluid viscosity coefficient, $\text{Pa}\cdot\text{s}$,
$s_{ij}=\frac{1}{2}\left(\frac{\partial u_i}{\partial x_j}+\frac{\partial u_j}{\partial x_i}\right)$ is the strain rate tensor;
$\rho\overline{\langle u_i' u_j'\rangle}$ are Reynolds stresses,
ρ is the averaged pressure,
u_i' is the pulsation velocity component.

To close this system of equations, the $k-\omega$ SST (Shear Stress Transport) turbulence model was used, combining the advantages of the $k-\varepsilon$ and $k-\omega$ models. The model is complemented by two additional equations for the transfer of the kinetic energy of turbulence and the relative dissipation rate of this energy:

$$\frac{\partial}{\partial t}(\rho k)+\frac{\partial}{\partial x_j}(\rho u_j k)=P_k-\rho\beta^* k\omega+\frac{\partial}{\partial x_j}\left[(\mu+\sigma_k\mu_t)\frac{\partial k}{\partial x_j}\right]; (3)$$

$$\frac{\partial}{\partial t}(\rho\omega)+\frac{\partial}{\partial x_j}(\rho u_j\omega)=\alpha\frac{\omega}{k}P_k-\rho\beta\omega^2+\frac{\partial}{\partial x_j}\left[(\mu+\sigma_\omega\mu_t)\frac{\partial\omega}{\partial x_j}\right]+2\cdot(1-F_i)\sigma_{\omega 2}\cdot\frac{1}{\omega}\frac{\partial k}{\partial x_i}\frac{\partial\omega}{\partial x_i}. (4)$$

To account for the compressibility of a fluid, this system was supplemented with the equation of state (isothermal):

$$p-p_0=a^2\cdot(\rho-\rho_0), (5)$$

where
p_0 is the initial pressure;
ρ_0 the initial density, a is the sound velocity in fluid.

As the boundary conditions when calculating the flow in the pump, the following combination was used.

1. Total inlet pressure, i.e.:

$$p+\frac{\rho v^2}{2}=\text{const}.$$

2. Outlet speed.

Modeling of the flow processes was carried out in a non-stationary formulation with a time step of 10^{-4} s. The criterion for the transition to the next step was the achievement of 10 iterations.

Two basic water hammer equations obtained by N.E. Zhukovsky. The expression for calculating the increase in pressure during water hammer:

$$\Delta p=\rho\cdot a\cdot v, (6)$$

where
ρ is the fluid density;
a is the shock wave velocity;
v is the average velocity of the fluid in the pipeline before the water hammer occurs.

Water hammer phase is defined as

$$\tau = \frac{2L}{a}, \quad (7)$$

where
L is the pipeline length.

If the time t_{cls} of the pipeline overlap (valve closing) is less than the impact phase value τ, the water hammer will be due to the loss of all speed and the overpressure will be maximum. Such a water hammer is called a complete or direct.

Closing the valve was simulated by setting the function of the speed change from time:

$$\begin{cases} v_{\text{out}}(t) = v_0, \ (t < t_s) \\ v_{\text{out}}(t) = v_0 \cdot \left(1 - \dfrac{t - t_c}{T_3}\right)^n, \ (t_s \le t \le t_s + T_{cls}), \\ v_{\text{out}}(t) = 0, \ (t > t_s + T_{cls}), \end{cases}$$

where
t_s is the start time of valve closing;
T_{cls} is the period for which the valve should close;
n is the index number;
v_0 is the initial speed.

In this work, the linear law of variation was used: $n = 1$. The speed of sound was assumed to be constant and equal to 1000 m/s. The fluid is water.

The computational grid has a different topology. In the flow core, cells are polyhedral of various shapes and sizes. Cells near solid walls are multi-sided prisms, elongated in the direction perpendicular to the wall. The number of prismatic layers is assumed to be 5, with a stretch ratio of 1.3. The thickness relative to the base size of a polyhedral cell is 33%. The total number of cells was 840,000. The flow part of the pump HM 7000-210 (figure 1) was taken as the object of study.

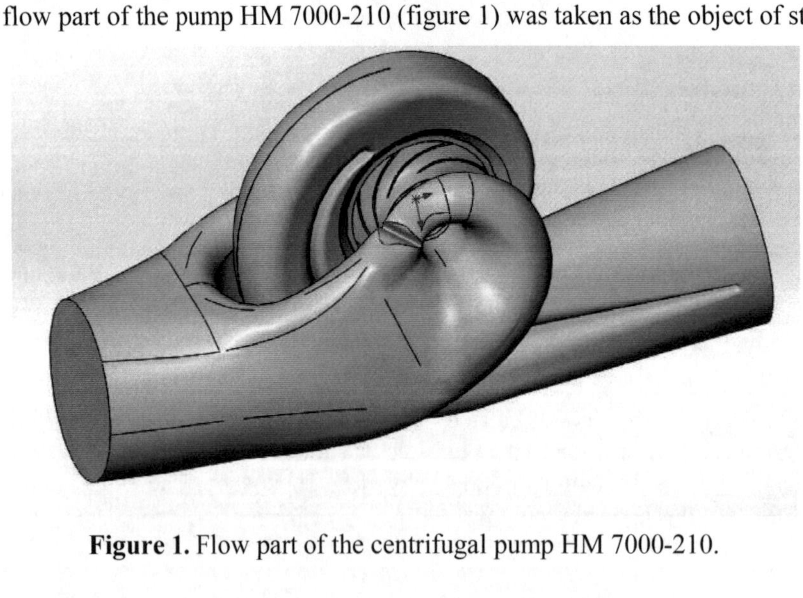

Figure 1. Flow part of the centrifugal pump HM 7000-210.

IOP Conf. Series: Materials Science and Engineering **492** (2019) 012038 doi:10.1088/1757-899X/492/1/012038

To create a wave process that occurs when a water hammer, you can attach the pipelines to the discharge and suction nozzles to the flow part of the pump. At the same time, the length of the pipes does not necessarily have to be equal to the actual length, since this will significantly increase the computational grid, and, accordingly, large computational resources will be required. According to (6), the amplitude of the pressure jump upon a water hammer does not depend on the length of the pipes, but is determined by the flow rate of the medium. The length of the tube affects the period of propagation and reflection of the waves should be sufficient for the appearance of the wave process. In this case, the calculation in a non-stationary formulation must be carried out with a corresponding time step, namely, less than the magnitude of the impact phase (7).

We consider the following problem statement: one pumping unit of the type HM 7000-210 with a pipeline connected to the outlet nozzle is installed at the pump station. At the other end of the pipe a valve is installed, which regulates the mode of operation of the pump. Overpressure at the inlet is provided approximately 70 m (figure 2).

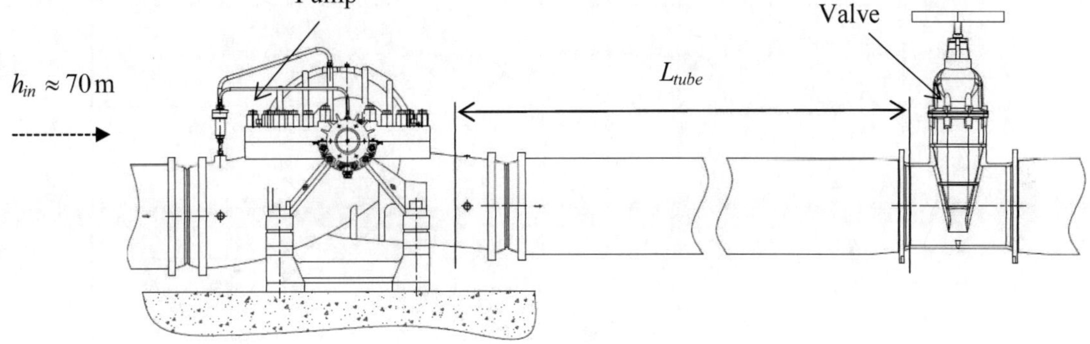

Figure 2. Calculation scheme.

The diameter of the pipeline is 600 mm. The length was chosen to be 14 m. With this value, the size of the computational domain remains acceptable for the calculation on a laptop.

In accordance with the calculation scheme (figure 2), the total inlet pressure was assumed to be 700,000 Pa. To set the output velocity function, we define the phase of the water hammer for this system. The total length of the pipeline for the propagating wave in this problem:

$$L = L_{tube} + L_{pump} = 14 + 5 = 19\,\text{m},$$

where

$L_{pump} = 5\,\text{m}$ is the path length for a wave inside the flow part of the pump (retraction, impeller, supply).

Then the phase of the water hammer will be:

$$\tau = \frac{2L}{a} = \frac{2 \cdot 19}{1000} = 0{,}038\,\text{s}.$$

Finally, we take the closing time to be

$$T_{cls} = 0{,}01\,\text{s}.$$

We understand that in a real situation there are no conditions for such a speed of closure, and water hammer occur in longer pipes and at longer closure times. However, in this case, the pipeline area will occupy most of the computational grid compared to the pump, which means that the computing power will be spent on modeling the processes in the pipe, and not in the pump, which is irrational, since it is the main object of study. First of all, the task of such a calculation is to understand what loads the pump unit will experience when a shock wave passes through it.

IOP Conf. Series: Materials Science and Engineering **492** (2019) 012038 doi:10.1088/1757-899X/492/1/012038

3. Results

As a result of simulating the impact of a water hammer on a centrifugal pump, the following data were obtained: pressure in the pipeline, radial force and torque on the rotor of the pump unit.

According to the obtained results, the oscillation period corresponds to the theoretical:

$$T = \frac{4L}{a} = \frac{4 \cdot 19}{1000} = 0.0766\,\text{s}.$$

The pressure around the valve at the moment of impact, taking into account the initial pressure in the pipe of about 2.7 MPa, should be equal to (6):

$$p_{out} = p_{out}^{0} + \Delta p = \rho \cdot a \cdot v = 2.7 + 1000 \cdot 1000 \cdot 6.88 = 9.58\,\text{MPa}.$$

That is confirmed on the graph (figure 3).

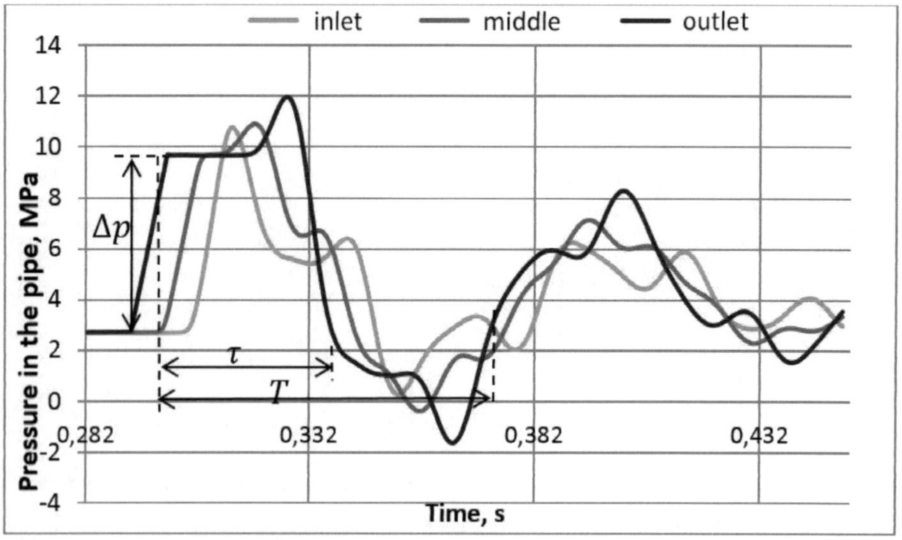

Figure 3. Pressure in the pipe with water hammer.

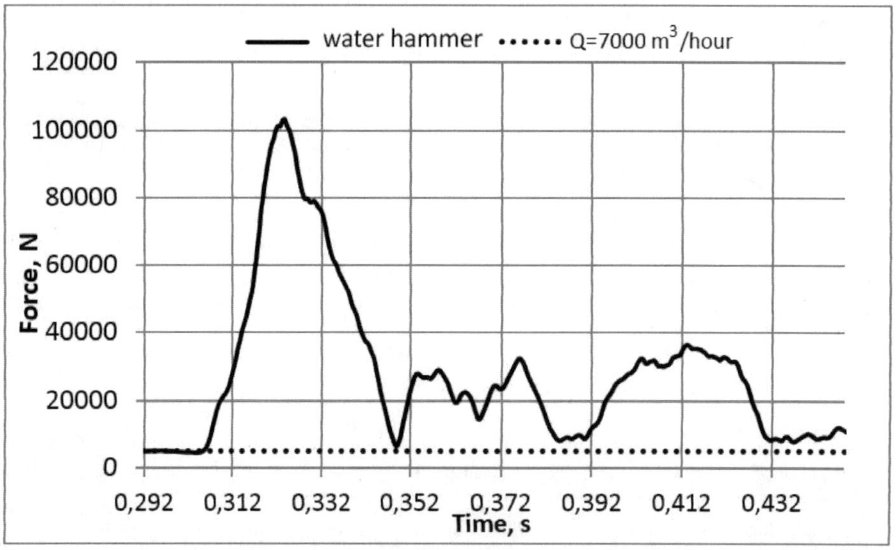

Figure 4. Radial force on the rotor of the pump HM 7000-210 with water hammer.

IOP Conf. Series: Materials Science and Engineering **492** (2019) 012038 doi:10.1088/1757-899X/492/1/012038

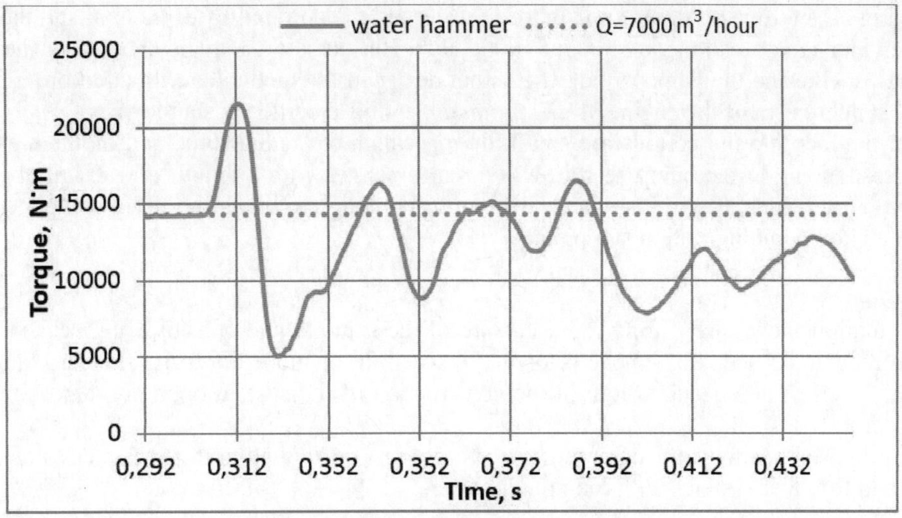

Figure 5. The moment on the rotor.

Figure 4 shows a graph of the radial force acting on the pump rotor during the water hammer period. Compared with the nominal mode, the force quickly increased to 100,000 N in the first period of the impact, i.e. increased by about 20 times compared with the nominal mode. Then, according to the obtained graph, the radial force fluctuated around 20,000 N, reaching maximum values of 30,000…36,000 N.

This pump unit is equipped with sliding bearings, and such shock loads can lead to the contact of the shaft neck and the liner, which will adversely affect their service life. Also in this case, it is possible to touch the gap seals, but this is less likely, since the gap between the wheel and hub is larger than in the supports. If rolling bearings act as supports on the pump, then such loads can lead to their failure, since they are not designed to absorb shock overloads.

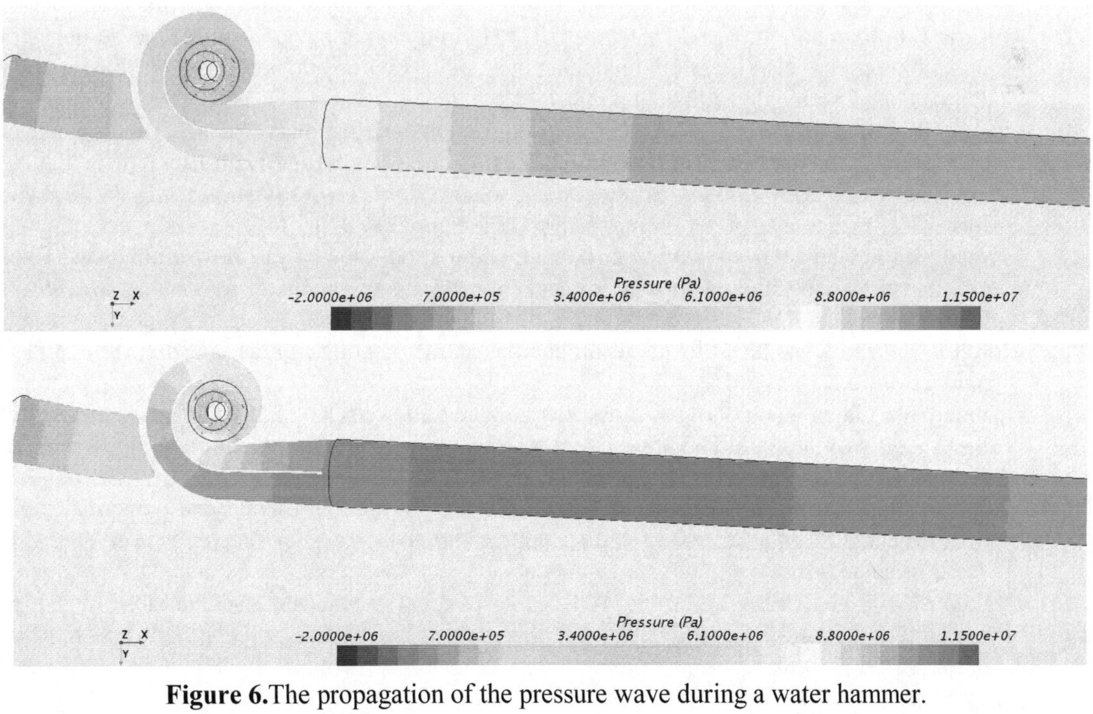

Figure 6. The propagation of the pressure wave during a water hammer.

In addition, the torque on the rotor (Figure 5) exceeded the nominal by 5000 N·m and then did not exceed the value in nominal mode of about 16000 N·m. But this blow may be enough for the coupling between the engine and the pump to fail, if it is not designed for such a load. In addition, it can cause loss of the synchronism of the engine of the pump unit and its emergency shutdown.

Despite the fact that the periods are small, during which the radial force and moment exceed the values corresponding to the pump zero flow, on real pipelines with a length of 100 km or more, the impact phases can reach 80…100 s, which is more than enough to break the listed components of the pump unit and oil pumping station stopping.

4. Discussion

1. The simulation shows that during rapid closure of the shut-off and control equipment in the main pumps the magnitude of forces on the rotor may exceed 20 or more times the standard values. This fact should be taken into account both in the construction of the system operation algorithms and in the pump design;

2. The proposed method for calculating such forces can be applied both for this class of hydraulic machines and for other systems with long pipelines.

References

[1] Lomakin V O, Kuleshova M S and Bozh'eva S M 2016 Numerical Modeling of Liquid Flow in a Pump Station *Power Technology and Engineering* (5) pp 324-327

[2] Lomakin V O, Kuleshova M S and Kraeva E A 2015 Fluid Flow in the Throttle Channel in the Presence of Cavitation *Procedia Engineering* **106** pp 27-35

[3] Kraposhin M V, Banholzer M, Pfitzner M and Marchevsky I K 2018 A hybrid pressure-based solver for non-ideal single-phase fluid flows at all speeds *International Journal for Numerical Methods in Fluids* **88** (2) pp 79-99

[4] Min'kov L L, Krokhina A V and Dueckc J 2011 Hydrodynamic mechanisms of the influence of injection on the classification characteristics of a hydrocyclone *Journal of Engineering Physics and Thermophysics* **84** (4) pp 807-819

[5] Ionaitis R R and Chain P L 2009 Principles for engineering active-passive hydrodynamic control-and-safety systems for nuclear reactors *Atomic Energy* **106** (3) pp 175-184

[6] Krapivtsev V G 2017 Model Studies of Coolant Flow Hydrodynamics in VVER-1000 In-Reactor Pressure Channel *Atomic Energy* **122** (5) pp 304-310

[7] Marchevskii I K and Puzikova V V 2016 Numerical simulation of the flow around two fixed circular airfoils positioned in tandem using the LS-STAG method *Journal of Machinery Manufacture and Reliability* **45** (2) pp 130-136

[8] Morch K A 1976 Dynamics of Cavitation bubbles and Cavitating liquids *Treatise on Materials Science and Technology* **16** pp 309-335

[9] Dazhuan Wu, Shuai Yang, Peng Wu and Leqin Wang 2015 MOC-CFD Coupled Approach for the Analysis of the Fluid Dynamic Interaction between Water Hammer and Pump *Journal of Hydraulic Engineering* February

[10] Dazhuan Wu, Shuai Yang, Peng Wu and Leqin Wang 2011 Experimental and numerical study of transient flow in a centrifugal pump during startup *Journal of Mechanical Science and Technology* March

[11] Dazhuan Wu, Shuai Yang and Peng Wu 2014 Transient characterisitcs of a close-loop pipe system during pump stopping periods *Journal of Pressure Vessel Technology, Transactions of the ASME* February

[12] Saemi S D, Raisee M, Cervantes M J and Nourbakhsh A 2014 Computation of laminar and turbulent water hammer flows *6th European Conference on Computational Fluid Dynamics* (ECFD VI) Barcelona, Spain

[13] Isaev S A, Kornev N V, Leontiev A I and Hassel E 2010 Influence of the Reynolds number and the spherical dimple depth on turbulent heat transfer and hydraulic loss in a narrow channel *International Journal of Heat and Mass Transfer***53** (1-3) pp 178-197

[14] Menter F R 1994 Two-Equation Eddy-Viscosity Turbulence Models for Engineering Applications *AIAA Journal***32** (8) pp 1598-1605

[15] SahilLahane, RachanaPatil, Rashmi Mahajan and KavitaPalve Analysis of Water Hammering in Pipeline and its CFD Simulation. International *Journal of Engineering Technology, Management and Applied Sciences***3** (5)

[16] Ismaier A and Schlücker E 2009 Fluid dynamic interaction between water hammer and centrifugal pumps. *Nuclear Engineering and Design***239** pp 3151–3154

[17] Patankar S 1980 *Numerical Heat Transfer and Fluid Flow* (NY: Hemisphere Publishing Corporation) 214 p

[18] Lomakin V O, PetrovA I 2012 Verifikatsiyarezul'tatovrascheta v paketegidrodinamicheskogomodelirovaniya STAR-CCM+ protochnojchastitsentrobezhnogonasosa AX 50-32-200 *Izvestiyavysshikhuchebnykhzavedenij. Mashinostroenie — Proceedings of Higher Educational Institutions. Machine Building* (5) pp 6-9

[19] Zimnitskiy V A and Umova V A red 1986 *Lopastnyenasosy: Spravochnik*[Vane pumps. Handbook] (Leningrad: Mashinostroenie) 334 p

[20] Zhukovsky N E 1949 O gidravlicheskom udare v vodoprovodnykh trubakh [On water hammer in hydrogen pipes] (Moscow: Gos. izdatel'stvotekhn.-teor. literatury Publ.) 105 p

IOP Conf. Series: Materials Science and Engineering **492** (2019) 012039 doi:10.1088/1757-899X/492/1/012039

Development and research of hydraulic disk pump

V Cheremushkin[1] and V Lomakin[1,2]

[1]Bauman Moscow State Technical University, 5 Second Baumanskaya Street, Moscow, 105005, Russian Federation

[2]E-mail: lomakin@bmstu.ru

Abstract. The article is devoted to the development of the disk pump driven by a hydraulic motor at Bauman Moscow State Technical University Department of Hydraulics. The relevance of the dynamic pumps of this type has been described, and the necessity to investigate the mechanism of disk impellers operation using the tools of numerical hydrodynamic modeling, has been shown. The article presents the mathematical model used in computer simulation of the flow and the description of the testing bench. A comparison of numerical and full-scale experiments to confirm the adequacy of the mathematical model is introduced. A research showing the influence of some basic parameters of disk impellers on their characteristics (head, efficiency) has been performed. The figures of the scalar physical quantities distribution in the flow part of the pump and graphs of the pressure and efficiency dependences on the studied parameters are presented for a visual illustration of the results.

1. Introduction

The dynamic pumps for various purposes: oil, water, chemical, food, slurry systems and other are widely used nowadays. It should be noted that most of these pumps are centrifugal. Their widespread use is caused by the relatively high efficiency. However, the design features of the impeller, as well as the mechanism of the working fluid flow in the inter-blade channel, limit their ability to pump viscous and highly contaminated liquids.

It is advisable to use disk impellers, the efficiency of which does not decrease with the viscosity increase [1] (figure 1), for viscous polluted liquids. Thus, in some enterprises, disc pumps have replaced successfully centrifugal slurry, screw pumps for pumping thickened sediment from the sumps of a water purification system. Some works are devoted to the use of disk pumps for blood pumping, including ones with imitation of the real heart pulsations. Such a field of application is justified by the ability of these pumps for gentle pumping, without introducing significant perturbations into the flow structure [2, 3, 4]. This property provided a niche for disc pumps in the food industry.

It should be noted, however, that there are few works devoted to the development of the theory of disk pumps at present days. Compared to the works that have been published quite a long time ago [5, 6, 7, 8], practically no changes were made to this theory. There are also very few works related to the numerical hydrodynamic simulation of disk pumps, which in its turn can significantly expand the understanding of the capabilities of these hydraulic machines.

This work is related to the research being conducted at the Department of Hydraulics of the Moscow Bauman State Technical University. Two variants of the design of a hydraulic disk drive

IOP Conf. Series: Materials Science and Engineering **492** (2019) 012039 doi:10.1088/1757-899X/492/1/012039

pump, shown in figure 2, were developed, manufactured and tested on the base of the department. In order to check the compliance of the flow part with the declared characteristics, a numerical hydrodynamic simulation of fluid flow in the pump was performed. According to the tests results, experimental characteristics of pumping units were compiled and compared with the results of computer analysis.

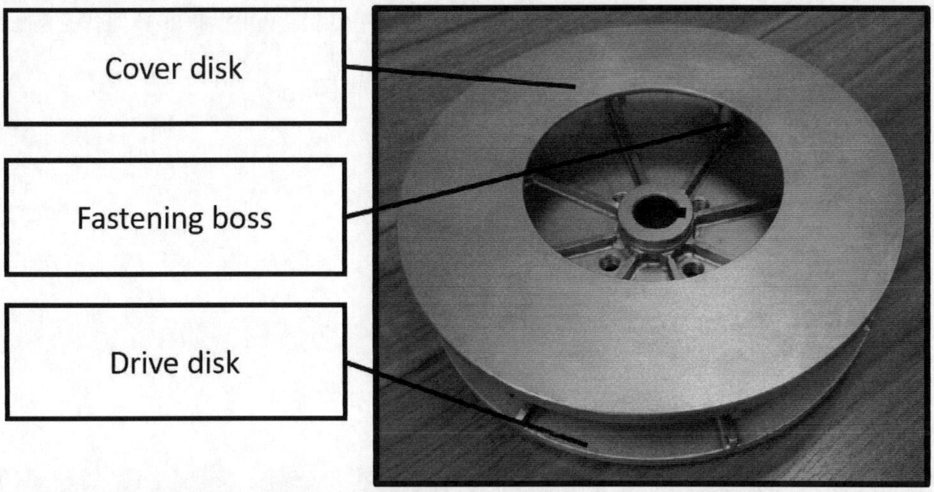

Figure 1. The disk impeller with ribs.

Figure 2. Variants of the disk hydraulic drive pump installed on the testing bench.

Also, numerical analysis has been carried out aimed at the research of the general influence of the disk impellers geometrical parameters on their characteristics.

2. Mathematical model and methods.

The method of numerical hydrodynamic modeling used in the work is based on solving discrete analogs of the equations of the continuous viscous medium dynamics. In the case of an incompressible fluid model (ρ = const), it can be found in [9, 10, 11]:

Mass conservation equation (continuity equation)

$$\frac{\partial \tilde{u}_j}{\partial x_j} = 0,$$

where
$\tilde{u}_j$ – the averaged value of the fluid velocity in the projection on the j-th axis (j = 1, 2, 3).

The equation of the momentum change (Reynolds averaging) in a nonstationary statement

$$\rho\left[\frac{\partial \mathrm{U}_i}{\partial t} + \mathrm{U}_\mathrm{j}\frac{\partial \mathrm{U}_\mathrm{i}}{\partial x_j}\right] = -\frac{\partial \mathrm{P}}{\partial \mathrm{x}_\mathrm{i}} + \frac{\partial}{\partial \mathrm{x}_\mathrm{i}}\left[T_{ij}^{(v)} - \langle \rho u_i u_j \rangle\right],$$

where
U, P – averaged speed and pressure;
$\tilde{T}_{ij}^{(v)} = 2\mu\tilde{s}_{ij}$ – viscous stress tensor for incompressible fluid;
$\tilde{\mathrm{s}}_{\mathrm{ij}} = \frac{1}{2}\left[\frac{\partial \tilde{\mathrm{u}}_\mathrm{i}}{\partial \mathrm{x}_\mathrm{j}} + \frac{\partial \tilde{\mathrm{u}}_\mathrm{j}}{\partial \mathrm{x}_\mathrm{i}}\right]$ – instant strain rate tensor;
$\langle \rho u_i u_j \rangle$ – Reynolds stresses;
ρ – fluid density;
μ – fluid viscosity coefficient.

The Reynolds stresses were simulates using the k—ω SST turbulence model, which was approved during dynamic pumps analysis [12, 13, 14, 15].

For the numerical solution by the finite element method in such software complexes, a mesh is used. In this case, the cells are polyhedral in the flow core and prismatic - near the walls (figure 3). The thickness of the prismatic layers which is necessary to ensure the value of Y + <100, which is recommended for high-Reynolds models like k—ω SST.

Tests of pumping units were carried out at the testing bench at the Hydraulics department, the scheme of which is shown in figure 4.

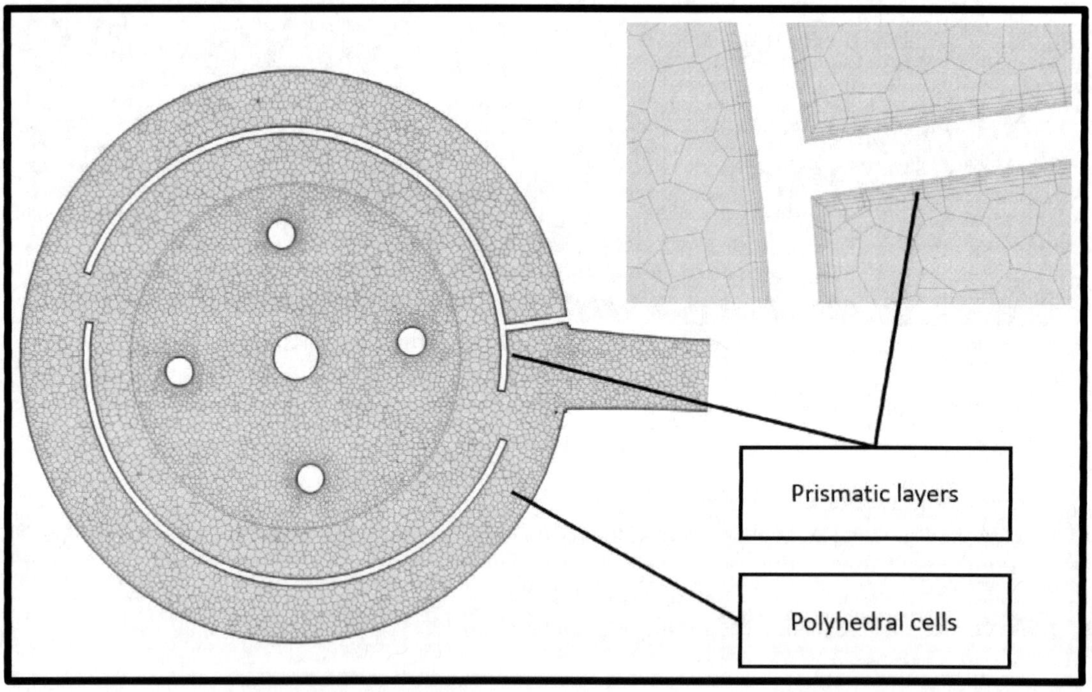

IOP Conf. Series: Materials Science and Engineering **492** (2019) 012039 doi:10.1088/1757-899X/492/1/012039

Figure3. Mesh.

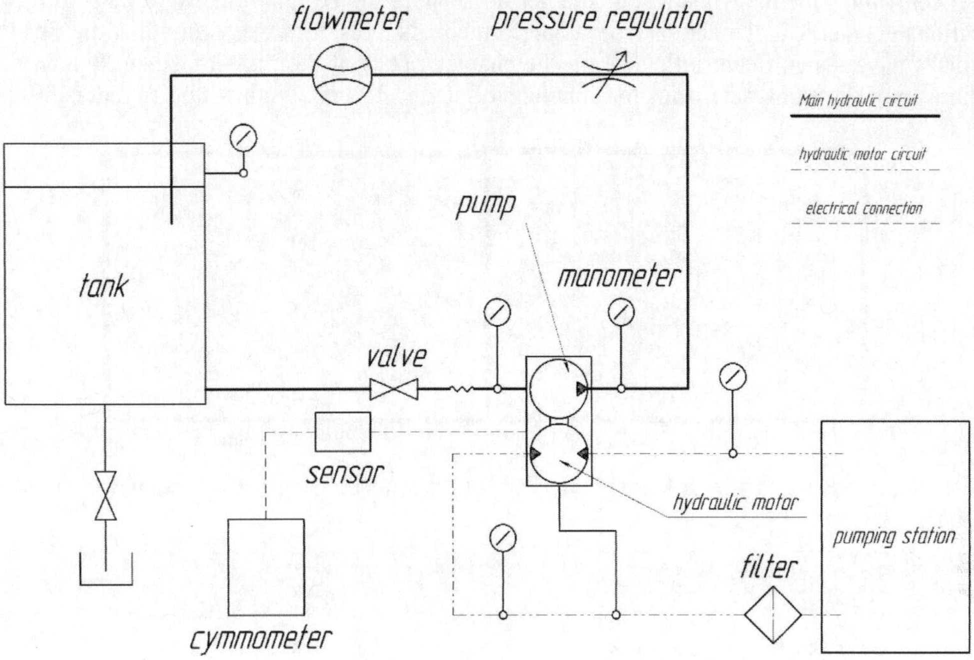

Figure 4. The scheme of the testing bench.

3. Results and analysis

The comparison of the results of flow part numerical simulation of the developed pumps and the results of bench tests are shown in figure 5. A fairly accurate coincidence of the characteristics of the head and efficiency shows the adequacy of the applied mathematical model with respect to disk pumps.

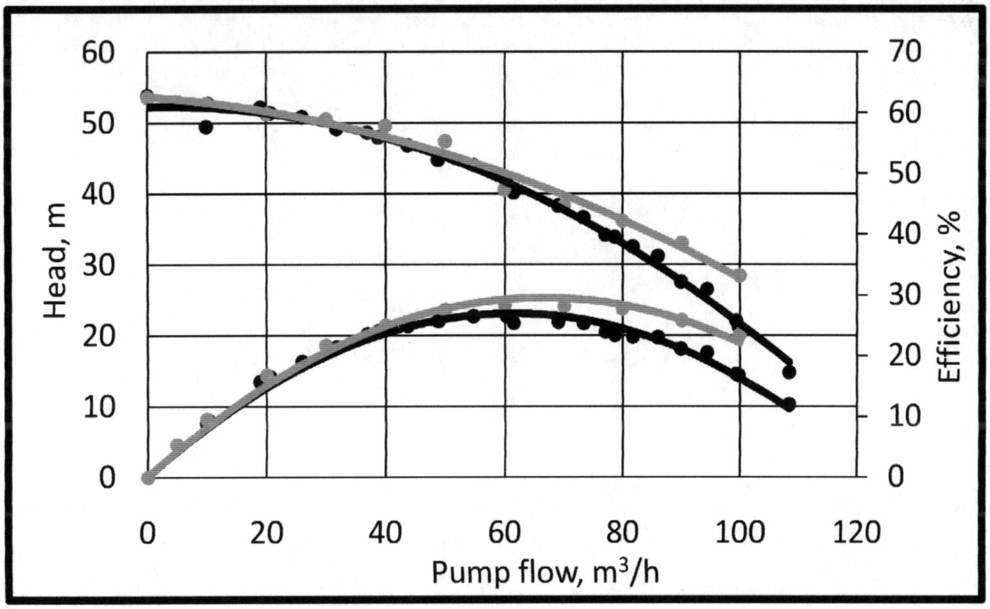

IOP Conf. Series: Materials Science and Engineering **492** (2019) 012039 doi:10.1088/1757-899X/492/1/012039

Figure 5. The graphs of the dependences of the head and the efficiency of the pump performance obtained experimentally (black) and by simulation (green).

After confirming the mathematical model performance, a simulation of many variants of flow parts with various geometrical characteristics was performed. As a result, it was found that the ribs located on its disks have a significant influence on the characteristics of the disk impellers. Figure 6 shows a comparison of these characteristics for impellers of 10 and 15 mm width with different configuration of the ribs.

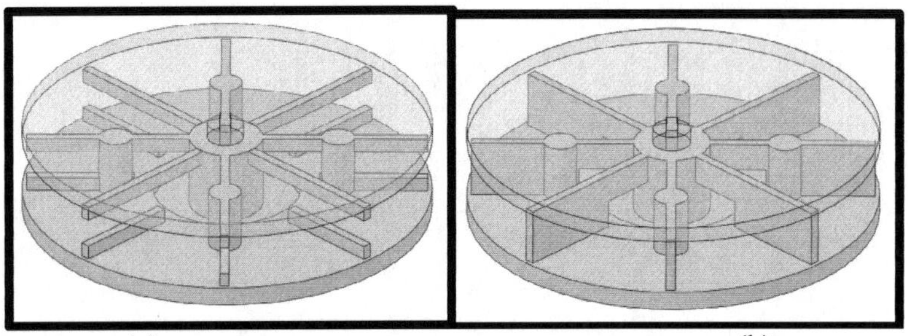

(a) (b)

Figure 6. The variants of the disk impeller ribs with a total height as a percentage of the distance between the disks: a) 50%; b) 100%.

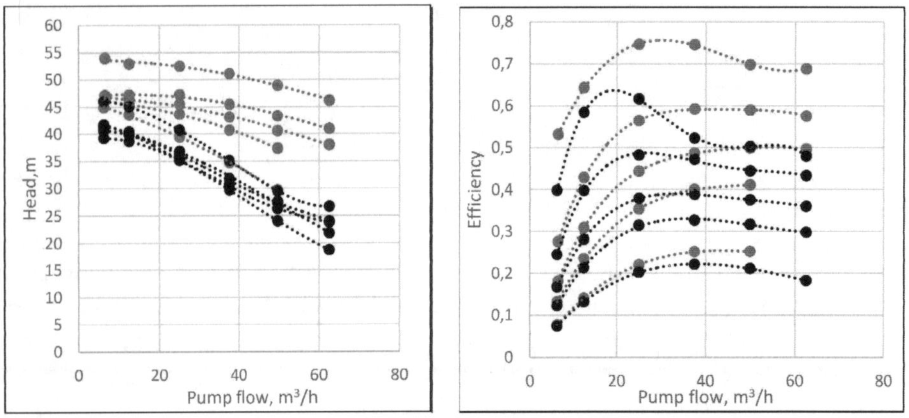

(a)

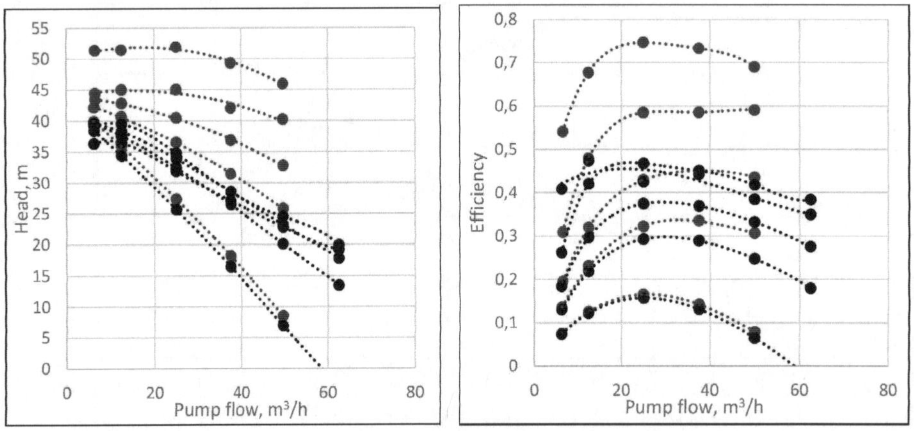

(b)

Figure 7. Dependencies of head and efficiency at distances between the disks: a) 15mm; b) 10mm. Black color — the height of the edges is 50%, red and purple colors — 100% of the distance between the discs.

In each figure, each impeller configuration corresponds to 5 graphs for five different viscosities: 1, 1000, 3000, 5000 and 10,000 cSt (from top to bottom). The presented graphs show the working capacity of disk impellers with ribs, including the full width of the impeller, at high viscosities.

4. Conclusions

As a result of the conducted research, the applicability of the numerical simulation method with respect to disk pumps was shown. According to the results of the study of disk impellers with different geometrical parameters at several viscosities, it can be concluded that the addition of radial ribs to the disks leads to an increase in pump's head and efficiency. At low viscosities, this increase is up to 20 m and 30%, respectively, and the blades retain their effectiveness up to 10,000 cSt. The obtained results require experimental studies, as well as further development of the numerical modelling.

References

[1] ZehevTadmor, Pradip S. Mehta, Lefteris N. Valsamis and Jan-Chin Yang. *Corotating Disk Pumps for viscous Liquids*Farrel Company, Division of USM, Ansonia, Connecticut, 06401.

[2] Chernyavskii A M, Medvedev A E et al. 2017 Flow – rate – pressure characteristics of a disk blood pump *Journal of Engineering Physics and Thermodynamics***90** (6) November

[3] Miller G E and Sidhu Amrita 1992 Multiple disk centrifugal pump as an artificial ventricle *Proceedings of Engineering Mechanics* pp 976–9

[4] Miller G E, Etter B D and Dorsi J M 1990 A multiple Disk Centrifugal Pump as a Blood Flow Device *IEEE Transactions on Biomedical Engineering***37** (2) pp 157–163 DOI: 10.1109/10.46255

[5] Naumov I E, Prikhodko Yu M, Chekhov V P and Fomichev V P 2012 On dimensionless parameters for generalization pressure head and flow rate characteristics of centrifugal disk pumps *Thermophysics and Aeromechanics***19** (1)

[6] Misyura V I, Ovsyannikov B V and Prisnyakov V F 1986 *Diskovyenasosy* [Disk pumps] (Moscow, Mashinostroenie Publ.) 112 p

[7] Crawford M E and Rice W 1974 Calculated Design Data For The Multiple-Disk Pump Using Incompressible Fluid *J Eng Power Trans ASME***96**Ser A (3) pp 274–82

[8] Ma X, Zhang R and Xiao C 2006 Application of disk friction pump in principal axis seal of hydroelectric generator ShuiliFadianXuebao*Journal of Hydroelectric Engineering***25** (1) зp 124–126+113

[9] Lomakin V, CheremushkinV and Chaburko P 2018 Investigation of vortex and hysteresis effects in the inlet device of a centrifugal pump *2018 Global Fluid Power Society PhD Symposium, GFPS*, no 8472374 DOI: 10.1109/GFPS.2018.8472374

[10] Lomakin V O, Chaburko P S and Kuleshova M S 2017 Multi-criteria Optimization of the Flow of a Centrifugal Pump on Energy and Vibroacoustic Characteristics *Procedia Engineering***176** pp 476–82 DOI: 10.1016/j.proeng 2017.02.347

[11] Lomakin V O 2015 Investigation of two-phase flow in axial-centrifugal impeller by hydrodynamic modeling methods *Proceedings of 2015 International Conference on Fluid Power and Mechatronics*, FPM 2015 no 7337302 pp 1204–6

DOI: 10.1109/FPM.2015.7337302

[12] Gouskov A M, Sorokin F D and Banin E P 2016 Simulation of an Inlet Structure of an Implantable Axial Blood Pump *Biomedical Engineering***50** pp 15–19
DOI: 10.1007/s10527-016-9578-2

[13] Gouskov A M, Lomakin V O, Banin E P and Kuleshova M S 2017 Minimization of Hemolysis and Improvement of the Hydrodynamic Efficiency of a Circulatory Support Pump by Optimizing the Pump Flow path *Biomedical Engineering***51** pp 229–33
DOI: 10.1007/s10527-017-9720-9

[14] Baturin O, Popov G, Kolmakova D, Zubanov V, Novikova J and Korneeva A 2018 Optimization of Powerful Two-stage Screw Centrifugal Pump *2nd International Conference on Mechanical, System and Control Engineering, ICMSC 2018* (Moscow) **220**
DOI: 10.1051/matecconf/201822003009

[15] Tesch K, and Kaczorowska-Ditrich K 2018 Discrete-continuous optimization of an axial flow blood pump *23rd Fluid Mechanics Conference, FMC 2018* (Zawiercie, Poland) **1101**
DOI: 10.1088/1742-6596/1101/1/012044

IOP Conf. Series: Materials Science and Engineering **492** (2019) 012040 doi:10.1088/1757-899X/492/1/012040

Optimal design of inlet device of multistage centrifugal pump

A Petrov[1,2], A Lysenko[1], T Valievand[1] and N Isaev[1]

Bauman Moscow State Technical University, 5 Second Baumanskaya Street, Moscow, 105005, Russian Federation

[2]E-mail: alex_i_petrov@mail.ru

Abstract: In the article the authors present a new type of inlet device for multistage centrifugal pumps. An overview of the existing types of inlet devices is given, their advantages and disadvantages are observed. The form of the proposed type of inlet device is described, which makes it possible to combine good cavitational properties of the pump with low losses in the inlet device. The results of hydrodynamic modeling of the proposed type of inlet device and its comparison with the existing ones are given.

1. Introduction

An important factor within the process of creation of modern multistage pumps is the designing of inlet chambers. The geometry and nature of fluid flow in these chambers directly affect pump characteristics such as NPSHR [1-3], the efficiency rate, the steepness of the pressure characteristics in the area of low flow rates [4-9]. Moreover, steeply decreasing characteristic is especially important for the pumps used in thermal power engineering and nuclear power engineering [10].

Now the following types of inlet devices are used for these pumps:

- annular chambers;
- annular chambers with a separating edge;
- semi-spiral chambers;
- channel inlets (deflectors) with annular chamber.

All these devices have their advantages and disadvantages. Thus, the pumps with an annular chamber have a higher efficiency rate due to its relatively small losses in the inlet devices. But when operating in the low feed rate area, the pump head decreases due to the reverse currents in the inlet devices, which leads to the characteristic's falling-off.

The integration of a separating edge of sufficient size to the inlet device allows to minimize this phenomenon, but also reduces the efficiency rate of the pump, as well as its cavitation properties [11–13].

The appliance of semi-spiral inlet devices, in turn, allows to improve the cavitation property of the pump, which is especially important for the multistage pumps designed for large feeds, but at the same time reduces the pump head (due to appearance of the moment of speed at the inlet to the first stage) and its efficiency rate. Channel inlet devices are used, on the contrary, for the pumps with low feed rates, but the losses in them are large compared to the other types of inlet devices. Figure 1 shows the lateral semi-spiral and annular inlet devices.

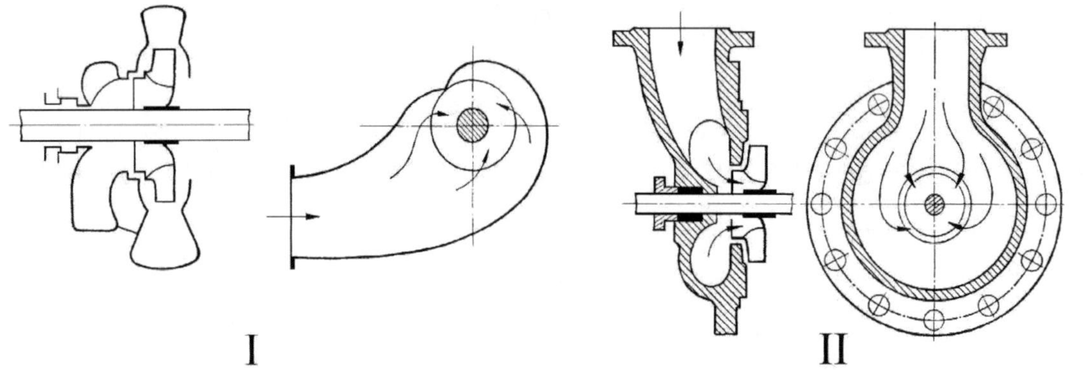

Figure 1. Lateral semi-spiral (I) and (II) annular inlet devices.

The purpose of this study was to create such a geometry of the inlet device of a multistage pump in order to ensure the smallest losses at the inlet to the first pump stage.

2. Methods

To determine the losses in the flow part of the inlet device, the 3D models of the original intake system and the optimized one were built. For these models, hydrodynamic modeling of fluid flow was carried out [14]. As the initial model an annular inlet of centrifugal pump type CNS 300-120 was taken.

Considered inlet systems are shown in figure 2. The distinguishing feature of the suggested inlet device lies in the fact that in addition to the reduced volume, it does not have thin-walled edges, and a massive streamlined separator with a small curvature is used instead. It is assumed that this solution will reduce the volume of stagnant zones and improve the shaft streamlining as a whole.

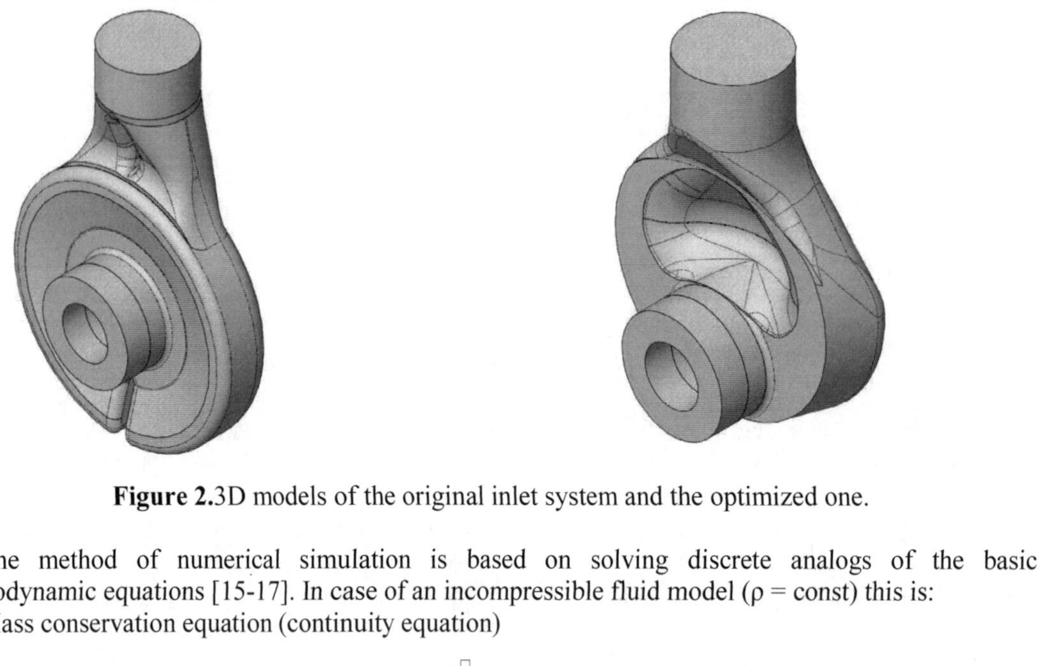

Figure 2.3D models of the original inlet system and the optimized one.

The method of numerical simulation is based on solving discrete analogs of the basic hydrodynamic equations [15-17]. In case of an incompressible fluid model (ρ = const) this is:

Mass conservation equation (continuity equation)

$$\frac{\partial u_j}{\partial x_j} = 0,$$

where

$\tilde{u}_j$ – the average value of the fluid velocity in the projection on the *j*-th axis (j = 1, 2, 3).

The equation of conservation of momentum (Reynolds averaging):

$$\rho\left[\frac{\partial U_i}{\partial t}+U_i\frac{\partial U_i}{\partial x_j}\right]=-\frac{\partial P}{\partial x_i}+\frac{\partial}{\partial x_i}\left[\tilde{T}_{ij}^{(v)}-\rho\langle u_i u_j\rangle\right]$$

where

U, P – the average velocity and pressure;

$\tilde{T}_{ij}^{(v)}=2\mu\tilde{S}_{ij}$ – viscous stress tensor for incompressible fluid;

$\tilde{S}_{ij}=\frac{1}{2}\left[\frac{\partial\tilde{u}_i}{\partial x_j}+\frac{\partial\tilde{u}_j}{\partial x_i}\right]$ – instant strain rate tensor;

$\rho\langle u_i u_j\rangle$ – Reynolds stress.

The introduction of the Navier—Stokes equation, averaged according to Reynolds, makes the system of equations not closed, since the additional unknown Reynolds stresses appear. To solve this system, in this problem a semiempirical model of k—ω SST turbulence was used, which introduced the additional equations needed: the equations of transfer of the kinetic energy of turbulence and the relative dissipation rate of this energy:

$$\frac{\partial k}{\partial t}+U_j\frac{\partial k}{\partial x_j}=P_k-\beta k\omega+\frac{\partial}{\partial x_j}\left[(\upsilon+\sigma_k\upsilon_T)\cdot\frac{\partial k}{\partial x_j}\right],$$

$$\frac{\partial\omega}{\partial t}+U_j\frac{\partial\omega}{\partial x_j}=\alpha\cdot S^2-\beta\cdot\omega^2+\frac{\partial}{\partial x_j}\left[(\upsilon+\sigma_\omega\upsilon_T)\cdot\frac{\partial\omega}{\partial x_j}\right]+2\cdot(1-F_l)\cdot\sigma_{\omega 2}\cdot\frac{1}{\omega}\cdot\frac{\partial k}{\partial x_j}\cdot\frac{\partial\omega}{\partial x_j}.$$

The flow part is divided into a set of finite cells, for each of which the discrete analogues of continuous equations are composed. The complex of all discrete analogs forms a closed system of algebraic equations.

As the boundary conditions, the velocity values at the inlet device and the pressure at the outlet from it were taken. The speed was set in such a way as to ensure the flow through the intake system equal to 300 m^3/h.

The computational grid for each model consisted of about 200 thousand cells. Figure 3 shows the computational grids for the original model and the optimized one.

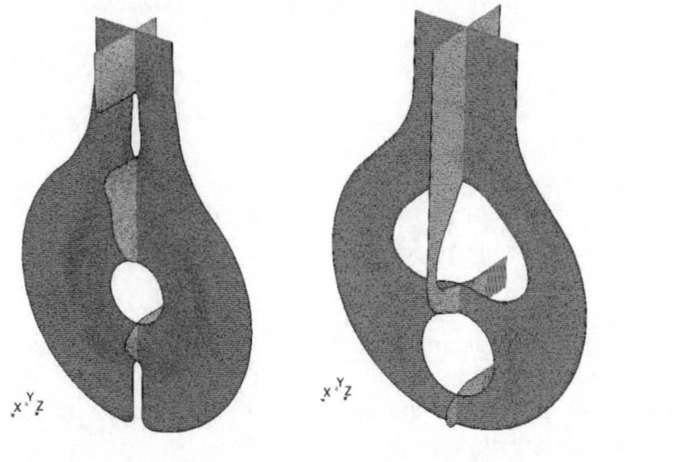

Figure 3. Computational grids for the original model and the optimized one.

3. Results

Based on the above modeling methods, a comparative numerical study of the flow in the original and proposed inlet devices was performed. The comparison criteria were the magnitude of the losses at the pump inlet, the even distribution of the absolute velocity at the inlet to the pump impeller.

Figure 4 shows the vector fields of the absolute velocity distribution in the meridional section of the original inlet device and the optimized one.

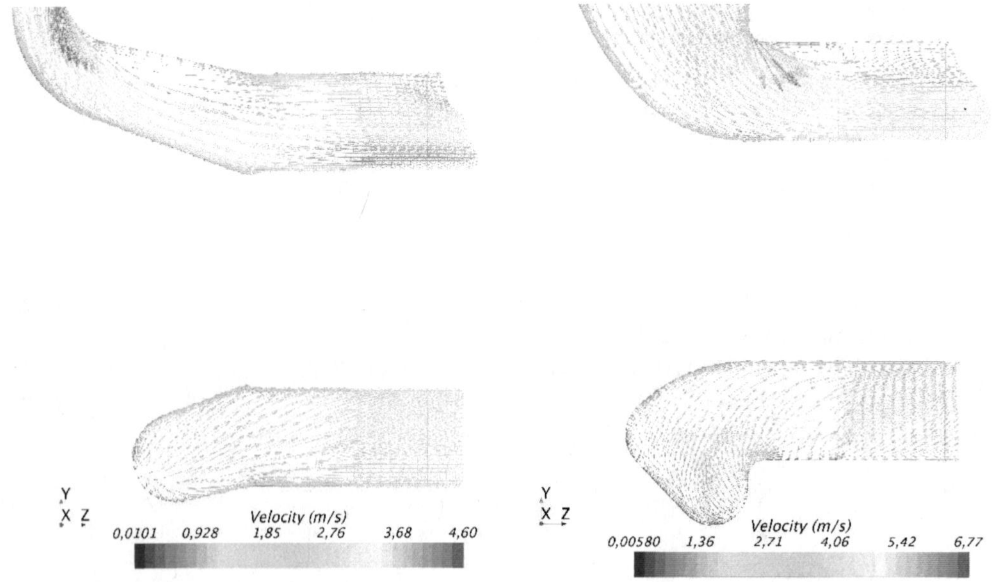

Figure 4. Vector fields of the absolute velocity distribution in the meridional section of the compared devices.

Then, after processing the results of the numerical experiment, the hydraulic losses of both devices were determined. With fluid flowing through the original intake system, the losses, expressed in terms of the differential pressure, were about 5000 Pa. The optimized model showed a result of about 3000 Pa.

The coefficient of evenness of the velocity distribution (Coriolis coefficient) allows to quantify the real distribution of velocities over the flow cross section:

$$\alpha = \frac{\int V^3 \mathrm{d}S}{V_{av}^3 S},$$

where

$V_{av} = \frac{Q}{S}$ – average flow velocity;

S – sectional calculated area;

V – real velocity within limits of dS.

For the annular inlet device $\alpha = 1.2814$, for the optimized one $\alpha = 1.1137$.

Figure 5 shows the fields of the velocities at the cross-sections, where the Coriolis coefficient is determined.

From the data analysis it can be seen that the proposed configuration of the inlet system allows a 30-40% reduction in losses at the pump inlet compared to an annular chamber with a separating edge. And also it allows to reduce the Coriolis coefficient at the inlet to the first stage of sectional pump with $\alpha = 1.2814$ до $\alpha = 1.1137$, which improves the conditions of flow impingement on the impeller blades and improves the cavitation property of the pump.

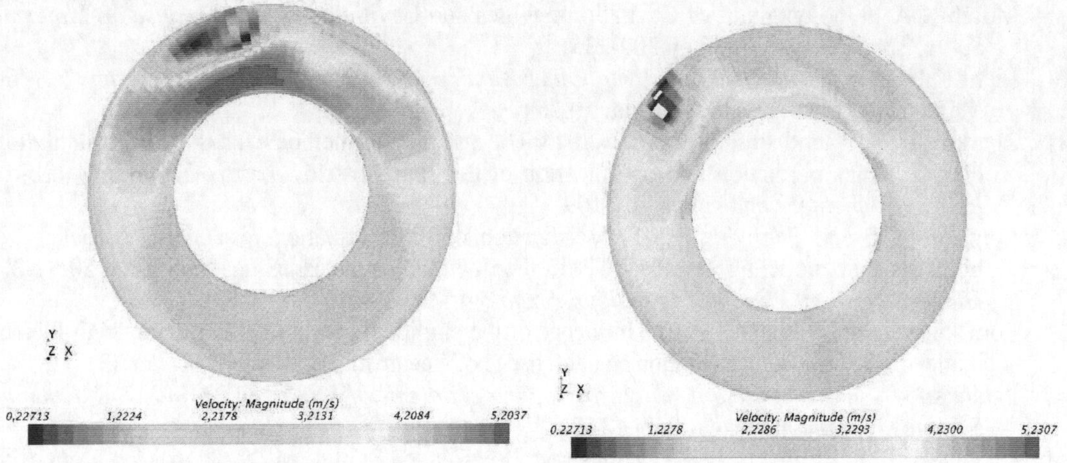

Figure 5. The distribution of the velocity field at the cross-section at the pump inlet.

4. Conclusions

1. As the numerical study has shown, the proposed design of the inlet system can significantly improve the flow structure at the inlet to the first stage of a multistage pump and increase its efficiency rate;

2. Currently, work is underway to manufacture an experimental pump with the geometry of the inlet device given, for carrying out its comprehensive tests and determining the prospects for the proposed design use.

References

[1] Lomakin V O, Chaburko P S and Kuleshova M S 2017 Multi-criteria Optimization of the Flow of a Centrifugal Pump on Energy and Vibroacoustic Characteristics *Procedia Engineering***176** pp 476-482

[2] Lomakin V O, Kuleshova M S and Bozh'eva S M 2016 Numerical Modeling of Liquid Flow in a Pump Station *Power Technology and Engineering***5** pp 324-327

[3] Lomakin V O, Kuleshova M S and Kraeva E A 2015 Fluid Flow in the Throttle Channel in the Presence of Cavitation *Procedia Engineering***106** pp 27-35

[4] Kraposhin M V, Banholzer M, Pfitzner M and Marchevsky I K 2018 A hybrid pressure-based solver for non-ideal single-phase fluid flows at all speeds *International Journal for Numerical Methods in Fluids***88** (2) pp 79-99

[5] Min'kov L L, Krokhina A V, Dueckc J 2011 Hydrodynamic mechanisms of the influence of injection on the classification characteristics of a hydrocyclone*Journal of Engineering Physics and Thermophysics***84** (4) pp 807-819

[6] Ionaitis R R and Chain P L 2009 Principles for engineering active-passive hydrodynamic control-and-safety systems for nuclear reactors *Atomic Energy***106** (3) pp 175-184

[7] Krapivtsev V G 2017 Model Studies of Coolant Flow Hydrodynamics in VVER-1000 In-Reactor Pressure Channel *Atomic Energy***122** (5) pp 304-310

[8] Marchevskii I K and Puzikova V V 2016 Numerical simulation of the flow around two fixed circular airfoils positioned in tandem using the LS-STAG method *Journal of Machinery Manufacture and Reliability***45** (2) pp 130-136

IOP Conf. Series: Materials Science and Engineering **492** (2019) 012040 doi:10.1088/1757-899X/492/1/012040

[9] Morch K A 1976 Dynamics of Cavitation bubbles and Cavitating liquids *Treatise on Materials Science and Technology* **16** pp 309-335
[10] Light K H 2005 *Development of cavitation erosion resistant advanced material system* (Orono, Maine USA: The University of Maine) 76 p
[11] Zharkovsky A A and Pospelov A Yu 2014 The use of 3D methods for the flow calculation, characteristics prediction and optimization of the shape of flow parts of hydraulic turbines *Hydraulic engineering* (11) pp 104-109
[12] Lomakin V O and Petrov A I 2012 Verification of the calculation results in the package of hydrodynamic modeling STAR-CCM+ of the flow part of the centrifugal pump AX 50-32-200 *HigherEducational Institutions Bulletin. Sociology. Economics. Politics* p 6
[13] Lomakin V O and Bibik O Yu 2017 Influence of the empirical coefficients in the Rayleigh-Plessett model on the calculated cavitation characteristics of a centrifugal pump *Hydraulics* (3)
[14] Patankar S 1984 *Numerical methods for solving problems of heat transfer and fluid dynamics* (Moscow: Energoatomizdat Publ.) 152 p
[15] Chaburko P S, Lomakin V O, Kuleshova M S and Baulin M N 2016 Comprehensive optimization of the flow part of the hermetic pump using the LP τ method *Pumps. Turbins. Systems* (1) pp 55-61
[16] Lomakin V O and Chaburko P S 2015 Influence of flow twist on the hydraulic efficiency rate of pump *BMSTU Engineering Bulletin. E-Magazine* (10)
[17] Karelin V Ya 1976 Cavitational phenomena in centrifugal and axial pumps (Moscow: Mashinostroenie Publ.) p 325

IOP Conf. Series: Materials Science and Engineering **492** (2019) 012041 doi:10.1088/1757-899X/492/1/012041

Electrical submersible centrifugal pump units of the new generation for the operation of marginal and inactive wells with a high content of free gas and mechanical impurities

A Trulev[1, 5], V Verbitsky[2], S Timushev[3] and P Chaburko[4, 6]

[1]Rimera Group of companies, Moscow, Russian Federation
[2]Gubkin Russian State University of Oil and Gas, Moscow, Russian Federation
[3]Moscow Aviation Institute, Moscow, Russian Federation
[4] Bauman Moscow State Technical University, 5 Second Baumanskaya Street, Moscow, 105005, Russian Federation

[5]e-mail: aleksey.trulev@rimera.com
[6]e-mail: chaburko@bmstu.ru

Abstract. The paper is devoted to the development of innovative solutions for oil production in complicated conditions using submersible pumping systems. The conditions in which these systems should operate are described. A mathematical model used for computational fluid dynamic simulation is presented. Technical solutions, as well as their key features and advantages over existing systems are presented. The main conclusions are made.

1. Introduction

In recent years, there has been an emerging steadytendency towards the deterioration of the operating conditions of wells in the oil fields of the Russian Federation. This is due to the fact that the fields with favorable geological and field parameters enter the late stage of development, with a corresponding increase in the relative share of oil production from marginal fields with hard to recover reserves.

Active work is underway to intensify the inflow from the bottom hole zone, to increase oil withdrawals, with the result that most of the wells were transferred to operating modes with bottom hole pressures of 3–5 MPa. The growth rate of the mean time between failures for equipment has begun to decline. It became clear that at a large number of wells, electrical centrifugal pumps with the applicable endurance group are no longer able to ensure a further significant increase in parameter of time between failures, since they do not correspond to the changed operating conditions [1].

Submersible pumping equipment of serial production for well operation is often simply unable to work in such complicated conditions.

The share of marginal wells in the Russian Federation is about 30-40%, and most of them have complicating factors: high content of free gas, mechanical impurities, corrosion, scale. A pressing issue is the introduction of an effective technology for their operation with nonfailure operation time of at least 600 days [3].

In line with this, the development, research and pilot implementation of innovative technologies for oil production in complicated conditions using submersible pumping systems becomes an actual problem.

2. Mathematical model

The calculation model is based on the Navier—Stokes (1) and continuity (2) equations:

$$\frac{\partial \mathbf{V}}{\partial t} + \mathbf{V}\nabla(\mathbf{V}) = -\frac{\nabla P}{\rho} + \frac{1}{\rho}\nabla((\mu + \mu_t)(\nabla \mathbf{V})) \quad (1)$$

$$\nabla \cdot \mathbf{V} = 0 \quad (2)$$

where

$\mathbf{V}$ – velocity vector;

t – physical time;

P – pressure;

ρ – density;

μ – molecular viscosity;

μ_t – turbulent viscosity.

These equations are supplemented with the corresponding equations of the turbulence model. The calculations were performed using the standard $k-\varepsilon$ (turbulent energy – dissipation rate) turbulence model [9], in which the turbulent viscosity is determined from relation

$$\mu_t = 0.09 \cdot \rho \cdot \frac{k^2}{\varepsilon}. \quad (3)$$

The presented mathematical model was successfully used in [16–20].

3. Proposed solutions

The new complete set of equipment is proposed: a combination of a special design of stages, pump and elements of a system. New technical solutions are protected by patents for inventions, for example [8] – the conceptual development of the Rimera Group of companies and specialists of the Gubkin Russian State University of Oil and Gas (Gubkin University) under the "White well" project [10].

Impellers have an additional blade ring of a special design, which allows to optimize the axial force acting on the impeller, significantly reducing it, almost three times. Due to the combination of cells and channels in the additional blade ring, the gas-liquid mixture is effectively dispersed, therefore the permissible content of free gas is increased by 10% [14].

The stages are made of cast Ni-Resist, which in terms of combination of the corrosion resistance and wear resistance characteristics is superior to ZhGr (powder steel based on iron (Zh) and graphite (Gr)). Materials testing in the abrasive environment showed that with only abrasive exposed to the surface, the wear rate of materials' samples had similar values. The wear rate increased almost three times for the sample of guide apparatus from ZhGr, which was previously tested in a corrosion-active environment. That is, if the working elements were previously exposed to an acid solution, their wear resistance in an abrasive environment is reduced. Adding 1.5% hydrochloric acid to the model fluid increases the wear rate of materials many times. In this case, the advantages of Ni-Resist as a material more resistant to the effects of "washouts" are fully manifested [4]. In many wells there is a corrosion complication. According to the unified technical requirements (UTR) of the oil company Rosneft (UTR 6), it is not allowed to use pumps with stages made of powder steels for equipment of an abrasion-resistant version (group N3).

Radial-stabilized pumps with a compression assembly configuration belong to the highest-class equipment for wear resistance [1]. This is the most reliable equipment among electrical submersible centrifugal pump (ESP) units. Traditionally they are made for high pump delivery flow rate starting from 200 m^3/day.

The design has been improved. Today, JSC Alnas manufactures such pumps in series for delivery flow rate from 15 m^3/day; it has become expedient to manufacture low-flow pumps in compression version. The high service life of compression pumps allows them to work continuously for three to five years without dismantling, thereby significantly reducing the cost of repair and installation of ESP units [13].

One may note the reliable operation of pumps for 15, 25 m^3/day; for example, at OJSC "Surgutneftegas" the nonfailure operation time is over 4 years. On average, compared to traditional equipment, the operation time increases by more than three times [10], [12].

After bedding-in the support washers, all the axial force is absorbed by the axial support in the protector. As a result, the performance factor of compression pumps can be 10–15% higher than that of traditionally designed ESP units [2].

Compression pumps work reliably in the left area, with a high content of free gas, since the axial load is absorbed by the axial support in the protector. Floating up is not observed in the right area.

The assembly of the compression pump is extremely simplified by design, eliminating the complicated operation of shimming, when plates are installed between the shafts. For accurate alignment of the shafts, a fixture is used that eliminates the need for a measuring tool and helps to avoid human-caused errors.

In addition to this, the specialists of R&D center of the Rimera Group of companies developed a stage of the 5th overall dimensions with an optimum delivery flow rate of 20 with the width of the channels of the flow part equal to ones in the stage with a performance of 125 m^3/day [11], [12].

An analogue is an offer of Schlumberger – pumps with a wide operating range. The operation of wells with a production rate of 25 m^3/day is proposed to be driven by pumps with an optimal flow rate of 125 m^3/day. In this case, the nominal flow rate is 0.2 of the optimal one. According to API 610, the reliability and safety standard developed by the American Petroleum Institute for pumps with face seals for the petroleum and petroleum refining industry, the recommended operating mode of the pump ranges from 0.7 to 1.1 of the optimum delivery flow rate. Only short-term operation at the flow rate of 0.5–0.7 from the optimal one is allowed; when working outside the recommended range, reversed currents occur that lead to pressure pulsations and vibration, an increase in axial and radial force that acts on the impeller of each stage. Moreover, when operating at a delivery flow rate of less than 0.5 from optimal, the pressure characteristic degrades and a possibility of surging, gas lock formation and pump starvation emerges even if there is a small amount of free gas at the pump inlet [6].

New pumping unit elements intended for efficient operation at low flow rates include gas separators with interchangeable inducers and multiphase modules with a new work concept.

Gas separator is considered an unreliable element of a pumping unit due to hydroabrasive cutting of the body. The reason is the reversed currents that occur if the pump delivery flow rate is more than two times less than the calculated gas separator flow rate, the optimum flow rate of the inducer installed at the inlet [5]. For example, the flow part of the gas separator is designed for a delivery flow rate of 250 m^3/day, and the most demanded pump delivery flow rate ranges from 15 to 80 m^3/day. Reversed currents are a trap for mechanical impurities, the concentration of which increases rapidly. A rotating abrasive ring can cut through the body and other elements of the gas separator.

In cooperation with the specialists of Gubkin University, vortex and centrifugal gas separators with replaceable elements of the flow part for delivery flow rate ranges up to 50, 100 and 250 m^3/day were developed [11].

The traditional solution aimed to reduce the angle of the blades at the inducer inlet is not advisable, since it unnecessarily increases the diffusivity of the flow part and reduces the efficiency of the inducer operation [5].

IOP Conf. Series: Materials Science and Engineering **492** (2019) 012041 doi:10.1088/1757-899X/492/1/012041

It was proposed to reduce the diameter at the inducer inlet, while maintaining the optimal flow part, the angles at the inlet, and at the same time maintaining the component parts list.

It was also decided to split the supply part of the gas separator, in which the axial stage is installed, and the separation chamber by a special conical sleeve. In the separation chamber, a separation process takes place; on the periphery a rotating ring with partially separated liquid is formed. In the design this ring presses on the fixed axial support in the form of a conical sleeve, then a high pressure gradient occurs, due to which efficient gas separation is carried out. Previously, the specified ring pressed on the blades of the stage that supplies the gas-liquid mixture of lower density, which led to reversed currents, dispersion, etc.

For the commonality purpose, it is allowed to use one gas separator for the whole range of for delivery flow rates from 15 m^3/day to the maximum rate indicated above.

It should be noted that gas separators designed for low-flow units have less power, which saves energy.

Gas separators developed in cooperation with the specialists of Gubkin University exceed known analogues in terms of separation properties and wear resistance [12].

In the centrifugal forces field it is effective to separate not only free gas, but also mechanical impurities. For marginal wells with a gas-oil mixture, a special design has been developed that combines a gas and mechanical impurities separator within one module [15].

Conventional multiphase modules for low-flow units disperse and compress the liquid-gas mixture (LGM). But, starting with a certain content of free gas at the module inlet, the reverse process of coalescence, merging and enlargement of gas bubbles begins [6], which limits the functionality of the module.

A new concept of compressor dispersing stages (DPT) – the circulation of fluid in the impeller. During the operation, in each stage the LGM is dispersed at the impeller inlet, compressed in the flow part, and the gas is partially separated at the impeller outlet. Gas with the main flow is directed to the next stage inlet, and a part of the liquid is directed to the same impeller inlet. In the Moscow Aviation Institute (MAI)computational studies of three-dimensional turbulent unsteady flow in DPT of different designs were conducted.

Figure 1 shows the distribution diagrams of the total pressure in one of the planes of the flow part of the multiphase stage.

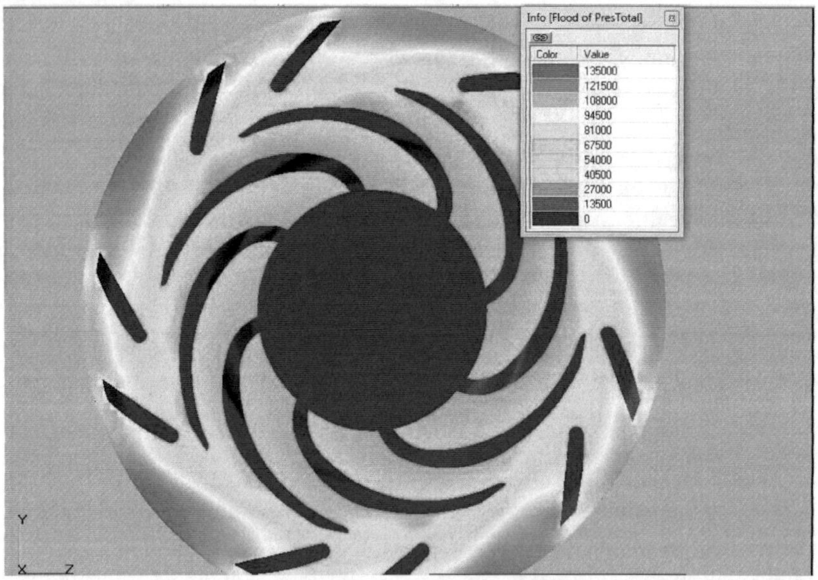

Figure 1. The distribution diagrams of the total pressure in one of the planes of the flow part of the multiphase stage.

IOP Conf. Series: Materials Science and Engineering **492** (2019) 012041 doi:10.1088/1757-899X/492/1/012041

This allows to more than twice reduce the content of free gas in the flow part of the impeller, to eliminate gas locks for off-design operating modes. The DPTs atomize and disperse bubbles of liquid-gas mixture, prepare a homogeneous, fine-dispersed structure of the flow, carry out its preliminary compression and delivery flow to the main stages of the pump. When a pump is operating together with a DPT, the maximum volumetric content of free gas at the inlet is allowed 15% higher compared to the use of a pump without DPT.

For the operation of wells with a high content of free gas and mechanical impurities, protectors of a new generation with a dynamic labyrinth have been developed. The compensating element may be of the diaphragm or piston type.

The design has two gas bells (figure 2).

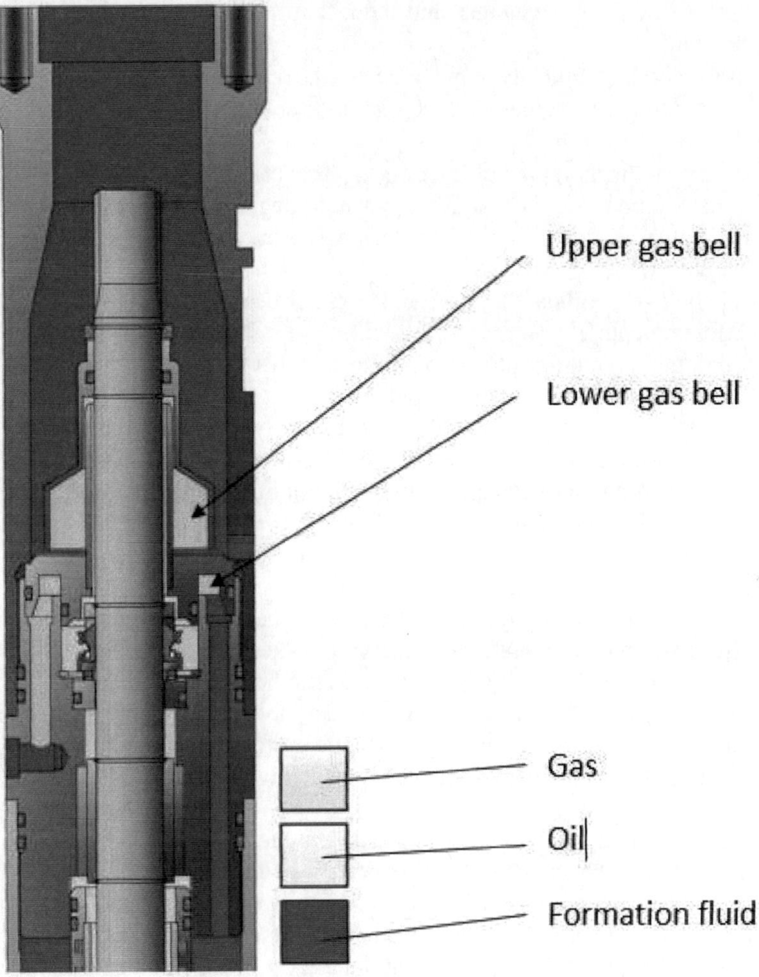

Figure 2. Sketch of the upper part of the protector with a dynamic labyrinth.

The advantages over the serial analogues:

- the gravity labyrinth to protect the upper face seal is replaced by a dynamic labyrinth (gas bell), which has absolute reliability, since moving face seals operating in adverse conditions are replaced with a fixed seal that is not subject to wear;
- the estimated separation efficiency of the dynamic labyrinth is 300 times higher than of the gravity one, since the centrifugal acceleration at the outer radius of the dynamic sleeve is corresponding number of times higher than the free fall acceleration;

- if there is even a small amount of free gas, which is always present in the reservoir fluid, a gas region forms under the bell, separating the face seal not only from mechanical impurities, but also from the reservoir fluid;
- the lower gas labyrinth in the nipple protects the piston module from mechanical impurities and scale, thereby eliminating the probability of piston seizure;
- a pumping device in the form of a labyrinth-screw pair provides that the oil pressure in the area of the face seal exceeds the pressure of the reservoir fluid by 7 meters of water column, which is necessary for the movement of the piston and reliable sealing [7];
- the mounting height of the piston type protector is reduced by 40% compared to the analogue, the cost of repair is reduced due to the smaller number of parts;
- the presence of a filter-heat exchanger and inducer in the area of the bearing assembly allows cooling the thrust bearing;
- the face seal is not hydraulically connected to the compensating element, therefore, when lowering, starting or stopping the engine, the face seal does not contact the formation fluid, it works in an enclosed oil area;
- in comparison with diaphragm type analogues, the effect of "thermos" is eliminated, heat removal is better, penetration of dissolved gas, which degrades the physical properties of the submersible electric motor oil and forms foam during the transition to the free state, through the diaphragm is impossible.

About 70% of all fields in Russia have two or more oil-bearing reservoirs, which are located in parallel planes at different depths.

A unit for dual production, taking into account the features of marginal reservoirs operation, has been developed. The unit is one-string, with one pump.

A reservoir switching valve is used, which allows alternate production from each reservoir.

The unit alternately pumps out the reservoir fluid from the upper and lower reservoir, for a relatively small amount of pressure change, while maintaining a given pressure drawdown for each reservoir. The pump works in continuous operation mode.

Accounting of samplings and measurement of the main parameters for each of the developed reservoirs in accordance with the requirements of RD 153-39.0-109-01 in the field of drill-hole hydrodynamic research and downhole logging is performed when one of the reservoirs is shut off.

Schlumberger uses two units to operate two reservoirs. These units are installed in the same well. Accordingly, the radial overall dimension of each unit should be less than in the case of using one unit.

In comparison with the above described option, the use of one unit allows to significantly save on the cost of equipment, installation and electricity, since the performance factor of one unit is higher.

There is an opportunity to complete unit with serial centrifugal, sucker-rod or progressive cavity pumps.

The unit has a versatility of application, both for dual production and dual injection.

The opportunity of using the algorithm of operation with varying pressure amplitude in the reservoirs from 1 to 250 atmospheres according to a given program and the possibility of changing the of operation program, without removing the unit from the well, are implemented.

New innovative submersible equipment allows, in cooperation with scientists from GubkinUniversity and MAI, to develop a fundamentally new wave technique for fields exploitation with controlled impact on the reservoir, with the aim of increasing oil recovery for the subsequent introduction of new oil production technology by service departments of Rimera Group of companies and by Yuganskneftegeofizika company.

When the pump stops during operating in constant mode and especially in the short-term operation mode, with a backpressure valve installed at the pump outlet, there is a possibility that the gas will fill the pump through the inlets and the formation fluid will flow out into the annular space. This happens just in the same way that gas is collected in a glass with the bottom up-directed, if the glass is installed in a liquid-gas mixture. In this case, there will be considerable difficulties in starting the pump, since the pressure developed by it is proportional to the density of the working medium, and the gas density

IOP Conf. Series: Materials Science and Engineering **492** (2019) 012041 doi:10.1088/1757-899X/492/1/012041

is three orders of magnitude lower than the liquid density. The problem is particularly urgent for low-flow pumps. In cooperation with specialists from Gubkin University and RN-Uganskneftegas company, shut-off and gas exhaust modules are being developed that can successfully solve this problem.

Under the integration project "White well", a comprehensive selection of material for the entire suspension and electrical submersible equipment is carried out for specific operating conditions. A technique that intended for the selection of material taking into account the content of corrosive components in the formation fluid has been developed in the Pervouralsk New Pipe Plant (PNTZ).

4. Conclusions

The innovative solutions proposed in the paper increase the efficiency of oil production in complicated conditions, which is proved by computational fluid dynamic simulation, as well as by operational testing.

References

[1] Yakimov S B 2016 On the prospects for the use of radially stabilized compression centrifugal pumps to improve the efficiency of operation of wells of the AV field group of the Samotlor reservoir (O perspektivakh ispol'zovaniya radial'no stabilizirovannykh kompressionnykh elektrotsentrobezhnykh nasosov dlya povysheniya effektivnosti ekspluatatsii skvazhin plastov gruppy AV Samotlorskogo mestorozhdeniya) *Territory "NEFTEGAZ"* (6) pp 78–86

[2] Loginov V F 2016 Compression pumps for oil production in complicated conditions (Kompressionnyye nasosy dlya neftedobychi v oslozhnennykh usloviyakh) *Oil and gas journal* (6) pp 54–56

[3] Kosilov D A 2015 Improving the efficiency of the management of the mechanical stock of the well in the current macroeconomic conditions. Mechanized production (Povysheniye effektivnosti upravleniya mekhfondom skvazhin v tekushchikh makroekonomicheskikh usloviyakh. Mekhanizirovannaya dobycha) *Proceedings of the specialized section of the conference of Rosneft Oil Company* (Moscow) pp 8–11

[4] Smirnov N I, Smirnov N N and Gorlanov S F2010 Scientific basis for increasing the resource of ESP units for marginal wells (Nauchnyye osnovy povysheniya resursa UETSN dlya malodebitnykh skvazhin) *Engineering practice (Inzhenernaya praktika)* (7) pp 18–21

[5] Borovskiy B I and PetrovV I1975 *High speed vane pumps (Vysokooborotnyye lopatochnyye nasosy)* (Moscow, Mashinostroyeniye Publ.)

[6] Drozdov A N 1982 *Development of methods for calculating the characteristics of a submersible centrifugal pump for the operation of wells with low pressure at the entrance to the pump (Razrabotkametodikiraschetakharakteristikipogruzhnogotsentrobezhnogonasosapriekspluata tsiiskvazhin s nizkimdavleniyem u vkhoda v nasos)*. Thesis for the degree of PhD

[7] Golubev A I 1986 *Seals and sealing technology (Uplotneniya i uplotnitel'naya tekhnika) Handbook* 449 p

[8] *Patent of the Russian Federation No. 2449176 for an invention*: Step of a submersible multistage centrifugal pump Application filed on July 12, 2010 published on April 27 2012

[9] Wilcox D. C. 1994 *Turbulence modeling for CFD* (DCW Industries, Inc) 460 p

[10] Trulev A V 2015 New equipment of CJSC Rimera for wells with complicated operating conditions, Proceedings of the conference on mechanized production *Oil and Gas Vertical (Neftegazovaya Vertikal)* **19**

IOP Conf. Series: Materials Science and Engineering **492** (2019) 012041 doi:10.1088/1757-899X/492/1/012041

[11] Trulev A V 2015 New equipment of OJSC Alnas for wells with complicated operating conditions *Engineering Practice (Inzhenernaya praktika)* (12) pp 71–73
[12] Trulev A V, Sabirov A A, Sibirev S V, Verbitsky V S and Timushev S F 2017 Submersible ESP units with wide channels in the flow part for the production of formation fluid from marginal wells with a high content of mechanical impurities *Engineering Practice (Inzhenernaya praktika)* (1–2) pp 60–63
[13] Trulev A V 2017 Submersible ESP unit of new generation. Innovative solutions for combatting complicated conditions in case of the marginal field, *Oil and Gas Journal Russia* (4) pp 30, 34
[14] Trulev A V 2017 New submersible centrifugal pumps with gray cast iron stages and between bearing design *Engineering Practice (Inzhenernaya praktika)* (5) pp 74–76
[15] Trulev A V and Grachev V V 2018 Bench tests of a submersible gas and mechanical impurities separator in model mixtures *Gas Industry (Gazovayapromyshlennost')* (8) pp 20–24
[16] Lomakin V O 2015 Investigation of two-phase flow in axial-centrifugal impeller by hydrodynamic modeling methods *Proceedings of 2015 International Conference on Fluid Power and Mechatronics, FPM 2015, November 24* pp 1204–1206 DOI: 10.1109/FPM.2015.7337302
[17] Lomakin V O, Kuleshova M S and Bozh'eva S M 2016 Numerical Modeling of Liquid Flow in a Pump Station*Power Technology and Engineering* January 1 **49**(5) pp 324–327 DOI: 10.1007/s10749-016-0623-9
[18] Lomakin V, Cheremushkin V and Chaburko P 2018 Investigation of vortex and hysteresis effects in the inlet device of a centrifugal pump *Global Fluid Power Society PhD Symposium, GFPS 2018*, September 25 2018 DOI: 10.1109/GFPS.2018.8472374
[19] Lomakin V O, Kuleshova M S and Kraeva E A 2015 Fluid Flow in the Throttle Channel in the Presence of Cavitation*Procedia Engineering***106** pp 27–35 DOI: 10.1016/j.proeng.2015.06.005
[20] Gouskov A M, Lomakin V O, Banin E P and Kuleshova M S 2017 Minimization of Hemolysis and Improvement of the Hydrodynamic Efficiency of a Circulatory Support Pump by Optimizing the Pump Flow path *Biomedical Engineering* November 1 **51**(4) pp 229–233 DOI: 10.1007/s10527-017-9720-9

IOP Conf. Series: Materials Science and Engineering **492** (2019) 012042 doi:10.1088/1757-899X/492/1/012042

Mathematical modeling of the mechanisms of volumetric hydraulic machines

S Semenov [1, 2] and D Kulakov[1, 3]

[1] Bauman Moscow State Technical University, 5 Second Baumanskaya Street, Moscow, 105005, Russian Federation

[2] E-mail: a1e10@rambler.ru
[3] E-mail: mitkul@rambler.ru

Abstract. The paper is concerned with an approach for mathematical modeling of the dynamics of the mechanisms of volumetric hydraulic machines based on the representation of them as tree-like structures with superimposed additional external and internal connections. For these purposes, the authors adapted the method of mathematical modeling of tree-like actuators, developed for robotic systems. In this paper, the authors considered the formation features of relationships equations in the framework of the using method. The authors showed the application of the method for modeling the mechanism of an axial-piston hydraulic machine with the peculiarities of its kinematics.

1. Introduction

The scope of application of various hydraulic equipment is extremely wide, due to its exceptionally high performance. To develop more advanced hydraulic equipment, it is necessary to improve the methods of mathematical modeling of volumetric hydraulic machines, hydraulic drive systems, and corresponding hydraulic apparatus. The mathematical models created should be not only adequate, but also compact, well-structured, making it possible to obtain both convenient analytical expressions and efficient computational algorithms. On their basis, both hydraulic equipment control algorithms and, if necessary, computer programs for modeling the processes under investigation can be created.

Developing methods of mathematical modeling, it is necessary to emphasize devices and systems containing complex mechanisms equipped with volumetric hydraulic actuators. Such are traditionally the executive systems of the road construction and load-lifting equipment, multi-support transport vehicles, and hydroficated robotic means.

Many types of modern volumetric hydraulic machines (for example, axial and radial-piston) contain a large number of moving parts that interact through heavy-loaded kinematic pairs of various classes. Corresponding theory and design methods continue to develop [1, 2]. The task of ensuring the reliability and durability of bearing assemblies, friction devices, including distributors, plunger-cylinder block pairs, slipper-swash plate [3, 4, 5, 6] remains an urgent task. Also relevant is the task of improving the workflow of volumetric hydraulic machines [7, 8, 9, 10]. Special mention should be made of studies of processes in thin gaps between rubbing parts [11, 12], an adequate consideration of which is not possible without taking into account the micro-displacements of coupled parts, the

IOP Conf. Series: Materials Science and Engineering **492** (2019) 012042 doi:10.1088/1757-899X/492/1/012042

magnitudes of which are comparable with the values of the gaps. However, the number of papers devoted to the mechanics of hydraulic machines [13] is relatively small. It is necessary to develop methods for mathematical modeling of mechanisms for kinematically complex three-dimensional hydraulic machines in order to increase the efficiency of calculating loads and micro-displacements in tribo-conjugations.

Detailed modeling of the mechanisms of complex volumetric hydraulic machines makes it possible to make more precise the determination of loads on structural elements, analysis of working processes, and improve the analysis of the operation of friction pairs, which basically determine the quality of work, life and reliability of modern volumetric hydraulic machines.

The developed mathematical models of the mechanisms of hydraulic machines should be easily integrated into complex models of hydraulic machines, which take into account a complex set of physical processes occurring in them. They should also allow the construction of efficient computational algorithms with the aim of creating programs for computer simulation and control systems for hydraulic equipment.

2. Methods

When choosing a method of forming mathematical models of the mechanisms of the above-mentioned devices and systems, it is advisable to take into account that in almost all the cases listed, the mechanisms can be represented as tree-like multi-link mechanisms with superimposed additional constant or variable external and internal links. This allows you to develop a unified for a wide class of hydraulic equipment method for the formation of mathematical models of mechanisms (figure 1).

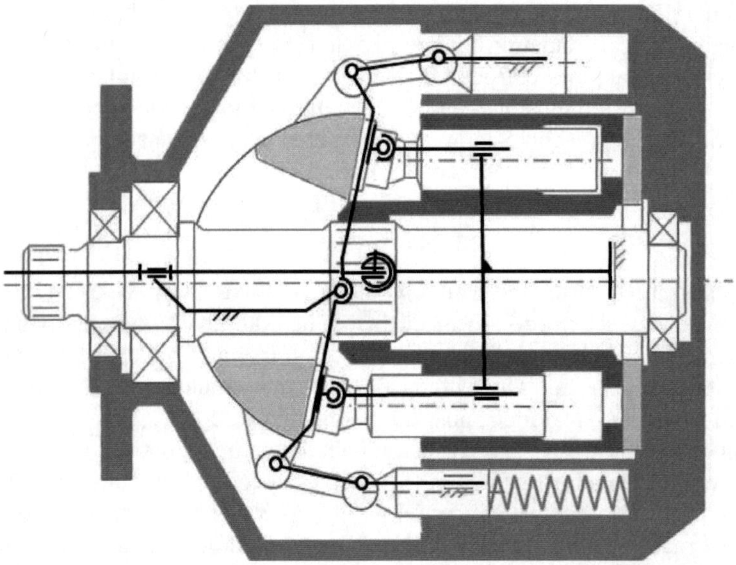

Figure 1. Design of an axial piston pump as a tree-form mechanism.

The form in which the mathematical model is written is also significant. The vector-matrix form is most convenient for the synthesis of control algorithms, it allows you to build compact and computationally efficient modeling and control computer programs.

Effective methods of modeling multi-link mechanisms for a long time developed in robotics. To describe the kinematics of robot mechanisms, Denavit—Hartenberg method [14] is most popular, suggesting certain rules for the assignment of coordinate systems associated with the links of the mechanism and linear (from rack to grips in incremental) order of assigning numbers of links. Based on this method, numerous researchers obtained visual analytical expressions and effective recursive dynamics modeling algorithms for mechanisms that represent simple kinematic chains, as well as robot control algorithms.

Recursive computations for mathematical models of mechanisms, which are simple kinematic chains, assume that the links "one by one" are searched in ascending or descending order of numbers. It is also very easy to refer to the previous or next links for a certain link *i*, simply by writing down, respectively, $i-1$ or $i+1$. However, in tree-like mechanisms this is not possible due to the presence of branching links. Moreover, the application of the Denavit—Hartenberg rules to tree-like mechanisms require that as many coordinate systems as the branches of the mechanism are attached to each branching link. Together, this makes it difficult to directly apply highly efficient mathematical methods of working with multilink mechanisms developed in robotics. A more advanced approach to modeling tree-like mechanisms is described in [15].

The authors propose a method suitable for solving most problems faced by developers in connection with the mathematical modeling of complex hydraulic equipment mechanisms. It is based on the presentation of the mechanisms of not only hydroficated machinery, but also mechanically complex hydraulic machines, as tree structures with superimposed additional external and internal connections. Using the basic principles adopted in robotics, the proposed method differs in the method of assigning the coordinate systems associated with the links and at the same time uses the method of organizing recursive calculations for tree structures described in [16, 17]. The result is a compact recording of the equations of mathematical models of tree-like mechanisms and the efficiency of the computational procedures created on their basis.

As an assumption, we assume that all links of the mechanism are absolutely rigid.

The structure of the mechanism will be described using **a tree-like root oriented graph**. The links of the mechanism correspond to the vertices of the graph, and the kinematic pairs of the fifth class correspond to the edges (figure 2).

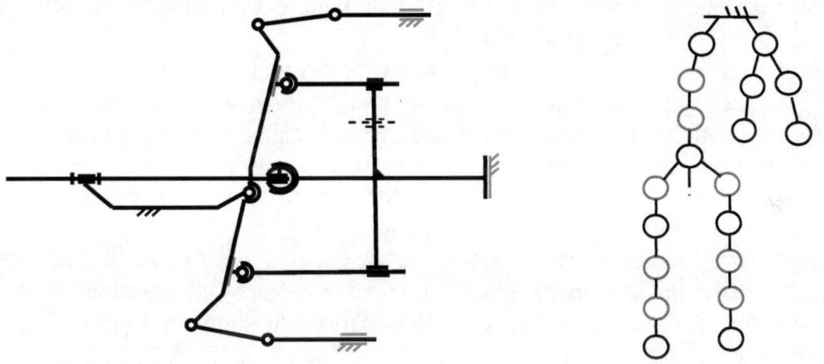

Figure 2. Design of the kinematic structure of the mechanism in the form of a graph.

In most cases, kinematic pairs of the fifth class are used in real mechanisms, however, if the class of the pair is lower, then the necessary number of fictitious weightless links and the corresponding kinematic pairs of the fifth class must be entered between the links connected by such a pair. Similarly, in case the mechanism is not fixed to a fixed base, a 6-step fictitious kinematic chain with zero mass and zero hinge forces must be additionally inserted between the conditional rack and any real link of the mechanism (for example, the machine body).

The links of the mechanism (the vertices of the corresponding graph), including the fictitious ones, are numbered without gaps in the numbers in any order. In the case of a mechanism of a system of equations that is not attached to a fixed base, it is desirable to assign the numbers from 1 to 6 to the links of the fictitious kinematic chain connecting the mechanism to the fixed base.

In the proposed method for the organization of recursive calculations, the following functions are used, known from graph theory (hereinafter N is the total number of links of the executive mechanism, including dummy):

$L = \{1, 2, \ldots, N\}$ – irregular set of graph vertices numbers (motion rod);

$d(i, j)$ – path length from vertex i to vertex j (number of edges on the way);

Dnum(i) – vertexdeep number *i,* which means its sequence number in the search for depth from the root of the tree ;

$\Gamma(i)$ – tuple of vertices numbers that are for vertex *i* children (directly connected be the following vertices);

$dg^{+}(i)$ – half degree of vertex outcome *i*(in terms of graph theory is the number of children).

Additionally added the following function:

$Dnum_k(i)$ – depth number of vertex *i* when in search for depth from vertex *k*;

In the notation of variables and parameters of links of the mechanism, it is proposed to use three index:

– the lower right shows the number of the link to which this value refers;

– upper right — any mathematical symbol, for example, the sign of transposition or exponentiation;

– top left — number of the coordinate system in which the value is expressed; if this index is absent or zero, then the value is set in the base (absolute) coordinate system; if some non-zero integer is specified *i*, then the rate is set in the link system *i*; if the index is *c*, then it is set in the system of the link to which it itself belongs (in the associated system).

In the case of simple kinematic chains, if necessary, designate the value of x related to the previous or subsequent links in relation to link*i* (in terms of graph theory to the parent or child), use the record xi-1 or xi + 1, respectively. However, in the case of a tree-like mechanism with arbitrarily numbered links, such a record loses its meaning. To overcome this difficulty, it is proposed to use the following index functions as indices:

s(i, k) – number of the link that is *k*-th next link of the graph G (*i*) (*k*-th vertex-child) for the link *i*;

f(i) – number of the link that is the previous (top-parent) for the link*i*;

ns(i) – determines the sequence number of the link i in graph G (f(i));

p(i, j, k) – number *k*-th summit on the path from *i* to *j*.

Then the values related to the previous (link parent) or the next (link child) in relation to link i links can be written, respectively — $\boldsymbol{x}_{f(i)}$and$\boldsymbol{x}_{s(i,j)}$. This approach allows mathematical expressions for tree-like mechanisms to be made as compact and descriptive as for simple kinematic chains.

The number of each joint (kinematic pair of the 5th class) in the mechanism coincides with the number of the link that this joint attaches to the father link. The type of articulation determines the multiplierσ_i: if σ_i= 1 – rotary joint; if σ_i=0, translational joint.

The proposed method assumes that, in contrast to the Denavit—Hartenberg coordinate systems, the links associated with the links are assigned so that the axis Z_i each *i*-th coordinate system is assigned lying on the joint axis of the link *i* with link *f(i)* (link parent), as shown in figure 3. This joint is also assigned number *i* [18].

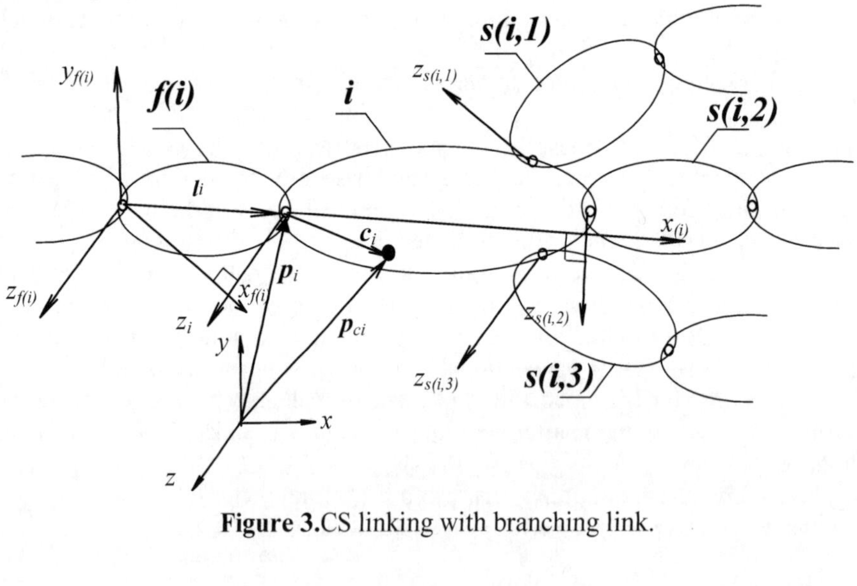

Figure 3.CS linking with branching link.

In general to transition from a path$i$ to system *f(i)* six elementary movements are required. Using the following combination of movements:

1) rotate the system *f(i)* through an angle β_i around the axis $Y_{f(i)}$, until the axis $X_{f(i)}$and the axis Z_i become perpendicular;

2) moving the system *f(i)* by the amount b_i along the axis $Y_{f(i)}$ to the intersection $X_{f(i)}$ and $Z_i$at an angle 90 degrees;

3) moving the system *f(i)* at an angelα_i around the axis $x_{f(i)}$ until the axes $z_{f(i)}$ and z_i become parallel;

4) displacement of the system *f(i)* at a distacea_i along the axis $x_{f(i)}$ to the aligment of the axes $z_{f(i)}$ and z_i,;

5) displacement of the system *f(i)* by the distance d_i along the axis $z_{f(i)}$ to the coincidence of the beginnings of the coordinate systems *i* and *f(i);*

6) system rotation *f(i)* at an angleθ_i around the axis $z_{f(i)}$ until all axes are aligned.

To convert the coordinates to each link *i*, it is traditional [14] to assign the transition matrix $A_i$from system *i* to system *f (i)* (father-link system) and T_i transition from the coordinate system associated with link *i* to the absolute coordinate system. When using the proposed method of assigning coordinate systems and the order of elementary motions, the expressions for calculating the elements of the matrices $A_i$are of the form (to shorten the writing of expressions, the function sin is designated as *S*, and the function cos is defined as *C*):

$$A_i = \begin{pmatrix} C\beta_i C\theta_i + S\alpha_i S\beta_i S\theta_i & -C\beta_i S\theta_i + S\alpha_i S\beta_i C\theta_i & C\alpha_i S\beta_i & a_i C\beta_i + d_i C\alpha_i S\beta_i \\ C\alpha_i S\theta_i & C\alpha_i C\theta_i & -S\alpha_i & b_i - d_i S\alpha_i \\ -S\beta_i C\theta_i + S\alpha_i C\beta_i S\theta_i & S\beta_i S\theta_i + S\alpha_i C\beta_i C\theta_i & C\alpha_i C\beta_i & -a_i S\beta_i + d_i C\alpha_i C\beta_i \\ 0 & 0 & 0 & 1 \end{pmatrix}$$

Relevant submatrices will be used below.

Traditionally, for the generalized coordinates of the mechanism are taken $q_i = \theta_i$ for rotating pairs, either $q_i = d_i$ for prismatic pair;

As for simple kinematic chains, the matrices $T_i$are calculated recurrently, and the inverse function of calculating depth numbers is used to determine the order of enumeration of the numbers of links [19]:

$$T_i = T_{f(i)} A_i; \qquad i = Dnum^{-1}(n); \quad n = 1,\ldots,N.$$

In block form, the Matrices A_i and T_i can be represented as follows:

$$A_i = \begin{pmatrix} {}^{f(i)}R_i & {}^{f(i)}\overline{l}_i \\ 0\ 0\ 0 & 1 \end{pmatrix}$$

$$T_i = \begin{pmatrix} R_i & \overline{p}_i \\ 0\ 0\ 0 & 1 \end{pmatrix} = \begin{pmatrix} \overline{x}_i & \overline{y}_i & \overline{z}_i & \overline{p}_i \\ 0 & 0 & 0 & 1 \end{pmatrix},$$

Their components will be used in the future.

The inverse problem of dynamics for a mechanism can be solved recursively [20]. Taking into account the proposed method of assigning coordinate systems, assumptions about the absolute rigidity of the links of the mechanism, immobility of the rack (real or, in the case of the mechanism not fixed to the fixed base, conditional), angular velocities and accelerations of the links, as well as linear velocities and accelerations of their characteristic points with links of coordinate systems are calculated as follows:

$${}^c\overline{\omega}_i = {}^{f(i)}R_i^T\, {}^c\overline{\omega}_{f(i)} + \sigma_i \overline{k}\dot{q}_i$$

$${}^c\overline{\varepsilon}_i = {}^{f(i)}R_i^T\, {}^c\overline{\varepsilon}_{f(i)} + \sigma_i(\overline{k}\ddot{q}_i + {}^c\overline{\omega}_i \times \overline{k}\dot{q}_i)$$

$$^c\overline{v}_i = {}^{f(i)}R_i^T({}^c\overline{v}_{f(i)} + {}^c\overline{\omega}_{f(i)} \times {}^{f(i)}\overline{l}_i) + (1-\sigma)\overline{k}\dot{q}_i)$$

$$^c\overline{a}_i = {}^{f(i)}R_i^T({}^c\overline{a}_{f(i)} + {}^c\overline{\varepsilon}_{f(i)} \times {}^{f(i)}\overline{l}_i + {}^c\overline{\omega}_{f(i)} \times ({}^c\overline{\omega}_{f(i)} \times {}^{f(i)}\overline{l}_i)) + (1-\sigma_i)(k\ddot{q}_i + 2\,{}^c\overline{\omega}_i \times k\dot{q}_i),$$

$$^c\overline{a}_{ci} = {}^c\overline{a}_i + {}^c\overline{\varepsilon}_i \times {}^c\overline{c}_i + {}^c\overline{\omega}_i \times ({}^c\overline{\omega}_i \times {}^c\overline{c}_i)$$

$$i = Dnun^{-1}(n), \quad n = 1,...,N$$

where

$\overline{\omega}_i$ – angular velocity of a link i;

$\overline{\varepsilon}_i$ – angular acceleration of a link i;

$\overline{v}_i$ – coordinate system speed of a link i;

$\dot{q}_i$ – i-th generalized speed equal to the speed of mutual displacement of link i and the previous link $f(i)$;

$\ddot{q}_i$ – i-th generalized acceleration;

$\overline{l}_i$ – vector connecting the beginning of the SC link of the parent for the$i$ link, with the beginning of the SC link i ;

$\overline{c}_i$ – mass position vector of a link i;

$\overline{a}_i$ – acceleration of the beginning of the coordinate system of the link i;

$\overline{a}_{ci}$ – acceleration of the center of mass of the link i.

To determine the order of iterating over the numbers of links in direct recursion, we use the inverse function to calculate the depth numbers.

In the reverse order of recursion, the interaction forces of the links directly connected by kinematic pairs are calculated (in related coordinate systems) $^c\overline{f}_i$ and $^c\overline{M}_i$ and accordingly the hinge efforts τ_i:

$$^c\overline{f}_i = \sum_{k=1}^{dg^+(i)} {}^iR_{s(i,k)} \cdot {}^c\overline{f}_{s(i,k)} + (R_i^T\overline{g} - {}^c\overline{a}_{ci})m_i + R_i^T\overline{F}_{вн i}$$

$$^c\overline{M}_i = \sum_{k=1}^{dg^+(i)} ({}^iR_{s(i,k)} \cdot {}^c\overline{M}_{s(i,k)} + {}^i\overline{l}_{s(i,k)} \times {}^iR_{s(i,k)} \cdot {}^c\overline{f}_{s(i,k)}) -$$

$$- {}^cJ_{ci}\,{}^c\overline{\varepsilon}_i - {}^c\overline{\omega}_i \times {}^cJ_{ci}\,{}^c\overline{\omega}_i + R_i^T\overline{M}_{вн i} + {}^c\overline{c}_i \times ((R_i^T\overline{g} - {}^c\overline{a}_{ci})m_i + R_i^T\overline{F}_{вн i})$$

$$i = Dnum^{-1}(n), \quad n = N,...,1$$

where

$\overline{F}_{вн}$ – sum of external forces acting on the link i;

$\overline{M}_{вн}$ – sum of moments of external forces acting on link i relative to its center of mass;

$^cJ_{ci}$ – inertia tensor of the i-th link, expressed in the system of this link;

$\overline{g}$ – gravity vector;

m_i – i-th link mass.

Expression for articulated force:

$$\tau_i = -\overline{k} \cdot (\sigma_i\,{}^c\overline{M}_i + (1-\sigma_i)\,{}^c\overline{f}_i) \quad i = Dnum^{-1}(n), \quad n = N,...,1,$$

where

$\overline{k} = (0,0,1)^T$.

Accordingly, the calculations are performed in the reverse order of listing the depth numbers.

To solve the direct problem of dynamics, the system of equations of the dynamics of a mechanism is often represented as:

$$I(\overline{q})\ddot{\overline{q}} + H(\overline{q},\dot{\overline{q}}) = \overline{\tau},$$

where
I – inertia matrix,
H – matrix column of mobility degrees of centrifugal and Coriolis forces, as well as external forces,
$\overline{\tau}$ – hinge force vector.

Adequately:

$$\ddot{\overline{q}} = I(\overline{q})^{-1}(\overline{\tau} - H(\overline{q},\dot{\overline{q}})).$$

Each column of the inertia matrix can be interpreted and calculated as a vector of hinge forces, causing a single generalized acceleration in the degree of mobility with the number of this column with the rest of the generalized accelerations equal to zero and all the generalized speeds equal to zero. Considering that the inertia matrix does not depend on the generalized velocities, external loads, reactions of additional constraints imposed on the mechanism, we take these values to be also zero when calculating. Then to calculate the i-th column of the inertia matrix in direct recursion order with the current values of the generalized coordinates:

$${}^{c}\overline{\varepsilon}_{j}^{**} = {}^{f(j)}R_{j}^{T}\,{}^{c}\overline{\varepsilon}_{f(j)}^{**} + \sigma_{i}e_{ij}\overline{k}$$

$${}^{c}\overline{a}_{j}^{**} = {}^{f(j)}R_{j}^{T}({}^{c}\overline{a}_{f(j)}^{**} + {}^{c}\overline{\varepsilon}_{f(j)}^{**} \times {}^{f(j)}\overline{l}_{j}) + (1-\sigma)e_{ij}\overline{k}$$

$${}^{c}\overline{a}_{cj}^{**} = {}^{c}\overline{a}_{j}^{**} + {}^{c}\overline{\varepsilon}_{j}^{**} \times {}^{c}\overline{c}_{j}$$

$$j = Dnum_{i}^{-1}(n); \quad n = 1,\ldots,\max Dnum_{i}$$

where
e_{ij}– element matrix E ($e_{ij} = 1$, if $I = j$, otherwise $e_{ij} = 0$).

The remaining accelerations are assigned zero values..

In reverse recursion order:

$${}^{c}\overline{f}_{j}^{**} = \sum_{k=1}^{dg^{+}(j)} {}^{j}R_{s(j,k)} \cdot {}^{c}\overline{f}_{s(j,k)}^{**} - {}^{c}\overline{a}_{cj}^{**}m_{j}$$

$${}^{c}\overline{M}_{j}^{**} = \sum_{k=1}^{dg^{+}(j)} ({}^{j}R_{s(j,k)} \cdot {}^{c}\overline{M}_{s(j,k)}^{**} + {}^{j}\overline{l}_{s(j,k)} \times {}^{j}R_{s(j,k)} \cdot {}^{c}\overline{f}_{s(j,k)}^{**}) -$$
$$- {}^{c}J_{cj}\,{}^{c}\overline{\varepsilon}_{j}^{**} - {}^{c}\overline{c}_{j} \times {}^{c}\overline{a}_{cj}^{**}m_{j}$$

$$j = Dnum^{-1}(n), \quad n = N,\ldots,1$$

Elements of the i-th column of the inertia matrix:

$$i_{ji} = -\overline{k} \cdot (\sigma_{j}\,{}^{c}\overline{M}_{j}^{**} + (1-\sigma_{j}){}^{c}\overline{f}_{j}^{**}), \qquad j = Dnum^{-1}(n), \quad n = N,\ldots,1$$

When calculating the elements of the matrix-column H of centrifugal, Coriolis and external forces, it is advisable to use the independence of its elements from generalized accelerations and reactions of superimposed external connections. We calculate its elements recursively as articulated efforts at zero generalized accelerations and in the absence of additional constraints imposed on the mechanism. In direct recursion:

$${}^{c}\overline{\omega}_{k}^{*} = {}^{f(k)}R_{k}^{T}\,{}^{c}\overline{\omega}_{f(k)}^{*} + \sigma_{k}\overline{k}\dot{q}_{k}$$

$${}^{c}\overline{\varepsilon}_{k}^{*} = {}^{f(k)}R_{k}^{T}\,{}^{c}\overline{\varepsilon}_{f(k)}^{*} + \sigma_{k}\,{}^{c}\overline{\omega}_{k}^{*} \times \overline{k}\dot{q}_{k}$$

$${}^{c}\overline{a}_{k}^{*} = {}^{f(k)}R_{k}^{T}({}^{c}\overline{a}_{f(k)}^{*} + {}^{c}\overline{\varepsilon}_{f(k)}^{*} \times {}^{f(k)}\overline{l}_{k} + {}^{c}\overline{\omega}_{f(k)}^{*} \times ({}^{c}\overline{\omega}_{f(k)}^{*} \times {}^{f(k)}\overline{l}_{k})) + 2(1-\sigma){}^{c}\overline{\omega}_{k}^{*} \times \overline{k}\dot{q}_{k}$$

$${}^{c}\overline{a}_{ck}^{*} = {}^{c}\overline{a}_{i}^{*} + {}^{c}\overline{\varepsilon}_{k}^{*} \times {}^{c}\overline{c}_{k} + {}^{c}\overline{\omega}_{k}^{*} \times ({}^{c}\overline{\omega}_{k}^{*} \times {}^{c}\overline{c}_{k})$$

$$k = Dnum^{-1}(n); \quad n = 1,\ldots,N.$$

In reverse recursion:

$${}^{c}\bar{f}_i^* = \sum_{k=1}^{dg^+(i)} {}^{i}R_{s(i,k)} \cdot {}^{c}\bar{f}_{s(i,k)}^* + {}^{i}\bar{F}_{вн} + (R_i^T \bar{g} - {}^{c}\bar{a}_{ci}^*)m_i$$

$${}^{c}\bar{M}_i^* = \sum_{k=1}^{dg^+(i)} ({}^{i}R_{s(i,k)} \cdot {}^{c}\bar{M}_{s(i,k)}^* + {}^{i}\bar{l}_{s(i,k)} \times {}^{i}R_{s(i,k)} \cdot {}^{c}\bar{f}_{s(i,k)}^*) +$$
$$+ {}^{c}\bar{M}_{вн\, i} - {}^{c}J_{ci}\, {}^{c}\bar{\varepsilon}_i^* - {}^{c}\bar{\omega}_i^* \times {}^{c}J_{ci}\, {}^{c}\bar{\omega}_i^* + {}^{c}\bar{c}_i \times ({}^{c}\bar{F}_{вн\, i} + (R_i^T \bar{g} - {}^{c}\bar{a}_{ci}^*)m_i)$$

$$i = Dnum^{-1}(n), \quad n = N,\ldots,1$$

Elements of matrix H

$$h_i = -\bar{k} \cdot (\sigma_i\, {}^{c}\bar{M}_i^* + (1-\sigma_i)\, {}^{c}\bar{f}_i^*),, \quad i = Dnum^{-1}(n), \quad n = N,\ldots,1$$

As a rule, for an adequate description of the kinematics and dynamics of the mechanisms, a certain number of both external and internal additional connections, which may be constant or variable in time, must be added to the tree structure. They can be elastic and rigid. The case where the bonds can be regarded as elastic is fairly simple. The structure of the equations does not change, and the coupling reactions are presented as functions of the movement of the mechanism and are taken into account as external forces in the H matrix. However, this is not always possible, and then additional equations of added constraints expressed relative to accelerations must be added to the system of equations obtained above. In matrix form:

$$J\ddot{\bar{q}} + P = 0,$$

where
J – matrix of partial derivatives (Jacobi matrix) of dimension $K \times N$, where K – number of links, P – matrix column dimension K.

When modeling the mechanisms of hydraulic equipment is convenient to use the equations of the form:

$$\ddot{\bar{r}}_i \cdot \bar{\theta} - \bar{a}_\theta(t) = 0,$$

$$\dot{\bar{\omega}}_i \cdot \bar{\theta} - \bar{\varepsilon}_\theta(t) = 0$$

corresponding situations when the projection of the linear vector (for some point) or angular acceleration of the link i to the direction θ equal to some time function ($a_\theta(t)$ or $\varepsilon_\theta(t)$). In case additional internal communication between the links of the mechanism is imposed, these equations will include the sums of accelerations of these links.

Linkage reactions $\bar{\lambda}$ (dimension of the vector is equal to the number of superimposed bonds), as well as other forces acting on the mechanism contribute to the hinge efforts:

$$\bar{\tau}_\lambda = -J^T \bar{\lambda}$$

Then, taking into account the constraints imposed on the accelerations, and the reactions of these constraints, the dynamics of the mechanism are written in the following block-matrix form:

$$\begin{pmatrix} I(\bar{q}) & -J^T(\bar{q}) \\ -J^T(\bar{q}) & 0 \end{pmatrix} \begin{pmatrix} \ddot{\bar{q}} \\ \bar{\lambda} \end{pmatrix} + \begin{pmatrix} H(\bar{q},\dot{\bar{q}}) \\ -P(\bar{q},\dot{\bar{q}}) \end{pmatrix} = \begin{pmatrix} \bar{\tau} \\ 0 \end{pmatrix}$$

where

$\overline{q}$ – vector of generalized coordinates of the mechanism without additional links;
$\overline{\lambda}$ – coupling reaction vector;
$\overline{\tau}$ – mechanism hinge vector;
$I(\overline{q})$ – symmetrical mechanism inertia matrix;
$J(\overline{q})$ – Jacobi matrix from the block of equations of additional constraints;
$H(\overline{q},\dot{\overline{q}})$ – matrix column of reduced centrifugal, Coriolis and external forces (without taking into account reactions of additional bonds);
$P(\overline{q},\dot{\overline{q}})$ – matrix column of the right sides of the equations of additional relations.

3. Specificity of mathematical modeling of axial-piston hydraulic machines

The proposed method can be successfully applied in modeling the most complex modern volumetric hydraulic machines: axial-piston, radial-piston and others, including those regulated with the possibility of detailed modeling of the motion of the regulator links with the definition of all the forces acting on the links of the mechanism. For this, the mechanism of the hydraulic machine is represented as a tree-like system of solids connected by kinematic pairs of the fifth class, with several additional external and internal connections. Along with the load on the links of the mechanism, determined by the working process of hydraulic machines, this method allows you to easily take into account other loads, such as weight forces, as well as inertial and vibration loads, hydrodynamic forces. Great prospects are opened up by the possibility of supplementing the mathematical and corresponding computer models with various models of physical processes that are strongly dependent on the micro-displacements of the links of the mechanism relative to each other. For example, these are friction processes in the most loaded moving mates of hydraulic machines: in sliding bearings, in distribution devices, in pairs of a plunger- cylinder block, a slipper-swash plate, and the like.

The proposed method makes it possible to simplify the overcoming of considerable difficulties arising in the course of hydrodynamic modeling of flows in the gaps of a cylinder block-distributor, plunger-cylinder block, slipper-swash plate. The difficulties are caused by the need to take into account in the numerical hydrodynamic modeling of relative micro-movements of the order of the magnitude of the gap between the parts forming the conjugation. For this, simultaneously with the simulation of fluid flow in the gap, it is required to simulate the movement of the mechanism, and very precisely and taking into account the instantaneous values of those loads that are created by the simulated fluid flow in the gap being studied.

The proposed method of modeling the mechanisms of volumetric hydraulic machines can be demonstrated on the example of an adjustable axial-piston pump with aswash plate. The pump mechanism should be presented in the form of a kinematic tree, branching directly from the root, for which it is advisable to take the case. The first branch corresponds to the shaft, the cylinder block and then the branches of the pistons with the slippers; the second branch corresponds to aswash plate with a variable angle of inclination with the drive mechanism of the regulator (figure 4).

In the first stage of the formation of a mathematical model of the mechanism, in accordance with the rules set forth above, it is necessary to assign the coordinate systems associated with the links, coordinate transformation parameters during the transition from system to system within the tree-like kinematic structure, set or calculate index functions, assign mass inertia characteristics of links of the mechanism.

The tree structure does not fully describe the kinematics of the pump mechanism; therefore, it is necessary to describe the interaction of the pump unit with the distributor, and the slippers - with aswash plate. For this, it is necessary to use, together with the mathematical model of the mechanism, mathematical models of the interaction of links with allowance for hydrodynamic and tribological processes in conjugations, taking into account the mutual displacement of parts, including on the scale of the gap values. The ability to form such detailed models of hydraulic machines is one of the main advantages of the proposed approach. But if the purpose of the research for which the model is being

IOP Conf. Series: Materials Science and Engineering **492** (2019) 012042 doi:10.1088/1757-899X/492/1/012042

formed allows us to limit ourselves to a detailed consideration of only one or a few mates, the rest should be described with additional external and internal rigid links from the following list:

1) three connections limiting the movement of the cylinder block relative to the distributor; for flat distributor:

- prohibition of cylinder block displacement along the axis of the distributor;
- prohibition of rotation of the cylinder block around the axes lying in the plane of the distributor;

2) a group of bonds holding slip planes of the slippers in the plane of the swash plate (three links for each slipper)

- prohibition of displacement of each slipper along an axis perpendicular to the working plane of the swash plate;
- the prohibition of rotation of each slipper around the axes lying in the working plane of the swash plate.

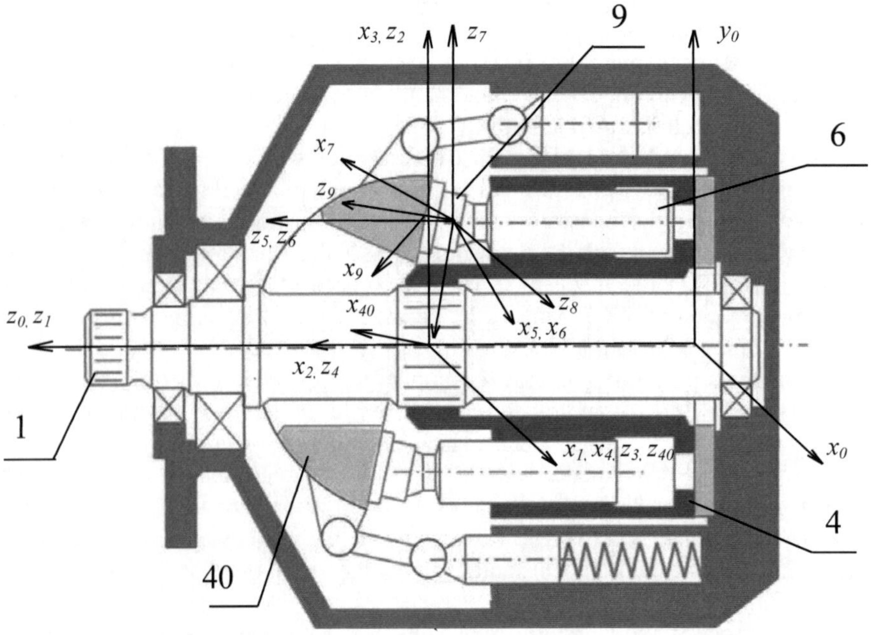

Figure 4. Axial piston pump mechanism and coordinate system associated with mechanism links.

To write equations of relations, it is necessary to assign in the coordinate systems of the corresponding links a point of overlapping relations. For the example in question, these may be:

${}^{c}\overline{r}_{c\,4}$ - cylinderblock radius vector;

${}^{c}\overline{r}_{c\,9}, {}^{c}\overline{r}_{c\,14}, {}^{c}\overline{r}_{c\,19}, {}^{c}\overline{r}_{c\,24}, {}^{c}\overline{r}_{c\,29}, {}^{c}\overline{r}_{c\,34}, {}^{c}\overline{r}_{c\,39}$ – radius vectors of slipper end centers.

For an axial-piston pump with aswash plate, the points of superposition of the links of the swash plate with the slippers are movable in the swash plate coordinate system; therefore their coordinates should be calculated during the simulation at each integration step.

In the case of imposition on any link i and its point with radius vector ${}^{c}\overline{r}_{c\,i}$ six constraints prohibiting both all rotational and all translational motions, the Jacobi matrix included in the corresponding block-matrix equation can be written as:

$$ {}^{i}J_{3T3R} = \begin{pmatrix} {}^{i}J_{3T} \\ {}^{i}J_{3R} \end{pmatrix} = \begin{pmatrix} {}^{i}j_{T1} & \cdots & {}^{i}j_{T\,k} & \cdots & {}^{i}j_{T\,N} \\ {}^{i}j_{R1} & \cdots & {}^{i}j_{R\,k} & \cdots & {}^{i}j_{R\,N} \end{pmatrix}, \quad k = 1,2,\cdots,N $$

where

${}^{i}J_{3T}$ and ${}^{i}J_{3R}$ – 3×N submatrices responsible, respectively, for blocking the translational and rotational motion;

${}^{i}j_{T\,k}$ and ${}^{i}j_{R\,k}$ – 3×1 submatrix of matrix column ${}^{i}J_{3T}$ and ${}^{i}J_{3R}$, corresponding k-th degree of mobility, calculated as follows:

$$ {}^{i}j_{T\,k} = \begin{cases} (0,0,0)^{T}; & k \notin P_{0,i} \\ \sigma_k \bar{z}_k \times (\bar{p}_i + R_i\,{}^{c}\bar{r}_{c\,i} - \bar{p}_k) + (1-\sigma_k)\bar{z}_k; & k \in P_{0,i} \end{cases} \qquad k = 1,\dots,N $$

$$ {}^{i}j_{R\,k} = \begin{cases} (0,0,0)^{T}; & k \notin P_{0,i} \\ \sigma_k \bar{z}_k; & k \in P_{0,i} \end{cases} \qquad k = 1,\dots,N $$

where

σ_k – a factor of 1 for the rotating pair and 0 is for the translational pair;

$\bar{z}_k$ – directional vector of the k-th articulation (5th grade);

$\bar{p}_k$ – pole vector of the kth link coordinate system;

R_i – orientation matrix in the absolute coordinate system of the kth link;

$P_{0,i}$ – many numbers of links that make up the path from the body to the link i.

If links are superimposed on links of less than six degrees, the dimension of the Jacobi matrices in the equations of bonds for these links decreases. With reference to the considered example with a seven-plunger pump, both the distributor and the slippers are superimposed, bonds blocking the translational motion along one axis, and rotation around two more. Then the dimension of the corresponding Jacobi matrices becomes 3xN. If the mobility of the link i is limited in the absolute system (as in the case of the cylinder block of the pump in this example), we denote the corresponding matrix ${}^{i}J_{1T2R}$. Its columns can be calculated as follows:

$$ {}^{i}j_{1T2R\;k} = \begin{pmatrix} {}^{i}j_{T\,k}^{T} \cdot \bar{n} \\ {}^{i}j_{R\,k}^{T} \cdot \bar{l} \\ {}^{i}j_{R\,k}^{T} \cdot \bar{m} \end{pmatrix} $$

where

i– number of the link to which the connection is imposed,

k – Jacobi matrix column number,

$\bar{n}$ – direction vector of the axis along which the link is imposed on the translational motion of the link point i,

$\bar{l}, \bar{m}$ – direction vectors of the axes around which the connection is imposed on the rotational motion of the link i.

If additional internal links are added between the links of the mechanism with the numbers a and b (as in the case of contact of the slippers with aswash plate), we denote the corresponding matrices ${}^{a,b}J_{1T2R}$. Their columns can be calculated as follows:

$$ {}^{a,b}j_{1T2R} = \begin{pmatrix} ({}^{a}j_{T\,k} - {}^{b}j_{T\,k})^{T} \cdot \bar{n} \\ ({}^{a}j_{R\,k} - {}^{b}j_{R\,k})^{T} \cdot \bar{l} \\ ({}^{a}j_{R\,k} - {}^{b}j_{R\,k})^{T} \cdot \bar{m} \end{pmatrix}, $$

where

a and b – numbers of interconnected links.

If the overlap point between two links is movable relative to one of the links (in this case, the connection point between the slipper and the swash plate is movable relative to the swash plate), then to calculate the corresponding elements of the matrix J and P of the connection equations for accelerations corresponding to this connection, it is necessary to determine its coordinates in which it is not fixed

$$ {}^{b}\overline{r}_{c\,b} = M T_b^{-1} T_a \, {}^{a}\overline{r}_{c\,a}, $$

where

${}^{a}\overline{r}_{c\,a}$ – radius-vector of the point of communication overlap in the link system, where it is fixed,

T_b^{-1} – link inverse matrix b,

$$ T_b^{-1} = \begin{pmatrix} & R_b^T & & -R_b^T p_b \\ 0 & 0 & 0 & 1 \end{pmatrix}, $$

M – a projecting matrix that allows you to “press” the point of imposition of the bond to the plane in which it is based on the physical meaning of the link, for example, in this case

$$ M = \begin{pmatrix} 0 & 0 & 0 & h_6 \\ 0 & 1 & 0 & 0 \\ 0 & 0 & 1 & 0 \\ 0 & 0 & 0 & 1 \end{pmatrix}, $$

where

h_6 – distance of the overlap point of the slipper connection from the working plane of the slipper.

The elements of the matrix P of the block-matrix equation of links are, in fact, taken with the opposite sign of the linear accelerations of the points of overlapping of the links, as well as the angular accelerations of the corresponding links projected onto the directions of the overlapping of the links and calculated at the current time values of the generalized coordinates and speeds accepted zero generalized accelerations. If the link is fixed in a fixed coordinate system along all 6 degrees of freedom, the full matrix P for link i and a point with a radius vector ${}^{i}\overline{r}_{c\,i}$ of link i is written as

$$ {}^{i}P_{3T3R} = \begin{pmatrix} {}^{i}P_T \\ {}^{i}P_R \end{pmatrix} = \begin{pmatrix} -\overline{a}^{*}_{rc\,i} \\ -\overline{\varepsilon}^{*}_i \end{pmatrix}. $$

If an additional link is imposed between the links of the mechanism, and the point of overlapping of the link on the link is movable in the coordinate system of this link, then the speed of this point in the connected system is determined by the absolute speed of the corresponding point fixed relative to the other link. In relation to the links of the slippers with aswash plate, the overlap point of the connection is fixed relative to the slipper (link a). The expression for the full speed of this point in the associated swash plate system (link b) we write as follows:

$$ {}^{b}\overline{v}_{rc\,b} = R_b^T R_a \, {}^{a}\overline{v}_{rc\,a}\,. $$

Then the relative velocity of the coupling point in the system of an swash plate (link *b),*

$$ {}^{b}\overline{v}_{rc\,b\,r} = R_b^T R_a \, {}^{a}\overline{v}_{rc\,a} - {}^{b}\overline{v}_b - {}^{b}\overline{\omega}_b \times {}^{b}\overline{r}_{c\,b}\,, $$

where

${}^{b}\overline{v}_b$ - pole speed of a coordinate system b;

${}^{b}\overline{\omega}_b$ - link angular velocity b.

This expression takes into account the instantaneous rate of rotation of the swash plate in the process of regulation and under the action of workloads.

Linear and angular accelerations of bond overlap points calculated at zero generalized accelerations and effective values of generalized coordinates and velocities taking into account Coriolis acceleration and expressed in absolute coordinate system

$$\overline{a}^*_{rc\,a} = R_a({}^a\overline{a}^*_a + {}^a\overline{\varepsilon}^*_a \times {}^a\overline{r}_{c\,a} + {}^a\overline{\omega}_a \times ({}^a\overline{\omega}_a \times {}^a\overline{r}_{c\,a}))$$
$$\overline{\varepsilon}^*_a = R_a\, {}^a\overline{\varepsilon}^*_a$$
$$\overline{a}^*_{rc\,b} = R_b({}^b\overline{a}^*_b + {}^b\overline{\varepsilon}^*_b \times {}^b\overline{r}_{c\,b} + {}^b\overline{\omega}_b \times ({}^b\overline{\omega}_b \times {}^b\overline{r}_{c\,b}) + 2\,{}^b\overline{\omega}_b \times {}^b\overline{v}_{rc\,b\,r})$$
$$\overline{\varepsilon}^*_b = R_b\, {}^b\overline{\varepsilon}^*_b$$

For the case of a connection, when for a link i, one linear movement of a certain point along the direction $\overline{n}$ and two turns around directions $\overline{l}$ and $\overline{m}$ (coupling mobile unit — fixed flat distributor), 3×1 submatrix ${}^iP_{1T2R}$ can be calculated as follows:

$${}^iP_{1T2R} = -\begin{pmatrix} \overline{a}^*_{rc\,i} \cdot \overline{n} \\ \overline{\varepsilon}^*_i \cdot \overline{l} \\ \overline{\varepsilon}^*_i \cdot \overline{m} \end{pmatrix}.$$

For the case of similar connections between the movable slipper (link a) and the movable swash plate (link b) 3×1 submatrix ${}^{a,b}P_{1T2R}$ can be calculated as follows:

$${}^{a,b}P_{1T2R} = {}^aP_{1T2R} - {}^bP_{1T2R} = \begin{pmatrix} (\overline{a}^*_{rc\,b} - \overline{a}^*_{rc\,a}) \cdot \overline{n} \\ (\overline{\varepsilon}^*_b - \overline{\varepsilon}^*_a) \cdot \overline{l} \\ (\overline{\varepsilon}^*_b - \overline{\varepsilon}^*_a) \cdot \overline{m} \end{pmatrix}.$$

Such 3×1 submatrices should be calculated for each slipper.

For the 7-plunger pump scheme considered in this example, the entire matrix J and P of the coupling equations is written as follows:

$$J = \begin{pmatrix} {}^4J_{1T2R} \\ {}^{9,40}J_{1T2R} \\ \cdots \\ {}^{i,40}J_{1T2R} \\ \cdots \\ {}^{39,40}J_{1T2R} \end{pmatrix}, \quad P = \begin{pmatrix} {}^4P_{1T2R} \\ {}^{9,40}P_{1T2R} \\ \cdots \\ {}^{i,40}P_{1T2R} \\ \cdots \\ {}^{39,40}P_{1T2R} \end{pmatrix}, \quad i = 14, 19, 24, 29, 34.$$

In the complex model of the pump, the hinge forces acting in the mechanism must be formed on the basis of their physical meaning. For example, in the example under consideration – a mathematical model of a seven-pump pump in the first degree of mobility, a moment acts from the side of the drive engine, as well as the friction force in bearings.

$$\tau_1 = M_{\partial s} - M_{mp}.$$

In degrees 5, 10, 15, 20, 25, 30, 35, corresponding to the translational motion of the pistons relative to the cylinder block, the joints are created by pressure in the cylinders, as well as by friction

$$\tau_i = p_i f_i - F_{mp}, \quad i = 5, 10, 15, 20, 25, 30, 35,$$

where

τ_i – instantaneous value of the force acting on the plunger, represented in the model of the mechanism as a link i;

p_i – pressure in the corresponding cylinder;

f_i – working area of this piston;
F_{mp} – friction force.

The pressures, in turn, are determined by equations describing the working process in the working cavities of the machine, including the process of throttling the main flow in the distributor and the influence of leaks. Their consideration is beyond the scope of this article.

Also, in all interfaces, it is advisable to add friction forces using mathematical models of the required degree of detail.

The proposed method gives the necessary flexibility in the simulation of volumetric hydraulic machines. For example, if the operation of the end distributor is investigated, then simplified models of friction in the remaining mates can be used, the kinematics of the mates themselves can be taken into account by introducing the corresponding rigid bonds using the above method. In this case, the contact of the unit with the distributor can be described not by hard link equations, but by a model of the required degree of complexity, taking into account the relative movement of the contacting parts in three corners and three directions, leakages in the distributor and the friction between the parts depending on the configuration of the resulting gap. In the same way, it is possible to investigate the processes of friction and flow in pairs of plunger- cylinderblock, slipper-swash plate and other.

4.Conclusions

The method presented in the paper allows:

– to formalize the formation of mathematical models of the mechanisms of the most diverse hydraulic equipment, hydraulic drive systems and hydraulic machines, the kinematic structure of which can be represented as a tree with superimposed additional external and internal connections, and the links considered absolutely rigid;

– to obtain a mathematical model of mechanisms in an analytical form, structured and visual;

– to simulate the work of volumetric hydraulic machines with complex mechanisms, while fully taking into account all the degrees of freedom of the hydraulic machine mechanism with all imposed connections without simplifications. A ready-made mathematical model and its computer implementation can be quite easily modified by increasing it with necessary submodels without "global" rework, while at the same time simplifying parts that are not essential for the process under study.

References

[1] Borisov B P 2018 *Volumetric hydraulic machines: study guide* (Moscow: BMSTU Publ.) 237 p

[2] Nikolenko I V and Zheglova V M 2018 *Design and calculation of axial piston hydraulic machines: monograph* (Simferopol: ARIAL Publ.) 300 p

[3] Fomin M V 2015 Seating selection for slip bearings with hydrodynamic lubrication *Russian Engineering Research* **35** (1) pp 25–9

[4] Grigoriev S N, Romanov R I, Fominski V Y, Volosova M A and Zhukova E A 2014 Structure and tribological behavior of nanocomposite C-Ti-WSex coatings *Journal of Friction and Wear* **35** (4) pp 263–9

[5] Kalashnikov I E, Bolotova L K, Bykov P A, Kobeleva L I, Katin I V, Mikheev R S and Kobernik N V 2016 Tribological properties of the babbit B83–based composite materials fabricated by powder metallurgy *Russian Metallurgy* (*Metally*) (7) pp 669–74

IOP Conf. Series: Materials Science and Engineering **492** (2019) 012042 doi:10.1088/1757-899X/492/1/012042

[6] Kobernik N V, Mikheev R S, Kalashnikov I E, Kobeleva L I and Bolotova LK 2017 Tribological properties of Babbitt alloy coatings modified with carbon nanotubes *Inorganic Materials: Applied Research* **8** (3) pp 428–33

[7] Berta G L, Casoli P and Vacca A 2002 *Proceedings of Fluid Power Systems and Technology Division (FPSTD) ASME International Mechanical Engineering Congress and Exposition* [*Conference*] Simulation mode od axial piston pumps inclusive of cavitation (New Orleans, Louisiana, USA : [s.n.], 2002, November 17–22)

[8] Hao M and Qi X Y2012 Modeling Analysis and Simulation of Hydraulic Axial Piston Pump *Advanced Materials Research* **430–432** pp 1532–35

[9] Perfilev A V and Mukhin A A 2018 Improving the energy efficiency of aircraft pumping installations *Hydraulics*(6) Available at: http://hydrojournal.ru/item/84-povyshenie-effektivnosti-aviatsionnykh-nasosnykh-ustanovok

[10] Zalogin O V, NoskovA S and CherevatovA G 2016 Experimental study of the dynamics of regulation of an axial piston pump with proportional electric control *Hydraulics* (2) Available at: http://hydrojournal.ru/item/43-eksperimentalnoe-issledovanie-dinamiki-regulirovaniya-aksialno-porshnevogo-nasosa-s-proportsionalnym-elektroupravleniem

[11] Deeken M 2003 Simulation of the tribological contacts in an axial piston machine *Ölhydraulik und Pneumatik* **47** Nr 11–12

[12] Zloto T 2017 Simulation of the Hydrostatic Load of the Valve Plate-cylinder Block System in an Axial Piston Pump*Procedia Engineering***177** pp 247–54

[13] Wayne D, Milestone S A and Olsen G. 1983 The kinematic analysis of axial piston pumps *Mechanism and Machine Theory***18** (6) pp 475–479

[14] Denavit J and Hartenberg R S 1955 Kinematic notation for lover pair mechanisms based on matrices *J. Appl. Mech.* **77** pp 215–21

[15] Wittenburg J 1977 *Dynamics of systems of rigid bodies* (B.G. Teubner Stuttgart) 223 p

[16] Kovalchuk A K, Kulakov D B and Semenov S E 2008 Block-matrix equations of robots executive mechanisms with tree-like kinematic structure. *Izvestiya vuzov.ysshikhuchebnykhzavedeniy. Mashinostroenie — Proceedings of Higher Educational Institutions. Machine Building* (12) pp 5–21 (in Russian)

[17] Kulakov D B, Semenov S E, Kulakov B B, Shcherbachev P V and Tarasov O I 2015 Hydraulic Bipedal Robots Locomotion Mathematical Modeling *Procedia Engineering***106** pp 62–70

[18] Semenov S E 2015 Mathematical modeling of the electro-hydraulic actuation systems of the machines with tree-like kinematic structure *Proceedings of 2015 International Conference on Fluid Power and Mechatronics, FPM 2015* article no 7337184 pp 583–92

[19] Semenov S E, Kovalchuk A K and Kulakov D B 2010 Mathematical modeling of mechanisms with a tree-arborescence kinematic structure *Drive Technology* (2)

[20] Yaskevich N 2014 Recursive algorithm for simulation of single chain manipulator dynamics *MESA 2014 10th IEEE/ASME International Conference on Mechatronic and Embedded Systems and Applications, Conference Proceedings*, article no 6935553